Leonardo Franco, David A. Elizondo, and José M. Jerez (Eds.)

Constructive Neural Networks

Studies in Computational Intelligence, Volume 258

Editor-in-Chief
Prof. Janusz Kacprzyk
Systems Research Institute
Polish Academy of Sciences
ul. Newelska 6
01-447 Warsaw
Poland
E-mail: kacprzyk@ibspan.waw.pl

Further volumes of this series can be found on our homepage: springer.com

Vol. 237. George A. Papadopoulos and Costin Badica (Eds.)
Intelligent Distributed Computing III, 2009
ISBN 978-3-642-03213-4

Vol. 238. Li Niu, Jie Lu, and Guangquan Zhang
Cognition-Driven Decision Support for Business Intelligence, 2009
ISBN 978-3-642-03207-3

Vol. 239. Zong Woo Geem (Ed.)
Harmony Search Algorithms for Structural Design Optimization, 2009
ISBN 978-3-642-03449-7

Vol. 240. Dimitri Plemenos and Georgios Miaoulis (Eds.)
Intelligent Computer Graphics 2009, 2009
ISBN 978-3-642-03451-0

Vol. 241. János Fodor and Janusz Kacprzyk (Eds.)
Aspects of Soft Computing, Intelligent Robotics and Control, 2009
ISBN 978-3-642-03632-3

Vol. 242. Carlos Artemio Coello Coello, Satchidananda Dehuri, and Susmita Ghosh (Eds.)
Swarm Intelligence for Multi-objective Problems in Data Mining, 2009
ISBN 978-3-642-03624-8

Vol. 243. Imre J. Rudas, János Fodor, and Janusz Kacprzyk (Eds.)
Towards Intelligent Engineering and Information Technology, 2009
ISBN 978-3-642-03736-8

Vol. 244. Ngoc Thanh Nguyen, Rados law Piotr Katarzyniak, and Adam Janiak (Eds.)
New Challenges in Computational Collective Intelligence, 2009
ISBN 978-3-642-03957-7

Vol. 245. Oleg Okun and Giorgio Valentini (Eds.)
Applications of Supervised and Unsupervised Ensemble Methods, 2009
ISBN 978-3-642-03998-0

Vol. 246. Thanasis Daradoumis, Santi Caballé, Joan Manuel Marquès, and Fatos Xhafa (Eds.)
Intelligent Collaborative e-Learning Systems and Applications, 2009
ISBN 978-3-642-04000-9

Vol. 247. Monica Bianchini, Marco Maggini, Franco Scarselli, and Lakhmi C. Jain (Eds.)
Innovations in Neural Information Paradigms and Applications, 2009
ISBN 978-3-642-04002-3

Vol. 248. Chee Peng Lim, Lakhmi C. Jain, and Satchidananda Dehuri (Eds.)
Innovations in Swarm Intelligence, 2009
ISBN 978-3-642-04224-9

Vol. 249. Wesam Ashour Barbakh, Ying Wu, and Colin Fyfe
Non-Standard Parameter Adaptation for Exploratory Data Analysis, 2009
ISBN 978-3-642-04004-7

Vol. 250. Raymond Chiong and Sandeep Dhakal (Eds.)
Natural Intelligence for Scheduling, Planning and Packing Problems, 2009
ISBN 978-3-642-04038-2

Vol. 251. Zbigniew W. Ras and William Ribarsky (Eds.)
Advances in Information and Intelligent Systems, 2009
ISBN 978-3-642-04140-2

Vol. 252. Ngoc Thanh Nguyen and Edward Szczerbicki (Eds.)
Intelligent Systems for Knowledge Management, 2009
ISBN 978-3-642-04169-3

Vol. 253. Akitoshi Hanazawa, Tsutom Miki, and Keiichi Horio (Eds.)
Brain-Inspired Information Technology, 2009
ISBN 978-3-642-04024-5

Vol. 254. Kyandoghere Kyamakya, Wolfgang A. Halang, Herwig Unger, Jean Chamberlain Chedjou, Nikolai F. Rulkov, and Zhong Li (Eds.)
Recent Advances in Nonlinear Dynamics and Synchronization, 2009
ISBN 978-3-642-04226-3

Vol. 255. Catarina Silva and Bernardete Ribeiro
Inductive Inference for Large Scale Text Classification, 2009
ISBN 978-3-642-04532-5

Vol. 256. Patricia Melin, Janusz Kacprzyk, and Witold Pedrycz (Eds.)
Bio-inspired Hybrid Intelligent Systems for Image Analysis and Pattern Recognition, 2009
ISBN 978-3-642-04515-8

Vol. 257. Oscar Castillo, Witold Pedrycz, and Janusz Kacprzyk (Eds.)
Evolutionary Design of Intelligent Systems in Modeling, Simulation and Control, 2009
ISBN 978-3-642-04513-4

Vol. 258. Leonardo Franco, David A. Elizondo, and José M. Jerez (Eds.)
Constructive Neural Networks, 2009
ISBN 978-3-642-04511-0

Leonardo Franco, David A. Elizondo,
and José M. Jerez (Eds.)

Constructive Neural Networks

Leonardo Franco
Dept. of Computer Science
University of Malaga
Campus de Teatinos S/N
Málaga 29071
Spain
E-mail: lfranco@lcc.uma.es

José M. Jerez
Dept. of Computer Science
University of Malaga
Campus de Teatinos S/N
Málaga 29071
Spain
E-mail: jja@lcc.uma.es

David A. Elizondo
Centre for Computational Intelligence
School of Computing
De Montfort University
The Gateway
Leicester LE1 9BH
UK
E-mail: elizondo@dmu.ac.uk

ISBN 978-3-642-04511-0 e-ISBN 978-3-642-04512-7

DOI 10.1007/978-3-642-04512-7

Studies in Computational Intelligence ISSN 1860-949X

Library of Congress Control Number: 2009937150

Typeset & Cover Design: Scientific Publishing Services Pvt. Ltd., Chennai, India.

Printed in acid-free paper

9 8 7 6 5 4 3 2 1

springer.com

Preface

This book presents a collection of invited works that consider constructive methods for neural networks, taken primarily from papers presented at a special session held during the 18th International Conference on Artificial Neural Networks (ICANN 2008) in September 2008 in Prague, Czech Republic.

The book is devoted to constructive neural networks and other incremental learning algorithms that constitute an alternative to the standard method of finding a correct neural architecture by trial-and-error. These algorithms provide an incremental way of building neural networks with reduced topologies for classification problems. Furthermore, these techniques produce not only the multilayer topologies but the value of the connecting synaptic weights that are determined automatically by the constructing algorithm, avoiding the risk of becoming trapped in local minima as might occur when using gradient descent algorithms such as the popular back-propagation. In most cases the convergence of the constructing algorithms is guaranteed by the method used.

Constructive methods for building neural networks can potentially create more compact and robust models which are easily implemented in hardware and used for embedded systems. Thus a growing amount of current research in neural networks is oriented towards this important topic.

The purpose of this book is to gather together some of the leading investigators and research groups in this growing area, and to provide an overview of the most recent advances in the techniques being developed for constructive neural networks and their applications.

The first chapter of the book presents a review of existing constructive neural network algorithms (M. Nicoletti, J. Bertini, D. Elizondo and L. Franco). Next, four different constructing approaches to solving classification problems are presented: Muselli and Ferrari introduce a constructing method for switching functions, Grochowski and Duck focus on a new method for highly complex functions, Anthony presents a constructive method based on decision lists and the SONN3 model is analyzed by Horzyk in chapter 5. Nguiph presents in chapter 6 concept lattice-based neural networks, followed in chapter 7 by the work of Sussner and Esmi who discuss the theory and experiments using morphological neural networks. Two extensions of constructive algorithms to multiclass problems are introduced in chapters 8 and 9 by Bertini and Nicoletti and by Elizondo and Ortiz de Lazcano respectively .

The application of constructive algorithms is used by Franco, Jerez, and Subirats for active learning in chapter 10, while Ollington,Vamplew and Swanson explore the use of constructive algorithms in a reinforcement learning framework in chapter 11. Chapter 12 by Huemer, Elizondo and Góngora shows the application of a constructive neural network for evolving a machine controller, followed by the contribution from Satizábal, Pérez-Uribe and Tomassini about avoiding prototype proliferation. The volume ends with two works on self organizing neural networks: chapter 14 by Barreto et al. where the parameter setting in a fuzzy growing network is analysed, while chapter 15 discusses the method of Inoue on Self-Organizing Neural Grove.

The editors wish to thank all the authors who contributed with their research in this volume and hope that the current snapshot of some of the latest work in the field of constructive neural network algorithms help in the further development of the field.

February, 2009

Leonardo Franco
Málaga, Spain

David A. Elizondo
Leicester, UK

José M. Jerez
Málaga, Spain

Contents

Constructive Neural Network Algorithms for Feedforward Architectures Suitable for Classification Tasks

Maria do Carmo Nicoletti, João R. Bertini Jr., David Elizondo, Leonardo Franco, and José M. Jerez

Abstract. This chapter presents and discusses several well-known constructive neural network algorithms suitable for constructing feedforward architectures aiming at classification tasks involving two classes. The algorithms are divided into two different groups: the ones directed by the minimization of classification errors and those based on a sequential model. In spite of the focus being on two-class classification algorithms, the chapter also briefly comments on the multiclass versions of several two-class algorithms, highlights some of the most popular constructive algorithms for regression problems and refers to several other alternative algorithms.

1 Introduction

Conventional neural network (NN) training algorithms (such as Backpropagation) require the definition of the NN architecture before learning starts. The common way for developing a neural network that suits a task consists of defining several different architectures, training and evaluating each of them, and then choosing the one most appropriate for the problem based on the error produced between the target and actual output values. Constructive neural network (CoNN) algorithms,

Maria do Carmo Nicoletti
CS Dept, UFSCar, S. Carlos, SP, Brazil
carmo@dc.ufscar.br

João R. Bertini Jr.
ICMC, USP, S. Carlos, SP, Brazil
bertini@icmc.usp.br

David Elizondo
De Monfort University, Leicester, UK
elizondo@dmu.ac.uk

Leonardo Franco and José M. Jerez
University of Málaga, Málaga, Spain
lfranco@lcc.uma.es,jja@lcc.uma.es

L. Franco et al. (Eds.): Constructive Neural Networks, SCI 258, pp. 1–23.
springerlink.com

however, define the architecture of the network along with the learning process. Ideally CoNN algorithms should efficiently construct small NNs that have good generalization performance. As commented in (Muselli, 1998), "...the possibility of adapting the network architecture to the given problem is one of the advantages of constructive techniques... [This] has also important effects on the convergence speed of the training process. In most constructive methods, the addition of a new hidden unit implies the updating of a small portion of weights, generally only those regarding the neuron to be added".

The automated design of appropriate neural network architectures can be approached by two different groups of techniques : evolutionary and non-evolutionary. In the evolutionary approach, a NN can be evolved by means of an evolutionary technique, i.e. a population-based stochastic search strategy such as a GA (see (Schaffer *et al.*, 1992) (Yao, 1999)). In the non-evolutionary approach, the NN is built not as a result of an evolutionary process, but rather as the result of a specific algorithm designed to automatically construct it, as is the case with a constructive algorithm.

CoNN algorithms, however, are not the only non-evolutionary approach to the problem of defining a suitable architecture for a neural network. The strategy implemented by the so called pruning methods can also be used and consists in training a larger than necessary network (which presumably is an easy task) and then, pruning it by removing some of its connections and/or nodes (see (Reed, 1993)).

As approached by Lahnajärvi and co-workers in (Lahnajärvi *et al.*, 2002), pruning algorithms can be divided into two main groups. Algorithms in the first group estimate the sensitivity of an error function to the removal of an element (neuron or connection); those with the least effect can be removed. Algorithms in the second group generally referred to as penalty-term as well as regularization algorithms, "add terms to the objective function that reward the network for choosing efficient and small solutions". The same group of authors also detected in the literature what they call combined algorithms that take advantage of the properties of both, constructive and pruning algorithms, in order to determine the network size in a flexible way (see (Fiesler, 1994), (Gosh & Tumer, 1994)).

There are many different kinds of neural network; new algorithms and variations of already known algorithms are constantly being published in the literature. Similarly to other machine learning techniques, neural network algorithms can also be characterized as supervised, when the target values are known and the algorithm uses the information, or as unsupervised, when such information is not given and/or used by the algorithm. The two main classes of NN architecture are feedforward, where the connections between neurons do not form cycles, and feedback (or recurrent), where the connections may form cycles. NNs may also differ in relation to the type of data they deal with; the two more popular being categorical and quantitative. Both, supervised and unsupervised learning with categorical targets are referred to as classification. Supervised learning with quantitative target values is known as regression. Classification problems can be considered a particular type of regression problems.

This chapter reviews some well-known CoNN algorithms for feedforward architectures, suitable for classification tasks, aiming to present the main ideas they are based upon as well as to stress their main similarities and differences. An empirical evaluation of several two-class and multiclass CoNN algorithms in a variety of knowledge domains can be seen in (Nicoletti & Bertini, 2007). Although focusing mainly on a particular class of constructive neural network algorithms, the chapter also aims at establishing the importance of using a constructive approach when working with neural networks, independent of the application problem at hand. For this reason towards the end of the chapter a few other approaches will be briefly presented.

The remainder of the chapter is organized as follows: Section 2 presents the main characteristics of constructive neural network algorithms. Section 3 approaches several two-class CoNN algorithms by grouping them into (3.1) algorithms directed by the minimization of classification errors and (3.2) algorithms based on the sequential model. Section 4 briefly describes the multiclass versions of a few two-class CoNN previously presented as well highlighting the main characteristics of a few other multiclass proposals. Section 5 introduces some CoNN algorithms for regression problem as well as other combined approaches. In Section 6 several CoNN algorithms that do not quite conform to the main focus of this chapter and others that do not qualify to be part of the groups characterized before are briefly presented and finally, in Section 7, the conclusions of this chapter are presented.

2 Main Characteristics of CoNN Algorithms

Several constructive algorithms that focus on feedforward architectures have been proposed in the literature. During the learning phase they all essentially repeat the same process: incrementally adding and training hidden neurons (generally Threshold Logic Units – TLUs) until a stopping criterion is satisfied. Generally they all begin the process having as the initial network only the input layer; output neuron(s) are then added and trained and, depending on their performance, the algorithm starts to add and connect hidden neurons to the current architecture and train them, aiming at improving the accuracy of the network performance.

The final result at the end of the constructive process implemented by these algorithms is a neural network that had its architecture defined along with its training. In spite of sharing the same basic mechanism, CoNN algorithms differ from each other in many different ways, such as:

1. Number of nodes they add per layer at each iteration;
2. Direction in which they grow the network:
    - forward, from input towards output nodes or
    - backward, from output towards input nodes;
3. Functionality of the added neurons (do they all play the same role?);
4. Stopping criteria;
5. Connectivity pattern of the newly added neuron;

6. Algorithm used for training individual neuron, such as
   - The Fisher Discriminant
   - Pocket algorithm (Gallant, 1986a)
   - Pocket with Ratchet Modification (PRM) (Gallant, 1986a, 1990)
   - MinOver (Krauth and Mézard, 1987)
   - Quickprop (Fahlman, 1988)
   - AdaTron (Anlauf & Biehl, 1989)
   - Thermal Perceptron algorithm (Frean, 1992)
   - Loss minimization (Hrycej, 1992)
   - Modified Thermal algorithm (Burgess, 1994)
   - Maxover (Wendemuth, 1995)
   - Barycentric Correction Procedure (BCP) (Poulard, 1995);
7. Type of input patterns they deal with: binary (or bipolar) valued, categorical or real valued attributes;
8. Type of problems they solve:
   - classification (two-class or multi-class), where the input is assigned to one of two or more classes
   - regression problems, characterized by a continuous mapping from inputs to an output or
   - clustering, where the patterns are grouped according to some similarity measure;
9. Topology of the connections among neurons (initially fixed or dynamically constructed);
10. 'Shape' of the feedforward architecture (e.g. tower-like, cascade-like, etc…).

Among the most well known CoNN algorithms for two-class classification problems are the Tower and the Pyramid (Gallant, 1986b), the Tiling (Mézard & Nadal, 1989), the Upstart (Frean, 1990), the Perceptron Cascade (Burgess, 1994), the PTI and the Shift (Amaldi & Guenin, 1997), the Irregular Partitioning Algorithm (IPA) (Marchand *et al.*, 1990; Marchand & Golea, 1993), the Target Switch (Campbell & Vicente, 1995), the Constraint Based Decomposition (CBD) algorithm (Drăghici, 2001) and the BabCoNN (Bertini Jr. & Nicoletti, 2008a).

Smieja, (1993), Gallant, (1994), Bishop, (1996), Mayoraz and Aviolat (1996), Campbell (1997) and Muselli (1998) discuss several CoNN algorithms in detail.

3 A Closer Look at Several Two-Class CoNN Algorithms

In order to review several of the two-class CoNN algorithms previously mentioned in a systematic way, this section divides them into two main categories: (3.1) those directed by the minimization of classification errors and (3.2) those based on the sequential learning model.

3.1 Algorithms Directed by the Minimization of Classification Errors

Since most of the CoNN algorithms can be characterized as belonging to this category, a few sub-categories have been adopted in this review aimed at grouping algorithms that share similar characteristics.

3.1.1 Growing NNs with Single-Neuron Hidden Layers

The Tower and the Pyramid (Gallant, 1986b, 1994) algorithms are two-class CoNN algorithms that can be viewed as incremental versions of the PRM algorithm; both can be characterized as forward methods since they grow the NN layer by layer, from the input towards the output layer. In the Tower algorithm each new hidden node added to the network is connected to all input nodes and to the last previously added hidden node, making the architecture of the network look like a tower.

The basic idea of the Tower algorithm is very simple. Initially the PRM algorithm is used for training the only node of the first hidden layer of the network which can be considered the output node. If p is the number of attributes that describe a training pattern, the first hidden node receives p + 1 inputs i.e., the input values associated with the p attributes plus the constant value associated with the bias. As is well known a single neuron can learn with 100% precision only from linearly separable training sets. If that is the case, the first neuron added to the network and subsequently trained will correctly learn to separate both classes and the Tower algorithm is reduced to the PRM algorithm. However, if that is not the case, the Tower algorithm continues to add hidden layers to the network (each containing only one TLU) until a stopping criterion is satisfied.

Generally three stopping criteria can be implemented: 1) the NN correctly classifies the training set; 2) adding a new hidden layer does not contribute to increasing the network accuracy and 3) a predefined maximum number of hidden neurons has been reached. Considering that the first step adds the first hidden node to the network, the kth step adds the kth hidden node. After the first hidden node is added, all the subsequent hidden nodes added will receive an extra input value, which corresponds to the output of the last added hidden node. This extra dimension, added after the first hidden node was created represents the behavior of the newly created hidden node.

The Pyramid algorithm is very similar to the Tower algorithm. The only difference between them is that each new hidden node created by the Pyramid method is connected to every hidden node previously added to the network, as well as to the input nodes, making the network look like a pyramid. Each step of the learning phase expands the training patterns in one dimension (which represents the last hidden neuron added). Considering that the first step adds the first hidden node to the network, the kth step adds the kth hidden node that receives the p+1 input values from the input layer plus bias, as well as the output of the (k–1) hidden nodes previously added.

Gallant (1994) presents a detailed description of both algorithms as well as the proof of their convergence, by stating and proving the following theorem: "with arbitrarily high probability, the Tower (Pyramid) algorithm will fit noncontradictory sets of training examples with input values restricted to {+1,–1}, provided enough cells are added and enough iterations are taken for each added cell. Furthermore each added cell will correctly classify a greater number of training examples than any prior cell."

3.1.2 Growing NNs with Hidden Neurons Performing Different Functions

The Tiling algorithm (Mézard & Nadal, 1989) is a CoNN algorithm originally proposed for Boolean domains that trains a multilayer feedforward NN where hidden nodes are added to a layer in a way comparable to the process of laying tiles. The neurons in each hidden layer in a Tiling NN perform one out of two different functions and their names reflect their functionality. Each layer has a *master neuron* that works as the output neuron for that layer. If the master neuron does not correctly classify all training patterns, however, the Tiling algorithm starts to add and train *ancillary neurons*, one at a time, aiming at obtaining a faithful representation of the training set. The output layer has only one master neuron. The *faithfulness criterion* employed by the Tiling algorithm establishes that no two training patterns, belonging to different classes, should produce the same outputs at any given layer.

As commented by Gallant (1994), "The role of these units (ancillary) is to increase the number of cells for layer L so that no two training examples with *different classifications* have the *same set of activations* in layer L. Thus each succeeding layer has a different representation for the inputs, and no two training examples with different classifications have the same representation in any layer. Layers with this property are termed *faithful layers*, and faithfulness of layers is clearly a necessary condition for a strictly layered network to correctly classify all training examples".

In order to construct the network, the first step of the Tiling method is to train the master neuron of the first hidden layer, using the original training set, aiming at minimizing the classification error. If the master neuron does not classify the training set correctly, ancillary neurons are added to this layer and subsequently trained, one at a time, in order to obtain a faithful representation. Tiling constructs an NN in successive layers such that each new layer has a smaller number of neurons than the previous layer and layer L only receives connections from hidden layer L – 1.

The way the Tiling method operates assures that the master neuron in layer L classifies the training set with higher accuracy than the master neuron in layer L – 1. Assuming a finite training set with non-contradictory patterns, the Tiling algorithm is guaranteed to converge to zero classification errors (under certain assumptions) (Gallant, 1994). Once the first layer is finished, the Tiling algorithm goes on adding layers (with one master and a few ancillary neurons) until one out of four stopping criteria is satisfied: 1) the network converges; 2) the master neuron added degrades the performance of the network; 3) a pre-defined maximum number of layers is reached or 4) a pre-defined number of ancillary neurons per layer is reached and the training set still does not have a faithful representation in this layer. The original Tiling algorithm as described in (Mézard & Nadal, 1989) uses the same TLU training algorithm for each neuron (master or ancillary) added to the NN; in the same article the authors state and prove a theorem that ensures its convergence.

The unnamed algorithm proposed in (Nadal, 1989) and described by its author as an algorithm similar in spirit to the Tiling algorithm was named by Smieja in (Smieja, 1993) as *Pointing*. The Pointing algorithm constructs a feedforward neural network with the input layer, one output neuron and an ordered sequence of single hidden neurons, each one connected to the input layer and to the previous hidden neuron in the sequence. Despite its author claiming that the algorithm is very similar to the Tiling one, although more constrained (which is true), actually the Pointing algorithm corresponds to the Tower algorithm as proposed by Gallant (Gallant, 1986b).

The Partial Target Inversion (PTI) algorithm is a CoNN algorithm proposed in (Amaldi & Guenin, 1997) that shares strong similarities with the Tiling algorithm. The PTI grows a multi-layer network where each layer has one master neuron and a few ancillary neurons. Following the Tiling strategy as well, the PTI adds ancillary neurons to a layer in order to satisfy the faithfulness criteria; the neurons in layer c are connected only to neurons in layer $c - 1$. If the training of the master neuron results in a weight vector that correctly classifies all training patterns or if the master neuron of layer c does not classify a larger number of patterns than the master neuron of layer $c - 1$, the algorithm stops. If a training pattern, however, was incorrectly classified by the master neuron, and the master neuron correctly classifies a greater number of patterns than the master of the previous layer, the algorithm starts adding ancillary neurons to the current layer aiming at its faithfulness. When the current c layer becomes faithful the algorithm adds a new layer, $c + 1$, initially only having the master neuron. The process continues until stopping criteria are met, such as when the number of master (or ancillary) neurons has reached a pre-defined threshold. The only noticeable difference between the Tiling and the PTI is the way the training set, used for training the ancillary neurons in the process of turning a layer faithful, is chosen.

For training the master neuron of layer c, the PTI uses all the outputs from the previous layer. Considering that the layer c needs to be faithful, the first ancillary neuron added to layer c will be trained with those patterns (used to train the master neuron) that made layer c unfaithful.

In addition, the patterns that activate the last added neuron (master, in this case) have their classes inverted. The authors justify this procedure by saying that "When trying to break several unfaithful classes simultaneously, it may be worthwhile to flip the target of the prototypes on layer c corresponding to several unfaithful classes. In fact, the specific targets are not important; the only requirement is that outputs corresponding to a different c–1 layer representation trigger a different output for at least one unit in the layer c". For clarification if after the addition of the first ancillary neuron the layer is not faithful yet, another ancillary neuron needs to be added to layer c. The second ancillary neuron will be trained with the c–1 outputs that provoked an unfaithful representation of layer c, this time, however, taking also into consideration the master and first ancillary neuron previously added. The patterns that activated the first ancillary neuron have their class inverted when they are included in the training set for the second ancillary neuron. The name PTI (Partial Target Inversion) refers to the fact that when constructing training sets only a partial number of patterns have their class (target) inverted.

3.1.3 Growing NNs Based on *Wrongly-On* and *Wrongly-Off* Errors

Generally CoNN algorithms based on discriminating between the two types of errors (*wrongly-on* and *wrongly-off*) tend to grow the neural network in a backward way, i.e. from the output towards the input layer.

The Upstart (Frean, 1990) is a constructive NN algorithm that dynamically grows a neural network whose structure resembles a binary tree. The algorithm starts the construction of the NN from the output layer towards the input layer and during the construction of an Upstart network, a neuron adds two other ancillary neurons in order to correct its misclassifications.

Let u_n be a neuron that classifies training patterns but produces *wrongly-off* errors (i.e. positive training patterns are misclassified by u_n as negative). The Upstart algorithm deals with *wrongly-off* errors by adding a *wrongly-off corrector* as a 'child' neuron u_{n+} which will try to correct the errors made by its parent u_n. The main tasks of neuron u_{n+} are 1) to correct the classification of positive training patterns that have been misclassified by u_n as negative and 2) to keep unchanged the other classifications made by u_n (i.e. u_{n+} should be inactive for any other pattern). The neuron u_{n+} is trained with the subset of training patterns that were *wrongly-off* plus the set of negative patterns.

Similarly, to deal with *wrongly-on* errors (i.e. negative training patterns that are misclassified by u_n as positive) the Upstart algorithm creates a *wrongly-on corrector* as a child neuron u_{n-} aiming: 1) to correct the classification of *wrongly-on* training patterns and 2) keep unchanged the other classifications made by neuron u_n (i.e. u_{n-} should stay inactive for any other pattern). Neuron u_{n-} is trained with the subset of training patterns that were *wrongly-on* plus the set of positive patterns.

When a neuron u_n does not correctly classify all training patterns, it gives rise to *wrongly-off* errors, *wrongly-on* errors or both. A hidden neuron in an Upstart network, therefore, can have up to two children. As can be seen in (Frean, 1990), two useful results follow from this training method because the children neurons have a simpler problem to solve than their parents. Children neurons can always make fewer misclassifications than their parents and connecting a child neuron to its parent with the appropriate weight will always reduce the misclassifications made by the parent .

For a child neuron (either a *wrongly-on* or a *wrongly-off corrector*) that has been added in order to correct misclassifications made by its parent, it is mandatory that this child only changes the activations of patterns that provoked the error; for this reason, the inactive output of the neuron should be 0. For a detailed description of Upstart, see also (Kubat, 2000).

Another algorithm that adopts the same error-correction strategy of the Upstart is the Perceptron Cascade (PC) algorithm (Burgess, 1994); PC also borrows the architecture of neural networks created by the Cascade Correlation algorithm (Fahlman & Lebiere, 1990).

PC begins the construction of the network by training the output neuron. If this neuron does not classify all training patterns correctly, the algorithm begins to add hidden neurons to the network. Each new added hidden neuron is connected to all previous hidden neurons as well as to the input neurons. The new hidden neuron is then connected to the output neuron; each time a hidden neuron is added, the output neuron needs to be retrained. The addition of a new hidden neuron enlarges the space in one dimension. The algorithm has three stopping criteria: 1) the network converges i.e. correctly classifies all training patterns; 2) a pre-defined maximum number of hidden neurons has been achieved and 3) the most common, the addition of a new hidden neuron degrades the network's performance.

Similarly to the Upstart algorithm, hidden neurons are added to the network in order to correct *wrongly-on* and *wrongly-off* errors. Following the same strategy employed by Upstart, what distinguishes a neuron created for correcting *wrongly-on* or for correcting *wrongly-off* errors caused by the output neuron is the training set used for training the neuron. To correct *wrongly-on* errors the training set used should have all negative patterns plus the patterns which produce *wrongly-on* errors. For correcting *wrongly-off* errors, the training set should have all positive patterns plus the negative patterns which produce *wrongly-off* patterns. Unlike Upstart, however, the PC algorithm only adds and trains one neuron at a time in order to correct the most frequent *wrongly-on* and *wrongly-off* errors produced by the output neuron.

Like the Upstart and Perceptron Cascade algorithms, the Shift algorithm (Amaldi & Guenin, 1997) also constructs the network beginning with the output neuron. This algorithm, however, creates only one hidden layer, iteratively adding neurons to it; each added neuron is connected to the input neurons and to the output neuron. The error correcting procedure used by the Shift method is similar to the one used by the Upstart method, in the sense that the algorithm also identifies *wrongly-on* and *wrongly-off* errors. However, it adds and trains a hidden neuron to correct the most frequent between these two types of errors. Also, the training set used for training a *wrongly-off* (or *wrongly-on*) corrector differs slightly from the training sets used by the Upstart algorithm.

3.2 Algorithms Based on the Sequential Model

The general CoNN model identified as sequential learning was proposed in (Marchand *et al.*, 1990) for Boolean domains and is an improved version of a previous proposal described in (Ruján & Marchand, 1989). In (Marchand & Golea, 1993) the algorithm is extended to deal with real data. Contrary to many of the CoNN algorithms, the sequential learning model is not directed by the minimization of the classification error.

Typically, a sequential learning algorithm adds to the network hidden neurons that are partial classifiers i.e. they separate a group of training patterns belonging to the same class from the remaining patterns in the training set; this was formally stated in (Muselli, 1998) as: Let Q^+ and Q^- be two subsets of the input space; a neuron will be called a *partial classifier* if it provides an output of +1 for at least one pattern of Q^+ and output of −1 for all elements of Q^-.

Any algorithm that implements the sequential learning model should use an efficient strategy for training individual TLUs as partial classifiers. Constructive algorithms based on the sequential learning model produce networks having partial classifiers as hidden nodes. In the paper (Marchand *et al.*, 1990) the authors describe a mechanism that uses the Perceptron as an auxiliary process for identifying the weight vector. However, as pointed out in (Poulard & Hernandez, 1997), this mechanism is time consuming and becomes prohibitive for large training sets and the authors suggest the BCPMax instead (Poulard & Labreche, 1995), which associates the BCP (Poulard, 1995) with a Pattern Exclusion Procedure (PEP); initially designed for the BCP the PEP finds the best value of the neuron's threshold for a fixed value of the weight vector: the one maximizing the number of excluded patterns.

The sequential learning model has been implemented by a few algorithms, namely Marchand *et al.*´s own proposal known in the literature as the Irregular Partitioning Algorithm (IPA), the Carve algorithm (Young & Downs, 1998), the Target Switch algorithm (Campbell & Vicente, 1995), the Oil Spot Algorithm (Mascioli & Martinelli, 1995), the Constraint Based Decomposition (CBD) algorithm, proposed in (Drăghici, 2001) and the Decomposition Algorithm for Synthesis and Generalization (DASG) recently introduced in (Subirats, Jerez and Franco, 2008).

The IPA algorithm creates a neural network with an input layer, a single hidden layer and the output neuron. The connections between the hidden layer and the output neuron have weights and the output neuron has a bias. The hidden layer is created by sequentially adding neurons to it; each added neuron represents a hyperplane that separates the greatest number of patterns belonging to the same class from the rest of the training set. Once the hyperplane is found, the identified patterns belonging to the same class are removed from the training set and the procedure is repeated. The process ends when the training set only has patterns belonging to the same class.

The Target Switch algorithm was originally designed to deal with binary patterns. The algorithm can induce two different network structures, namely a cascade that uses linear neurons and a tree-like structure that uses threshold neurons. The algorithm is based on the concept of *dichotomy* which, for a classification problem with a training set $E = E^- \cup E^+$ can be summarised as: "A set of weights and thresholds which correctly store all the E^+ patterns and some of the E^- will be said to induce a *(+)dichotomy* while a *(−)dichotomy* will correspond to correct storage of all the E^- patterns and some of the E^+" (Campbell, 1997).

For growing either type of structure, neurons are always added in pairs, one for inducing a *(+)dichotomy* and the other for inducing a *(−)dichotomy*. The patterns belonging to E^- that are correctly stored by the *(+)dichotomy* and those belonging to E^+ that are correctly stored by the *(–)dichotomy* are the patterns that will be correctly separated by the neuron introduced in order to connect the above mentioned pair.

When the architecture is a cascade type, the introduced neuron is a linear neuron that implements a summation function on the pair outputs. If the result is positive or negative the current pattern is correctly classified otherwise a misclassification is produced, which will be dealt with by the next pair of neurons to be added. For growing neural networks with a tree structure the introduced neuron is a threshold neuron that implements a threshold function on the pair outputs. Considering that the first iteration adds one threshold neuron (the output), each following iteration will add two more threshold neurons to those already added in the previous iteration.

To obtain the dichotomies the authors propose the use of any Perceptron-like TLU training algorithm. Roughly speaking, the idea is to run the TLU training algorithm and then shift the resulting hyperplane in order to correctly classify all patterns of a given class.

The Constraint Based Decomposition (CBD) is another algorithm that follows the sequential model. The algorithm builds an architecture with four layers which are named input, hyperplane, AND and OR layers respectively. The whole training set is used for training the first hidden neuron in the hyperplane layer. The next hidden neuron to be added will be trained with those training patterns that were misclassified by the first hidden neuron. The algorithm goes on adding neurons to the first layer until no pattern is left in the training set. For training a neuron u_i, one pattern from each class is randomly chosen and removed from the training set E. These patterns are put in the training set E_{u_i}. After u_i has been trained with E_{u_i}, the algorithm starts to add patterns to E_{u_i}, one at a time, in a random manner. Each time a pattern is added to the set, u_i is retrained with the updated E_{u_i}. However, if the addition of a new pattern to E_{u_i} results in misclassification, the last pattern added is removed from E_{u_i} and marked as ´used´ by the neuron. Before adding a new hidden neuron, the algorithm considers all patterns in E for the current neuron. A new neuron will be added when all training patterns left have been tried for the current neuron. The neurons of the AND layer are connected only to relevant neurons from the hyperplane layer and in the OR layer the output neurons are connected only to neurons from the AND layer which are turned on for the given class.

The recently introduced DASG algorithm belongs also to the class of sequential learning algorithm. It works with binary inputs by decomposing the original Boolean function (or partially defined Boolean function) into two new lower complexity functions, which in turn are decomposed until all obtained functions are threshold functions that can be implemented by a single neuron. The final solution incorporates all functions in a single hidden layer architecture with an output neuron that computes and OR or AND Boolean function.

The BabCoNN (Barycentric-based CoNN) (Bertini Jr. & Nicoletti, 2008a) is a new two-class CoNN that borrows some of the ideas of the BCP (Barycentric Correction Procedure, see (Poulard, 1995), (Poulard & Labreche 1995)) and can be considered a representative of the sequential model. Like the Upstart,

Table 1 Overview of fourteen two-class CoNN algorithm characteristics

Algorithm	Group	# HL Growth direction	New Neuron Connected to	Special Feature	Stopping criteria
Tower	One HN per HL	Various Forward	Previously added HN and INs	Weight update	CON AD NHL
Pyramid	One HN per HL	Various Forward	All previously added HNs and INs	Dimension of weight space increases	CON AD NHL
Tiling	Neurons perform different functions	Various Forward	Previous layer	Faithful layers – divide and conquer	CON AD NHL NHN
PTI	Neurons perform different functions	Various Forward	Previous layer	Faithful layers – inversion of classes	CON AD NHL NHN
Upstart	Wrongly-on/off correctors	Binary tree Backward	Parent neuron	Children correct the father's mistakes	CON AD NHL
Shift	Wrongly-on/off correctors	One Backward	INs	Weighted connections are used to correct the output error	CON AD NHL
Perceptron cascade	Wrongly-on/off correctors	Cascade-like Backward	Previously added HNs and INs	Output increases the dimension of its weight space every time a neuron is added	CON AD NHL
Cascade correlation	Wrongly-on/off correctors	Cascade-like Backward	Previously added HNs and INs	Suitable for regression tasks	CON AD NHL
Offset	Neurons perform different functions	Two Forward	Previous layer	Parity machine	CON AD NHL
IPA	Sequential	One Forward	INs	Sequentially classifies the training set	TSC
Target switch	Sequential	Cascade (tree-like) Backward	Previously added HNs and INs	(+) and (–) dichotomies	TSC
CBD	Sequential	Three Forward	Previous layer	AND/OR layers	TSC
BabCoNN	Sequential	One Backward	Input	HN fires –1, 0 or 1	TSC
DASG	Sequential	One Forward	Input	AND/OR output function	TSC

Perceptron Cascade (PC) and Shift, the BabCoNN also constructs the network, beginning with the output neuron. However, it creates only one hidden layer; each hidden neuron is connected to the input layer as well as to the output neuron, like with the Shift algorithm.

Although the Upstart, PC and Shift construct the network by adding new hidden neurons specialized in correcting *wrongly-on* and *wrongly-off* errors, the BabCoNN employs a different strategy. The BCP is used for constructing a hyperplane in the hyperspace defined by the training set; the classified patterns are removed from the set and the process is repeated again with the updated training set. Due to the way the algorithm works a certain degree of redundancy is inserted in the process, in the sense of a pattern being correctly classified by more than one hidden neuron. This has been fixed by the BabCoNN classification process, where hidden neurons have a particular way of firing their output.

Table 1 summarizes the main characteristics of fourteen two-class algorithms previously discussed. For presenting the table the following abbreviations were adopted: Forward (the NN is grown from input towards output layer); Backward (the NN is grown from output towards input layer); INs: all neurons in the input layer; HN: a hidden neuron; HL: a hidden layer; #HL: number of hidden layers. The following abbreviations were adopted for stopping criteria: CON (convergence); AD (accuracy decay); NHL (number of hidden layers exceeds a given threshold); NHN (number of hidden neurons per hidden layer exceeds a given threshold); TSC (all training patterns have been correctly classified).

4 A Brief Approach to Multiclass Classification Using CoNN

A multiclass classification problem is a classification problem involving *m* (> 2) classes usually treated as *m* two-class problems. Generally multiclass CoNN start by training as many output neurons as there are classes in the training set, using one of two strategies: individual (I) and winner-takes-all (WTA).

The multiclass versions of a few two-class algorithms have been proposed in (Parekh *et al.*, 1995), (Yang *et al.*, 1996), (Parekh *et al.*, 1997a), (Parekh *et al.*, 1997b), (Parekh *et al.*, 2000) and they are the MTower, MPyramid, MTiling, MUpstart and MPerceptron Cascade, which can be considered extensions of their two-class counterparts.

The MTower algorithm deals with an m-class problem by adding and training *m* hidden neurons per hidden layer at each iteration. In an MTower architecture each of the *m* neurons in a certain hidden layer has connections with all the neurons of the input layer as well as with all the *m* neurons of the previous hidden layer. The MPyramid also deals with an m-class problem by adding and training *m* hidden neurons per hidden layer at each iteration. The *m* hidden neurons in each hidden layer, however, are connected to all the hidden neurons of all the hidden layers as well as to the input neurons.

Although the two-class Upstart algorithm constructs the neural network as a binary tree of TLUs starting with the output neuron, its multiclass version, the MUpstart, creates a network with a single hidden layer where each single hidden neuron is directly connected to every neuron in the output layer. The input layer is fully connected to the hidden neurons as well as to the output neurons. As mentioned before, each neuron added to a hidden layer by the Tiling algorithm can be a master (one per hidden layer) or an ancillary neuron (a few per layer). The MTiling also constructs a multi layer neural network where the first hidden layer has connections to the input layer and each subsequent hidden layer has connections only to the previous hidden layer. For training data containing m classes (> 2), MTiling adds m master neurons and as many ancillary neurons as necessary to make the layer faithful. The output layer has exactly m neurons.

The multiclass MPerceptron-Cascade is very similar to the MUpstart. Their main difference is the architecture of the neural network they induce. While the MUpstart adds the new hidden neurons in a single layer, the MPerceptron-Cascade adds the new hidden neurons in new layers. The MBabCoNN (Bertini Jr. & Nicoletti, 2008b) is the multiclass version of BabCoNN and constructs a network beginning with the output layer containing as many neurons as there are classes in the training set (each output neuron is associated to a class). The algorithm allows the neurons to be trained using any TLU algorithm combined with either strategy, individual or WTA. After adding m output neurons, the algorithm starts to add neurons to its single hidden layer in order to correct the classification mistakes made by the output neurons. A detailed description of MBabCoNN and an empirical evaluation of its performance *versus* the performance of several multiclass CoNN algorithms is described in one chapter of this book and is an extended version of the paper (Bertini Jr. & Nicoletti, 2008b).

5 CoNN Algorithms for Regression Problems and Combined Approaches

In spite of their strong focus on classification tasks, many CoNN proposals specifically aim at regression problems (see (Kwok & Yeung, 1997a), (Ma & Khorasani, 2003), (Ma & Khorasani, 2004)). A review of the CoNN algorithms for regression problems, approached from the perspective of a state-space search can be seen in (Kwok & Yeung, 1997a).

In their proposed taxonomy Kwok & Yeung group the algorithms into six different categories, each named after its most representative algorithm (1) Dynamic node creation (DNC) (Ash, 1989); (2) Projection pursuit regression, based on the statistical technique proposed in (Friedman & Stuetzle, 1981); (3) Cascade-Correlation, that mostly groups variants of the cascade-correlation architecture proposed in (Fahlman & Lebiere, 1990); (4) Resource-allocating networks (RAN) (Platt, 1991); (5) Group methods of data handling, a class of algorithms inspired by the GMDH proposed by Ivakhnenko and described in (Farlow, 1984) and (6) Miscellaneous, a category that groups CoNN that have 'multivaluated state transition mappings while still retraining the whole network upon hidden unit addition'. In the last category, however, the authors only talk about a hybrid algorithm, proposed in (Nabhan & Zomaya, 1994) that employs both, a constructive and a pruning strategy.

Kwok & Yeung in (Kwok & Yeung, 1997b) conducted a very careful investigation on the objective functions for training hidden neurons in CoNN for multilayer feedforward networks for regression problems, aiming at deriving a class of objective functions whose value and the corresponding weight updates could be computed in O(N) time, for a training set with N patterns.

In spite of the many CoNN algorithms surveyed in (Kwok & Yeung, 1997a), the most popular for regression problems is no doubt the Cascade Correlation algorithm (CasCor) and maybe the second most popular is the DNC. While the DNC algorithm constructs neural networks with a single hidden layer, the CasCor creates them with multiple hidden layers, where each hidden layer has one hidden neuron. The popularity of CasCor can be attested by the various ways this algorithm has inspired new variations and also has been used in the combined approaches between learning methods.

A similar approach to CasCor called Constructive Backpropagation (CBP) was proposed in (Lehtokangas, 1999). The RCC, a recurrent extension to CasCor is described in (Fahlman, 1991) and its limitations are presented and discussed in (Giles *et al.*, 1995). In (Kremer, 1996) the conclusions of Giles *et al.* in relation to RCC are extended. An investigation into problems and improvements in relation to the basic CasCor can be found in (Prechelt, 1997), where CasCor and five of its variations are empirically compared using 42 different datasets from the benchmark PROBEN1 (Prechelt, 1994).

CasCor has also inspired the proposal of the Fixed Cascade Error (FCE), described in (Lahnajärvi *et al.*, 1999c), (Lahnajärvi *et al.*, 2002), which is an enhanced version of a previous algorithm proposed by the same authors known as Cascade Error (CE) (see (Lahnajärvi *et al.*, 1999a), (Lahnajärvi *et al.*, 1999b)). While the general structure of both algorithms is the same, they differ in the way the hidden neurons are created.

The Rule-based Cascade-correlation (RBCC) proposed in (Thivierge *et al.*, 2004) is a collaborative symbolic-NN approach which is partially inspired by the KBANN (Knowledge-Based Artificial Neural Networks) model proposed in (Towel *et al.*, 1990), (Towel, 1991) where the NN used is a CasCor network. In the KBANN an initial set of rules is translated into a neural network which is then refined using a training set of patterns; the refined neural network can undergo a further step and be converted into a set of symbolic rules which could, again, be used as the starting point for constructing a neural network and the whole cycle would be repeated.

According to the authors the RBCC is a particular case of the Knowledge-based Cascade-correlation algorithm (KBCC) (Shultz & Rivest, 2000) (Shultz & Rivest, 2001). The KBCC extends the CasCor by allowing as hidden neurons during the growth of a NN not only single neurons, but previously learned networks as well. In (Thivierge et al., 2003) an algorithm that implements simultaneous growing and pruning of CasCor networks is described; the pruning is done by removing irrelevant connections using the Optimal Brain Damage (OBD) procedure (Le Cun *et al.*, 1990).

In (Islam & Murase, 2001) the authors propose the CNNDA (Cascade Neural Network Design Algorithm) for inducing two-hidden-layer NNs. The method automatically determines the number of nodes in each hidden layer and can also reduce a two-hidden-layer network to a single-layer network. It is based on the use of a temporary weight freezing technique. The Fast Constructive-Covering Algorithm (FCCA) for NN construction proposed in (Wang, 2008) is based on geometrical expansion. It has the advantage of each training example having to be learnt only once, which allows the algorithm to work faster than traditional training algorithms.

6 Miscellaneous

A few constructive approaches have also been devised for RBF (Radial Basis Function) networks, such the Orthogonal Least Squares (OLS) (Chen *et al.*, 1989) (Chen *et al.*, 1991) and the Growing Radial Basis Function (GRBF) networks (Karayiannis & Weiqun, 1997).

Although CoNN algorithms seem to have a lot of potential in relation to both the size of the induced network and its accuracy, it is really surprising that their use, particularly in the area of classification problems, is not as widespread as it should be, considering their many advantages. In regression problems, however, CoNNs have been very popular, particularly the Cascade-Correlation algorithm and many of its variations. In what follows some of the most recent works using CoNN are mentioned.

In (Lahnajärvi *et al.*, 2004) four CasCor-based CoNN algorithms, have been used for evaluating the movements of a robotic manipulator. In (Huemer *et al.*, 2008) the authors describe a method for controlling machines, such as mobile robots, using a very specific CoNN. The NN is grown based on a reward value given by a feedback function that analyses the on-line performance of a certain task. In fact since conventional NNs are commonly used in controlling tasks (Alnajjar & Murase, 2005), this is a potential application area for CoNN algorithms as well.

In (Giordano *et al.*, 2008), a committee of CasCor neural networks was implemented as a software filter, for the online filtering of CO_2 signals from a bioreactor gas outflow. The knowledge-based CasCor proposal (KBCC) previously mentioned has been used in a few knowledge domains, such as simulation of cognitive development (see e.g. (Mareschal & Schultz, 1999) and (Sirois & Shultz, 1998)), vowel recognition (Rivest & Shultz, 2002) and for gene-splice-junction determination (Thivierge & Shultz, 2002), a benchmark problem from the UCI Machine Learning Repository (Asuncion & Newman, 2007). A more in depth investigation into the use of the knowledge-based neural learning implemented by the KBCC in developmental robotics can be seen in (Shultz *et al.*, 2007).

A few other non-conventional approaches to CoNN can be found in recent works, such as the one described in (García-Pedrajas & Ortiz-Boyer, 2007), based on cooperative co-evolution, for the automatic induction of the structure of an NN for classification purposes; the method partially tries to avoid the problems of greedy approaches. In (Yang *et al.*, 2008) the authors combined the ridgelet function with feedforward neural networks in the ICRNN (Incremental Constructive Ridgelet Neural Network) model. The ridgelet function was chosen as the activation function due to its efficiency in describing linear-, curvilinear- and hyperplane-like structures in its hidden layer; the structure of the network is induced via a constructive method.

The CoNN classifier known as the Recursive Deterministic Perceptron (RDP) (Tajine & Elizondo, 1996) is a generalization of the Perceptron, capable of solving any two-class classification problem. It works by transforming any non-linearly separable two-class problem into a linearly one, which can be easily learnt by the Perceptron. Its multiclass version (Tajine *et al.* 1997) is a generalization that allows separation of the m-classes in a deterministic way. Results show that in certain domains, both the multiclass version and the backpropagation have similar performance (Elizondo *et al.*, 2008).

The Switching Neural Network (SNN) is a connectionist model recently proposed in (Muselli, 2006) suitable for classification problems. The first layer of an SNN contains converters, called latticizers that change the representation of the input vectors into binary strings. The two other layers of the SNN represent a Boolean function that solves, in the lattice domain, the original classification problem. As proposed in (Ferrari & Muselli, 2008) the construction of an SNN can be done by a constructive algorithm known as Switch Programming (SP) which is based on solutions of a linear programming problem. Good simulation results suggest that this proposal is worthy of a deeper investigation.

The constructive proposals CLANN and its multiclass version M-CLANN described in (Tsopzé *et al.*, 2007) and (Nguifo *et al.*, 2008) respectively are based on concept lattices and aim at a semantic interpretation of the involved neurons and consequently at an 'interpretable' (in the sense of comprehensibility) neural network. CLANN and M-CLANN can be approached as representation-translators, in the same sense as the KBANN model is (Towel *et al.*, 1990), (Towel, 1991).

A different approach to CoNN can be found in (Barreto-Sanz *et al.*, 2008), where the authors propose the FGHSON (Fuzzy Growing Hierarchical Self-Organizing Networks), an adaptive network method capable of representing the underlying structure of the data, in a hierarchical fuzzy way.

Transformation of original data features usually helps to find interesting low dimensional data that can reveal previously unseen structures. This process aims to ease the problem for a classifier. The simplest of these transformations is the linear projection. Many methods search for the optimal and the most informative linear projection. Friedman (Friedman, 1987) proposed a framework to find interesting data transformations by maximizing an index of projection pursuit. Grochowski and Duch in (Grochowski & Duch, 2008) proposed the QPC network, a constructive neural network that can implement this framework. The algorithm introduces a new index based on the quality of projected clusters that can be used to define specific representations for the hidden layer of a neural network and may help to construct the network.

The recently introduced C-Mantec algorithm (Subirats, Franco et al, 2008) that works by error correction using the thermal perceptron (Frean, 1992) incorporates competition between neurons in the hidden layer and it has been shown to lead to very compact architectures. The intrinsic dynamics of the algorithm has been applied for detecting and filtering noisy instances, reducing overfitting and improving the generalization ability.

7 Conclusions

This chapter presents an overview of several CoNN algorithms and highlighted some of their applications and contributions. Although focusing on feedforward architectures for classification tasks, the chapter also tries to present a broad view of the area, discussing several of the most recent contributions.

An interesting aspect of CoNN research is its chronological aspect. It may be noticeable that most of the CoNN algorithms for classification tasks were proposed in the nineties and since then not many new proposals have been published. Another point to consider also is the lack of real world applications involving the use of CoNN algorithms; this can be quite surprising, considering the many that are available and the fact that several have competitive performances in comparison to other more traditional approaches. The tendency in the area is for diversifying both the architecture and the constructive process itself, by means of including collaborative techniques. What has been surveyed in this chapter is just a part of the research work going on in the area of CoNN algorithms. As mentioned in the Introduction, there is a very promising area characterized as the group of evolutionary techniques that has been contributing a lot to the development of CoNNs and was not the subject of this chapter.

Acknowledgments. To FAPESP and CAPES for the support provided to M. C. Nicoletti and to J. R. Bertini Jr. respectively. L. Franco, D. Elizondo and J.M. Jerez acknowledge support from MICIIN through grant TIN2008-04985 and to Junta de Andalucía grants P06-TIC-01615 and P08-TIC-04026.

References

Alnajjar, F., Murase, K.: Self-organization of spiking neural network generating autonomous behavior in a real mobile robot. In: Proceedings of The International Conference on Computational Intelligence for Modeling, Control and Automation, vol. 1, pp. 1134–1139 (2005)

Amaldi, E., Guenin, B.: Two constructive methods for designing compact feedfoward networks of threshold units. International Journal of Neural System 8(5,6), 629–645 (1997)

Anlauf, J.K., Biehl, M.: The AdaTron: an adaptive perceptron algorithm. Europhysics Letters 10, 687–692 (1989)

Ash, T.: Dynamic node creation in backpropagation networks, Connection Science, vol. Connection Science 1(4), 365–375 (1989)

Asuncion, A., Newman, D.J.: UCI Machine Learning Repository. University of California, School of Information and Computer Science, Irvine (2007), `http://www.ics.uci.edu/~mlearn/MLRepository.html`

Barreto-Sanz, M., Pérez-Uribe, A., Peña-Reyes, C.-A., Tomassini, M.: Fuzzy growing hierarchical self-organizing networks. In: Kůrková, V., Neruda, R., Koutník, J. (eds.) ICANN 2008,, Part II. LNCS, vol. 5164, pp. 713–722. Springer, Heidelberg (2008)

Bertini Jr., J.R., Nicoletti, M.C.: A constructive neural network algorithm based on the geometric concept of barycenter of convex hull, Computational Intelligence: Methods and Applications. In: Rutkowski, R.L., Tadeusiewicz, R., Zadeh, L.A., Zurada, J. (eds.) Academic Publishing House Exit, pp. 1–12. IEEE Computational Intelligence Society, Poland (2008a)

Bertini Jr., J.R., Nicoletti, M.C.: MBabCoNN – A multiclass version of a constructive neural network algorithm based on linear separability and convex hull. In: Kůrková, V., Neruda, R., Koutník, J. (eds.) ICANN 2008,, Part II. LNCS, vol. 5164, pp. 723–733. Springer, Heidelberg (2008)

Bishop, C.M.: Neural networks for pattern recognition. Oxford University Press, USA (1996)

Burgess, N.: A constructive algorithm that converges for real-valued input patterns. Int. Journal of Neural Systems 5(1), 59–66 (1994)

Campbell, C.: Constructive learning techniques for designing neural network systems. In: Leondes, C. (ed.) Neural Network Systems Technologies and Applications, vol. 2. Academic Press, San Diego (1997)

Campbell, C., Vicente, C.P.: The target switch algorithm: a constructive learning procedure for feed-forward neural networks. Neural Computation 7(6), 1245–1264 (1995)

Chen, S., Billings, S., Luo, W.: Orthogonal least squares learning methods and their application to non-linear system identification. International Journal of Control 50, 1873–1896 (1989)

Chen, S., Cowan, C., Grant, P.M.: Orthogonal least squares learning algorithm for radial basis function networks. IEEE Transactions on Neural Networks 2, 302–309 (1991)

Drăghici, S.: The constraint based decomposition (CBD) training architecture. Neural Networks 14, 527–550 (2001)

Elizondo, D., Ortiz-de-Lazcano-Lobato, J.M., Birkenhead, R.: On the generalization of the m-class RDP neural network. In: Kůrková, V., Neruda, R., Koutník, J. (eds.) ICANN 2008, Part II. LNCS, vol. 5164, pp. 734–743. Springer, Heidelberg (2008)

Fahlman, S.E.: Faster-learning variations on backpropagation: an empirical study. In: Touretzky, D.S., Hinton, G.E., Sejnowski, T.J. (eds.) Proceedings of the 1988 Connectionist Models Summer School, pp. 38–51. Morgan Kaufmann, San Mateo (1988)

Fahlman, S., Lebiere, C.: The cascade correlation architecture, Advances in Neural Information Processing Systems, vol. 2, pp. 524–532. Morgan Kaufman, San Mateo (1990)

Fahlman, S.: The recurrent cascade-correlation architecture, Advances in Neural Information Processing Systems, vol. 3, pp. 190–196. Morgan Kaufman, San Mateo (1991)

Farlow, S.J. (ed.): Self-organizing methods in Modeling: GMDH Type Algorithms. In: Statistics: Textbooks and Monographs, vol. 54. Marcel Dekker, New York (1984)

Ferrari, E., Muselli, M.: A constructive technique based on linear programming for training switching neural networks. In: Kůrková, V., Neruda, R., Koutník, J. (eds.) ICANN 2008, Part II. LNCS, vol. 5164, pp. 744–753. Springer, Heidelberg (2008)

Fiesler, E.: Comparative bibliography of ontogenic neural networks. In: Proceedings of The International Conference on Artificial Neural Networks (ICANN), Sorrento, vol. 94, pp. 793–796 (1994)

Frean, M.: The upstart algorithm: a method for constructing and training feedforward neural networks. Neural Computation 2, 198–209 (1990)

Frean, M.: A thermal perceptron learning rule. Neural Computation 4, 946–957 (1992)

Friedman, J.H., Stuetzle, W.: Projection pursuit regression. Journal of the American Statistical Association 76(376), 817–823 (1981)

Friedman, J.: Exploratory projection pursuit. Journal of the American Statistical Association 82, 249–266 (1987)

Gallant, S.I.: Optimal linear discriminants. In: Proceedings of The Eighth International Conference on Pattern Recognition, pp. 849–852 (1986a)

Gallant, S.I.: Three constructive algorithms for network learning. In: Proceedings of The Eighth Annual Conference of the Cognitive Science Society, Amherst, Ma, pp. 652–660 (1986b)

Gallant, S.I.: Perceptron based learning algorithms. Proceedings of the IEEE Transactions on Neural Networks 1(2), 179–191 (1990)

Gallant, S.I.: Neural Network Learning & Expert Systems. The MIT Press, England (1994)

García-Pedrajas, N., Ortiz-Boyer, D.: A cooperative constructive method for neural networks for pattern recognition. Pattern Recognition 40, 80–98 (2007)

Ghosh, J., Tumer, K.: Structural adaptation and generalization in supervised feed-forward networks. Journal of Artificial Neural Networks 1(4), 431–458 (1994)

Giles, C.L., Chen, D., Sun, G.-Z., Chen, H.-H., Lee, Y.-C., Goudreau, M.W.: Constructive learning of recurrent neural network: limitations of recurrent cascade correlation and a simple solution. IEEE Transactions on Neural Networks 6(4), 829–836 (1995)

Giordano, R.C., Bertini Jr., J.R., Nicoletti, M.C., Giordano, R.L.C.: Online filtering of CO2 signals from a bioreactor gas outflow using a committee of constructive neural networks. Bioprocess and Biosystems Engineering 31(2), 101–109 (2008)

Grochowski, M., Duch, W.: Projection pursuit constructive neural networks based on quality of projected cluster. In: Kůrková, V., Neruda, R., Koutník, J. (eds.) ICANN 2008,, Part II. LNCS, vol. 5164, pp. 754–762. Springer, Heidelberg (2008)

Hrycej, T.: Modular Learning in Neural Networks. Addison Wiley, New York (1992)

Huemer, A., Elizondo, D., Gongora, M.: A reward-value based constructive method for the autonomous creation of machine controllers. In: Kůrková, V., Neruda, R., Koutník, J. (eds.) ICANN 2008,, Part II. LNCS, vol. 5164, pp. 773–782. Springer, Heidelberg (2008)

Islam, M.M., Murase, K.: A new algorithm to design compact two-hidden-layer artificial networks. Neural Networks 14, 1265–1278 (2001)

Karayiannis, N.B., Weiqun, G.: Growing radial basis neural networks: merging supervised and unsupervised learning with network growth techniques. IEEE Transactions on Neural Networks 8(6), 1492–1506 (1997)

Krauth, W., Mézard, M.: Learning algorithms with optimal stability in neural networks. Journal of Physics A 20, 745–752 (1987)

Kremer, S.: Comments on constructive learning of recurrent neural networks: limitations of recurrent cascade correlation and a simple solution. IEEE Transactions on Neural Networks 7(4), 1047–1049 (1996)

Kubat, M.: Designing neural network architectures for pattern recognition. The Knowledge Engineering Review 15(2), 151–170 (2000)

Kwok, T.-Y., Yeung, D.-Y.: Constructive algorithms for structure learning in feedforward neural networks for regression problems. IEEE Transactions on Neural Networks 8(3), 630–645 (1997a)

Kwok, T.-Y., Yeung, D.-Y.: Objective functions for training new hidden units in constructive neural networks. IEEE Transactions on Neural Networks 8(5), 1131–1148 (1997b)

Lahnajärvi, J.J.T., Lehtokangas, M.I., Saarinen, J.P.P.: Fast constructive methods for regression problems. In: Proceedings of the 18th IASTED International Conference on Modelling, Identification and Control (MIC 1999), Innsbruck, Austria, pp. 442–445 (1999a)

Lahnajärvi, J.J.T., Lehtokangas, M.I., Saarinen, J.P.P.: Comparison of constructive neural networks for structure learning. In: Proceedings of the 18th IASTED International Conference on Modelling, Identification and Control (MIC 1999), Innsbruck, Austria, pp. 446–449 (1999b)

Lahnajärvi, J.J.T., Lehtokangas, M.I., Saarinen, J.P.P.: Fixed cascade error – a novel constructive neural network for structure learning. In: Proceedings of the Artificial Neural Networks in Engineering Conference (ANNIE 1999), St. Louis, USA, pp. 25–30 (1999c)

Lahnajärvi, J.J.T., Lehtokangas, M.I., Saarinen, J.P.P.: Evaluation of constructive neural networks with cascaded architectures. Neurocomputing 48, 573–607 (2002)

Lahnajärvi, J.J.T., Lehtokangas, M.I., Saarinen, J.P.P.: Estimating movements of a robotic manipulator by hybrid constructive neural networks. Neurocomputing 56, 345–363 (2004)

Le Cun, T., Denker, J.S., Solla, S.A.: Optimal brain damage. In: Touretzky, D.S. (ed.) Advances in Neural Information Processing Systems, vol. 2, pp. 598–605. Morgan Kauffman, San Mateo (1990)

Lehtokangas, M.: Modelling with constructive backpropagation. Neural Networks 12, 707–716 (1999)

Ma, L., Khorasani, K.: A new strategy for adaptively constructing multilayer feedforward neural networks. Neurocomputing 51, 361–385 (2003)

Ma, L., Khorasani, K.: New training strategies for constructive neural networks with application to regression problems. Neural Networks 17, 589–609 (2004)

Marchand, M., Golea, M., Ruján, P.: A convergence theorem for sequential learning in two-layer perceptrons. Europhysics Letters 11(6), 487–492 (1990)

Marchand, M., Golea, M.: On learning simple neural concepts: from halfspace intersections to neural decision lists. Network: Computation in Neural Systems 4(1), 67–85 (1993)

Mareschal, D., Schultz, T.R.: Development of children´s seriation: a connectionist approach. Connection Science 11, 149–186 (1999)

Mascioli, F.M.F., Martinelli, G.: A constructive algorithm for binary neural networks: the oil-spot algorithm. IEEE Transaction on Neural Networks 6, 794–797 (1995)

Mayoraz, E., Aviolat, F.: Constructive training methods for feedforward neural networks with binary weights. International Journal of Neural Networks 7, 149–166 (1996)

Mézard, M., Nadal, J.: Learning feedforward networks: the tiling algorithm. J. Phys. A: Math. Gen. 22, 2191–2203 (1989)

Muselli, M.: Sequential constructive techniques. In: Leondes, C. (ed.) Neural Network Systems Techniques and Applications, vol. 2, pp. 81–144. Academic, San Diego (1998)

Muselli, M.: Switching neural networks: a new connectionist model for classification. In: Apolloni, B., Marinaro, M., Nicosia, G., Tagliaferri, R. (eds.) WIRN 2005 and NAIS 2005. LNCS, vol. 3931, pp. 23–30. Springer, Heidelberg (2006)

Nabhan, T.M., Zomaya, A.Y.: Toward generating neural network structures for function approximation. Neural Networks 7(1), 89–99 (1994)

Nadal, J.-P.: Study of a growth algorithm for a feedforward network. International Journal of Neural Systems 1(1), 55–59 (1989)
Nguifo, E.M., Tsopzé, N., Tindo, G.: M-CLANN: Multi-class concept lattice-based artificial neural network for supervised classification. In: Kůrková, V., Neruda, R., Koutník, J. (eds.) ICANN 2008,, Part II. LNCS, vol. 5164, pp. 812–821. Springer, Heidelberg (2008)
Nicoletti, M.C., Bertini Jr., J.R.: An empirical evaluation of constructive neural network algorithms in classification tasks. International Journal of Innovative Computing and Applications 1(1), 2–13 (2007)
Parekh, R.G., Yang, J., Honavar, V.: Constructive neural network learning algorithm for multi-category classification, TR ISU-CS-TR95-15a, Iowa State University, IA (1995)
Parekh, R.G., Yang, J., Honavar, V.: MUpstart a constructive neural network learning algorithm for multi-category pattern classification. In: Proceedings of the IEEE/INNS International Conference on Neural Networks (ICNN 1997), vol. 3, pp. 1924–1929 (1997a)
Parekh, R.G., Yang, J., Honavar, V.: Pruning strategies for the MTiling constructive learning algorithm. In: Proceedings of the IEEE/INNS International Conference on Neural Networks (ICNN 1997), 3rd edn., pp. 1960–1965 (1997b)
Parekh, R.G., Yang, J., Honavar, V.: Constructive neural-network learning algorithms for pattern classification. IEEE Transactions on Neural Networks 11(2), 436–451 (2000)
Platt, J.: A resource-allocating network for function interpolation. Neural Computation 3, 213–225 (1991)
Poulard, H.: Barycentric correction procedure: a fast method for learning threshold unit. In: Proceedings of WCNN 1995, vol. 1, pp. 710–713 (1995)
Poulard, H., Estève, D.: A convergence theorem for barycentric correction procedure, Technical Report 95180, LAAS-CNRS, Toulouse, France (1995)
Poulard, H., Labreche, S.: A new threshold unit learning algorithm, Technical Report 95504, LAAA-CNRS, Toulouse, France (1995)
Poulard, H., Hernandez, N.: Training a neuron in sequential learning. Neural Processing Letters 5, 91–95 (1997)
Prechelt, L.: PROBEN1 A set of neural-network benchmark problems and benchmarking rules, Fakultät für Informatik, Univ. Karlsruhe, Germany, Tech. Rep. 21/94 (1994)
Prechelt, L.: Investigation of the CasCor family of learning algorithms. Neural Networks 10(5), 885–896 (1997)
Ruján, P., Marchand, M.: Learning by minimizing resources in neural networks. Complex Systems 3, 229–241 (1989)
Reed, R.: Pruning algorithms – a survey. IEEE Transaction on Neural Networks 4(5), 740–747 (1993)
Rivest, F., Shultz, T.R.: Application of knowledge-based cascade-correlation to vowel recognition. In: Proceedings of The International Joint Conference on Neural Networks, pp. 53–58 (2002)
Schaffer, J.D., Whitely, D., Eshelman, L.J.: Combinations of genetic algorithms and neural networks: a survey of the state of the art. In: Proceedings of the International Workshop of Genetic Algorithms and Neural Networks, pp. 1–37 (1992)
Shultz, T.R., Rivest, F.: Knowledge-based cascade-correlation. In: Proceedings of The IEEE-INNS-ENNS International Joint Conference on Neural Networks, vol. 5, pp. 641–646 (2000)
Shultz, T.R., Rivest, F.: Knowledge-based cascade-correlation: using knowledge to speed learning. Connection Science 13, 1–30 (2001)

Shultz, T.R., Rivest, F., Egri, L., Thivierge, J.-P., Dandurand, F.: Could knowledge-based neural learning be useful in developmental robotics? The case of KBCC. International Journal of Humanoid Robotics 4(2), 245–279 (2007)

Sirois, S., Shultz, T.R.: Neural network modeling of developmental effects in discrimination shifts. Journal of Experimental Child Psychology 71, 235–274 (1998)

Smieja, F.J.: Neural network constructive algorithms: trading generalization for learning efficiency? Circuits, Systems and Signal Processing 12, 331–374 (1993)

Subirats, J.L., Jerez, J.M., Franco, L.: A new decomposition algorithm for threshold synthesis and generalization of Boolean Functions. IEEE Transactions on Circuits and Systems I 55, 3188–3196 (2008)

Subirats, J.L., Franco, L., Molina, I.A., Jerez, J.M.: Active learning using a constructive neural network algorithm. In: Kůrková, V., Neruda, R., Koutník, J. (eds.) ICANN 2008,, Part II. LNCS, vol. 5164, pp. 803–811. Springer, Heidelberg (2008)

Tajine, M., Elizondo, D.: Enhancing the perceptron neural network by using functional composition. Technical Report 96-07, Computer Science Department, Université Louis Pasteur, Strasbourg, France (1996)

Tajine, M., Elizondo, D., Fiesler, E., Korczak, J.: Adapting the 2-class recursive deterministic perceptron neural network to m-classes. In: Proceedings of The IEEE International Conference on Neural Networks (ICNN 1997), Los Alamitos, vol. 3, pp. 1542–1546 (1997)

Thivierge, J.-P., Dandurand, F., Shultz, T.R.: Transferring domain rules in a constructive network: introducing RBCC. In: Proceedings of The IEEE International Joint Conference on Neural Networks, vol. 2, pp. 1403–1408 (2004)

Thivierge, J.-P., Shultz, T.R.: Finding relevant knowledge: KBCC applied to DNA splice-junction determination. In: Proceedings of The IEEE International Joint Conference on Neural Networks, pp. 1401–1405 (2002)

Thivierge, J.-P., Rivest, F., Shultz, T.R.: A dual-phase technique for pruning constructive networks. In: Proceedings of The IEEE International Joint Conference on Neural Networks, vol. 1, pp. 559–564 (2003)

Towell, G.G., Shavlik, J.W., Noordewier, M.O.: Refinement of approximate domain theories by knowledge-based neural networks. In: Proceedings of the Eight National Conference on Artificial Intelligence, Boston, MA, pp. 861–866 (1990)

Tsopzé, N., Nguifo, E.M., Tindo, G.: Concept-lattice-based artificial neural network. In: Diatta, J., Eklund, P., Liquiére, M. (eds.) Proceedings of the Fifth International Conference on Concept Lattices and Applications (CLA 2007), Monpellier, France, pp. 157–168 (2007)

Towell, G.G.: Symbolic knowledge and neural networks: insertion, refinement and extraction. Doctoral dissertation, Madison, WI. University of Wisconsin, USA (1991)

Wang, D.: Fast constructive-covering algorithm for neural networks and its implement in classification. Applied Soft Computing 8, 166–173 (2008)

Yang, J., Parekh, R.G., Honavar, V.: MTiling – a constructive network learning algorithm for multi-category pattern classification. In: Proceedings of the World Congress on Neural Networks, pp. 182–187 (1996)

Yang, S., Wang, M., Jiao, L.: Incremental constructive ridgelet neural network. Neurocomputing 72, 367–377 (2008)

Yao, X.: Evolving neural networks. Proceedings of the IEEE 87(9), 1423–1447 (1999)

Young, S., Downs, T.: CARVE – a constructive algorithm for real-valued examples. IEEE Transactions on Neural Network 9(6), 1180–1190 (1998)

Wendmuth, A.: Learning the unlearnable. Journal of Physics A 28, 5423–5436 (1995)

Efficient Constructive Techniques for Training Switching Neural Networks

Enrico Ferrari and Marco Muselli

Abstract. In this paper a general constructive approach for training neural networks in classification problems is presented. This approach is used to construct a particular connectionist model, named Switching Neural Network (SNN), based on the conversion of the original problem in a Boolean lattice domain. The training of an SNN can be performed through a constructive algorithm, called *Switch Programming* (*SP*), based on the solution of a proper linear programming problem. Since the execution of SP may require excessive computational time, an approximate version of it, named *Approximate Switch Programming* (*ASP*) has been developed. Simulation results obtained on the StatLog benchmark show the good quality of the SNNs trained with SP and ASP.

Keywords: Constructive method, Switching Neural Network, Switch Programming, positive Boolean function synthesis, Statlog benchmark.

1 Introduction

Solving a classification problem consists of finding a function $g(\mathbf{x})$ capable of providing the most probable output in correspondence of any feasible input vector $\mathbf{x}$, when only a finite collection S of examples is available. Since the probability of misclassifying a pattern $\mathbf{x}$ is generally unknown, classification algorithms work by

Enrico Ferrari
Institute of Electronics, Computer and Telecommunication Engineering,
Italian National Research Council, Via De Marini, 6 - 16149 Genoa, Italy
e-mail: ferrari@ieiit.cnr.it

Marco Muselli
Institute of Electronics, Computer and Telecommunication Engineering,
Italian National Research Council, Via De Marini, 6 - 16149 Genoa, Italy
e-mail: muselli@ieiit.cnr.it

L. Franco et al. (Eds.): Constructive Neural Networks, SCI 258, pp. 25–48.
springerlink.com

minimizing at the same time the error on the available data and a measure of the complexity of g. As a matter of fact, according to the Occam razor principle and the main results in statistical learning theory [18] on equal training errors, a simpler function g has a higher probability of scoring a good level of accuracy in examples not belonging to S.

To pursue this double target any technique for the solution of classification problems must perform two different actions: choosing a class Γ of functions (*model definition*) and retrieving the best classifier $g \in \Gamma$ (*training phase*). These two tasks imply a trade-off between a correct description of the data and the generalization ability of the resulting classifier. In fact, if the set Γ is too large, it is likely to incur the problem of *overfitting*: the optimal classifier $g \in \Gamma$ has a good behavior in the examples of the training set, but scores a high number of misclassifications in the other points of the input domain. On the other hand, the choice of a small set Γ prevents retrieval of a function with a sufficient level of accuracy on the training set.

Backpropagation algorithms [17] have been widely used to train multilayer perceptrons: when these learning techniques are applied, the choice of Γ is performed by defining some topological properties of the net, such as the number of hidden layer and neurons. In most cases, this must be done without having any prior information about the problem at hand and several validation trials are needed to find a satisfying network architecture.

In order to avoid this problem, two different approaches have been introduced: pruning methods [16] and constructive techniques [10]. The former consider an initial trained neural network with a large number of neurons and adopt smart techniques to find and eliminate those connections and units which have a negligible influence on the accuracy of the classifier. However, training a large neural network may increase the computational time required to obtain a satisfactory classifier.

On the other hand, constructive methods initially consider a neural network including only the input and the output layers. Then, hidden neurons are added iteratively until a satisfactory description of the examples in the training set is reached. In most cases the connections between hidden and output neurons are decided before training, so that only a small part of the weight matrix has to be updated at each iteration. It has been shown [10] that constructive methods usually present a rapid convergence to a well-generalizing solution and allow also the treatment of complex training sets. Nevertheless, since the inclusion of a new hidden unit involves only a limited number of weights, it is possible that some correlations between the data in the training set may be missed.

Here, we will present a new connectionist model, called *Switching Neural Network (SNN)* [12], which can be trained in a constructive way while achieving generalization errors comparable to those of best machine learning techniques. An SNN includes a first layer containing a particular kind of A/D converter, called *latticizers*, which suitably transform input vectors into binary strings. Then, the subsequent two layers compute a positive Boolean function that solves in a lattice domain the original classification problem.

Since it has been shown [11] that positive Boolean functions can approximate any measurable function within any desired precision, the SNN model is sufficiently rich to treat any real-world classification problem. A constructive algorithm, called *Switch Programming* (*SP*) has been proposed [3] for training an SNN. It is based on integer linear programming and can lead to an excessive computational burden when a complex training set is analyzed. To allow a wider application of SNN, a suboptimal method, named *Approximate Switch Programming* (*ASP*) will be introduced here. Preliminary results on the Statlog benchmark [9] show that ASP is able to considerably reduce the execution time while keeping a high degree of accuracy in the resulting SNN.

The chapter is organized as follows. In Sec. 2 the considered classification problem is formalized, whereas in Sec. 3 a general schema for a wide class of constructive methods is presented. The SNN model is presented in Sec. 4 and in Sec. 5 the general schema introduced in Sec. 2 is employed to describe the SP and the ASP algorithms.

Sec. 6 shows how it is possible obtain a set of intelligible rules starting from any trained SNN, whereas Sec. 7 illustrates a very simple example with the purpose of making clear the functioning of an SNN. Finally, Sec. 8 presents the good results obtained with SP and ASP algorithms on the well-known datasets of the Statlog benchmark. Some concluding remarks end the chapter.

2 Problem Setting

Consider a general binary classification problem, where d-dimensional patterns $\mathbf{x}$ are to be assigned to one of two possible classes, labeled with the values of a Boolean output variable $y \in \{0,1\}$. According to possible situations in real world problems, the type of the components x_i, $i = 1,\ldots,d$, may be one of the following:

- **continuous ordered**: when x_i can assume values inside an uncountable subset U_i of the real domain $\mathbb{R}$; typically, U_i is an interval $[a_i,b_i]$ (possibly open at one end or at both the extremes) or the whole $\mathbb{R}$.
- **discrete ordered**: when x_i can assume values inside a countable set C_i, where a total ordering is defined; typically, C_i is a finite subset of $\mathbb{Z}$.
- **nominal**: when x_i can assume values inside a finite set H_i, where no ordering is defined; for example, H_i can be a set of colors or a collection of geometric shapes.

If x_i is a binary component, it can be viewed as a particular case of discrete ordered variable or as a nominal variable; to remove a possible source of ambiguity, a binary component will always be considered henceforth as a nominal variable.

Denote with I_m the set $\{1,2,\ldots,m\}$ of the first m positive integers; when the domain C_i of a discrete ordered variable is finite, it is isomorphic to I_m, with $m = |C_i|$, being $|A|$ the cardinality of the set A. On the other hand, if C_i is infinite, it can be shown to be isomorphic to the set $\mathbb{Z}$ of the integer numbers (if C_i is neither lower nor upper bounded), to the set $\mathbb{N}$ of the positive integers (if C_i is lower bounded), or to the set $\mathbb{Z} \setminus \mathbb{N}$ (if C_i is upper bounded).

For each of the above three types of components x_i a proper metric can be defined. In fact, many different distances have been proposed in the literature to characterize the subsets of $\mathbb{R}$. Also when the standard topology is assumed, different equivalent metrics can be adopted; throughout this paper, the *absolute metric* $d_a(v,v') = |v - v'|$ will be employed to measure the dissimilarity between two values $v, v' \in \mathbb{R}$ assumed by a continuous ordered component x_i.

In the same way, when x_i is a discrete ordered component, the above cited isomorphism between its domain C_i and a suitable subset K of $\mathbb{Z}$ makes it possible to adopt the distance $d_c(v,v')$ induced on C_i by the absolute metric $d_a(v,v') = |v - v'|$ on K. Henceforth we will use the term *counter metric* to denote the distance d_c.

Finally, if x_i is a nominal component no ordering relation exists between any pair of elements of its domain H_i; we can only assert that a value v assumed by x_i is equal or not to another value v'. Consequently, we can adopt in H_i the *flat metric* $d_f(v,v')$ defined as

$$d_f(v,v') = \begin{cases} 0 \text{ if } v = v' \\ 1 \text{ if } v \neq v' \end{cases}$$

for every $v, v' \in H_i$. Note that, by considering a counting function η which assigns a different positive integer in I_m to each element of the set H_i, being $m = |H_i|$, we can substitute the domain H_i of x_i with the set I_m without affecting the given classification problem. It is sufficient to employ the flat metric $d_f(v,v')$ also in I_m, which is no longer seen as an ordered set.

According to this framework, to simplify the exposition we suppose henceforth that the patterns $\mathbf{x}$ of our classification problem belong to a set $X = \prod_{i=1}^{d} X_i$, where each monodimensional domain X_i can be a subset of the real field $\mathbb{R}$ if x_i is a continuous ordered variable, a subset of integers in $\mathbb{Z}$ if x_i is a discrete ordered component, or the finite set I_m (for some positive integer m) without ordering on it if x_i is a nominal variable.

A proper metric $d_X(\mathbf{x},\mathbf{x}')$ on X can be simply defined by summing up the contributions $d_i(x_i,x_i')$ given by the different components

$$d_X(\mathbf{x},\mathbf{x}') = \sum_{i=1}^{d} d_i(x_i,x_i') \,, \quad \text{for any } \mathbf{x},\mathbf{x}' \in X$$

where d_i is the absolute metric d_a if x_i and x_i' are (continuous or discrete) ordered variables or the flat metric d_f if x_i and x_i' are nominal variables.

The target of a binary classification problem is to choose within a predetermined set Γ of decision functions the classifier $g : X \to \{0,1\}$ that minimizes the number of misclassifications on the whole set X. If Γ is equal to the collection $\mathscr{M}$ of all the measurable decision functions, this amounts to selecting the *Bayes classifier* $g_{\text{opt}}(\mathbf{x})$ [2]. On the other hand, if Γ is a proper subset of $\mathscr{M}$, the optimal decision function corresponds to the classifier $g^* \in \Gamma$ that best approximates g_{opt} according to a proper distance in $\mathscr{M}$.

Unfortunately, in real world problems we have only access to a *training set* S, i.e. a collection of s observations $(\mathbf{x}_k, y_k)$, $k = 1,\ldots,s$, for the problem at hand.

Thus, the solution to the binary classification problem is produced by adopting a *learning algorithm* $\mathscr{A}$ that employs the information contained in the training set to retrieve the best classifier g^* in Γ or a good approximation $\hat{g}$ to it.

This approach consists therefore of two different stages:

1. at first the class Γ of decision functions must be suitably determined (*model selection*);
2. then, the best classifier $g^* \in \Gamma$ (or a good approximation $\hat{g}$) is retrieved through the learning algorithm $\mathscr{A}$ (*training phase*).

In the next section we will introduce a general constructive model, which is sufficiently rich to approximate within an arbitrary precision any measurable function $g : X \to \{0,1\}$.

3 A General Structure for a Class of Constructive Methods

Denote with $S_1 = \{\mathbf{x}_k \mid y_k = 1\}$ the set of positive examples in the training set S and with $S_0 = \{\mathbf{x} \mid \; y_k = 0\}$ the set of negative examples. Moreover, let $s_1 = |S_1|$ and $s_0 = |S_0|$.

In many constructive methods the function $\hat{g}$ is realized by a two layer neural network; the hidden layer is built incrementally by adding a neuron at each iteration of the training procedure. In order to characterize the hidden neurons consider the following

Definition 1. A collection $\{\{L_h, \hat{y}_h\}, \, h = 1, \ldots, t+1\}$, where $L_h \subset X$ and $\hat{y}_h \in \{0,1\}$ for each $h = 1, \ldots, t$, will be called a *decision list* for a two class problem if $L_{t+1} = X$.

In [10] the decision list is used hierarchically: a pattern $\mathbf{x}$ is assigned to the class y_h, where h is the lower index such that $\mathbf{x} \in L_h$. It is possible to consider more general criteria in the output assignment: for example a weight $w_h > 0$ can be associated with each domain L_h, measuring the reliability of assigning the output value $\hat{y}_h$ to every point in L_h.

It is thus possible to associate with every pattern $\mathbf{x}$ a weight vector $\mathbf{u}$, whose h-th component is defined by

$$u_h = \begin{cases} w_h & \text{if } \mathbf{x} \in L_h \\ 0 & \text{otherwise} \end{cases}$$

for $h = 1, \ldots, t$. The weight u_h can be used to choose the output for the pattern $\mathbf{x}$. Without loss of generality suppose that the decision list is ordered so that $\hat{y}_h = 0$ for $h = 1, \ldots, t_0$, whereas $\hat{y}_h = 1$ for $h = t_0 + 1, \ldots, t_0 + t_1$, where $t_0 + t_1 = t$. The value of $\hat{y}_{t+1}$ is the *default* decision, i.e. the output assigned to $\mathbf{x}$ if $\mathbf{x} \notin L_h$ for each $h = 1, \ldots, t$.

In order to fix a criterion in the output assignment for an input vector $\mathbf{x}$ let us present the following

Definition 2. A function $\sigma(\mathbf{u}) \in \{0,1\}$ is called an *output decider* if

$$\sigma(\mathbf{u}) = \begin{cases} y_{t+1} & \text{if } u_1 = \ldots = u_{t_0} = u_{t_0+1} = \ldots = u_t = 0 \\ 1 & \text{if } u_1 = \ldots = u_{t_0} = 0 \text{ and some } u_h > 0 \text{ with } t_0 < h \le t \\ 0 & \text{if } u_{t_0+1} = \ldots = u_t = 0 \text{ and some } u_h > 0 \text{ with } 0 < h \le t_0 \end{cases}$$

This classifier can then be implemented in a two layer neural network: the first layer retrieves the weights u_h for $h = 1,\ldots,t$, whereas the second one realizes the output decider σ. The behavior of σ is usually chosen *a priori* so that the training phase consists of finding a proper decision list and the relative weight vector $\mathbf{w}$. For example, σ can be made equivalent to a comparison between the sum of the weights of the two classes:

$$\sigma(\mathbf{u}) = \begin{cases} 0 & \text{if } \sum_{h=1}^{t_0} u_h > \sum_{h=t_0+1}^{t} u_h \\ 1 & \text{if } \sum_{h=1}^{t_0} u_h < \sum_{h=t_0+1}^{t} u_h \\ \hat{y}_{t+1} & \text{otherwise} \end{cases}$$

The determination of the decision list $\{L_h, \hat{y}_h\}$, $h = 1,\ldots,t$, can be performed in a constructive way, by adding at each iteration h the best pair $\{L_h, \hat{y}_h\}$ according to a smart criterion. Each domain L_h corresponds to a neuron characterized through the function introduced by the following

Definition 3. Consider a subset $T \subset S_y$, $y \in \{0,1\}$. The function

$$\hat{g}_h(\mathbf{x}) = \begin{cases} 1 & \text{if } \mathbf{x} \in L_h \\ 0 & \text{otherwise} \end{cases}$$

is called a *partial classifier* for T if $T \cap L_h$ is not empty whereas $L_h \cap \backslash S_{1-y} = \emptyset$. If $\hat{g}_h(\mathbf{x}) = 1$ the h-th neuron will be said to *cover* $\mathbf{x}$.

The presence of noisy data can also be taken into account by allowing a small number of errors in the training set. To this aim Def. 3 can be generalized by the following

Definition 4. Consider a subset $T \subset S_y$, $y \in \{0,1\}$. The function

$$\hat{g}_h(\mathbf{x}) = \begin{cases} 1 & \text{if } \mathbf{x} \in L_h \\ 0 & \text{otherwise} \end{cases}$$

is called a *partial classifier with error* ε for T if $T \cap L_h$ is not empty whereas $|L_h \cap S_{1-y}| \le \varepsilon |S_{1-y}|$.

It is easy to notice that Def. 3 is recovered when setting $\varepsilon = 0$.

Since the target of the training phase is to find the simplest network satisfying the input-output relations in the training set, the patterns already covered by at least one neuron can be ignored when training further neurons having the same value of $\hat{y}_h$.

Fig. 1 shows a general constructive procedure for training a neural network in the case of binary output. At each iteration the set T contains the patterns belonging to

Constructive training for a two layer perceptron

For $y \in \{0,1\}$ do

1. Set $T = S_y$ and $h = 1$.
2. Find a partial classifier $\hat{g}_h$ for T.
3. Let $W = \{\{\mathbf{x}_k, y_k\} \in T \mid \hat{g}_h(\mathbf{x}_k) = 1\}$.
4. Set $T = T \setminus W$ and $h = h + 1$.
5. If T is nonempty go to step 2.
6. Prune redundant neurons and set $t_y = h$.

Fig. 1 General constructive procedure followed for neural network training.

the current output value not covered by the neurons already included in the network. Notice that removing elements from T allows a considerable reduction of the training time for each neuron since a lower number of examples has to be processed at each iteration.

A pruning phase is performed at the end of the training process in order to eliminate redundant overlaps among the sets L_h, $h = 1, \ldots, t$.

Without entering into details about the general theoretical properties of constructive techniques, which can be found in [10], in the following sections we will present the architecture of Switching Neural Networks and an appropriate training algorithm.

4 Switching Neural Networks

A promising connectionist model, called Switching Neural Network (SNN), has been developed recently[12]. According to this model, the input variables are transformed into n-dimensional Boolean strings by means of a particular mapping $\varphi : X \to \{0,1\}^n$, called *latticizer*.

Consider the Boolean lattice $\{0,1\}^n$, equipped with the well known binary operations '$+$' (*logical sum* or OR) and '$\cdot$' (*logical product* or AND). To improve readability, the elements of this Boolean lattice will be denoted henceforth as strings of bits: in this way, the element $(0,1,1,0) \in \{0,1\}^4$ will be written as 0110. The usual priority on the execution of the operators $+$ and $\cdot$ will be adopted; furthermore, when there is no possibility of misleading, the symbol $\cdot$ will be omitted, thus writing $\mathbf{v}\mathbf{v}'$ instead of $\mathbf{v} \cdot \mathbf{v}'$.

A standard partial ordering on $\{0,1\}^n$ can be defined by setting $\mathbf{v} \leq \mathbf{v}'$ if and only if $\mathbf{v} + \mathbf{v}' = \mathbf{v}'$; this definition is equivalent to writing $\mathbf{v} \leq \mathbf{v}'$ if and only if $v_i \leq v'_i$ for every $i = 1, \ldots, n$. According to this ordering, a Boolean function

$f : \{0,1\}^n \to \{0,1\}$ is called *positive* (resp. *negative*) if $\mathbf{v} \leq \mathbf{v}'$ implies $f(\mathbf{v}) \leq f(\mathbf{v}')$ (resp. $f(\mathbf{v}) \geq f(\mathbf{v}')$) for every $\mathbf{v}, \mathbf{v}' \in \{0,1\}^n$. Positive and negative Boolean functions form the class of *monotone Boolean functions*.

Since the Boolean lattice $\{0,1\}^n$ does not involve the complement operator NOT, Boolean expressions developed in this lattice (sometimes called *lattice expressions*) can only include the logical operations AND and OR. As a consequence, not every Boolean function can be written as a lattice expression. It can be shown that only positive Boolean functions are allowed to be put in the form of lattice expressions.

A recent theoretical result [11] asserts that positive Boolean functions are universal approximators, i.e. they can approximate every measurable function $g : X \to \{0,1\}$, being X the domain of a general binary classification problem, as defined in Sec. 2. Denote with Q_n^l the subset of $\{0,1\}^n$ containing the strings of n bits having exactly l values 1 inside them. A possible procedure for finding the positive Boolean function f that approximates a given g within a desired precision is based on the following three steps:

1. (*Discretization*) For every ordered input x_i, determine a finite partition $\mathscr{B}_i$ of the domain X_i such that a function $\hat{g}$ can be found, which approximates g on X within the desired precision and assumes a constant value on every set $B \in \mathscr{B}$, where $\mathscr{B} = \{\prod_{i=1}^d B_i : B_i \in \mathscr{B}_i, i = 1, \ldots, d\}$.
2. (*Latticization*) By employing a proper function φ, map the points of the domain X into the strings of Q_n^l, so that $\varphi(\mathbf{x}) = \varphi(\mathbf{x}')$ if $\mathbf{x}$ and $\mathbf{x}'$ belong to the same set $B \in \mathscr{B}$, whereas $\varphi(\mathbf{x}) \neq \varphi(\mathbf{x}')$ if $\mathbf{x} \in B$ and $\mathbf{x}' \in B'$, being B and B' two different sets in $\mathscr{B}$.
3. (*Positive Boolean function synthesis*) Select a positive Boolean function f.

If g is completely known these three steps can be easily performed; the higher the required precision is, the finer the partitions $\mathscr{B}_i$ for the domains X_i must be. This affects the length n of the binary strings in Q_n^l, which has to be big enough to allow the definition of the 1-1 mapping φ.

If $\mathbf{a} \in \{0,1\}^n$, let $P(\mathbf{a})$ be the subset of $I_n = \{1, \ldots, n\}$ including the indexes i for which $a_i = 1$. It can be shown [14] that a positive Boolean function can always be written as

$$f(\mathbf{z}) = \bigvee_{\mathbf{a} \in A} \bigwedge_{j \in P(\mathbf{a})} z_j \tag{1}$$

where A is an *antichain* of the Boolean lattice $\{0,1\}^n$, i.e. a set of Boolean strings such that neither $\mathbf{a} < \mathbf{a}'$ nor $\mathbf{a}' < \mathbf{a}$ holds for each $\mathbf{a}$, $\mathbf{a}' \in A$. It can be proved that a positive Boolean function is univocally specified by the antichain A, so that the task of retrieving f can be transformed into searching for a collection A of strings such that $\mathbf{a}' < \mathbf{a}$ for each $\mathbf{a}$, $\mathbf{a}' \in A$.

The symbol $\bigvee$ (resp. $\bigwedge$) in (1) denotes a logical sum (resp. product) among the terms identified by the subscript. The logical product $\bigwedge_{j \in P(\mathbf{a})} z_j$ is an *implicant* for the function f; however, when no confusion arises, the term implicant will also be used to denote the corresponding binary string $\mathbf{a} \in A$.

Preliminary tests have shown that a more robust method for classification problems consists of defining a positive Boolean function f_y (i.e. an antichain A_y to be

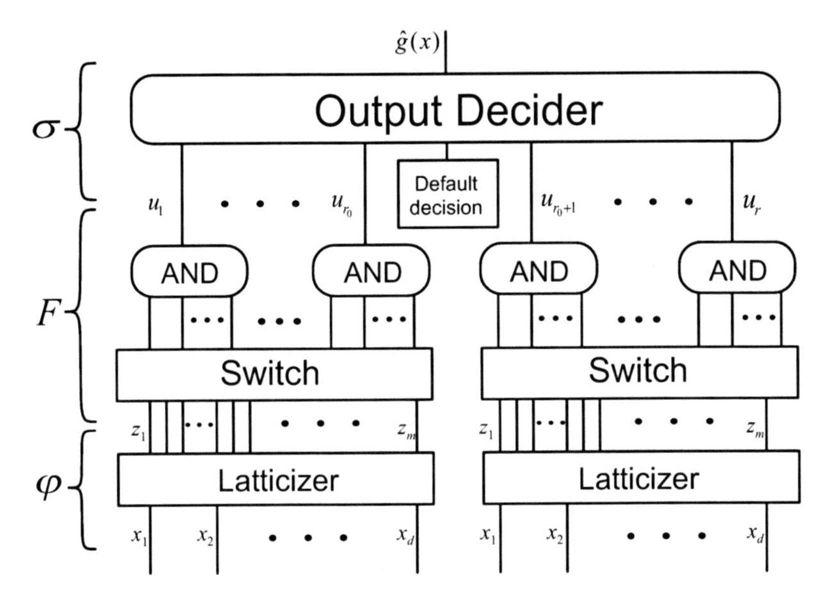

Fig. 2 The schema of a Switching Neural Network

inserted on (1)) for each output value y and in properly combining the functions relative to the different output classes. To this aim, each generated implicant can be characterized by a weight $w_h > 0$, which measures its significance level for the examples in the training set. Thus, to each Boolean string $\mathbf{z}$ can be assigned a weight vector $\mathbf{u}$ whose h-th component is

$$u_h = F_h(\mathbf{z}) = \begin{cases} w_h & \text{if } \bigwedge_{j \in P(\mathbf{a}_h)} z_j = 1 \\ 0 & \text{otherwise} \end{cases}$$

where $\mathbf{a}_h \in A_0 \cup A_1$ for $h = 1, \ldots, t$.

At the final step of the classification process, an output decider $\sigma(\mathbf{u})$ assigns the correct class to the pattern $\mathbf{z}$ according to a comparison between the weights u_h of the different classes. If no h exists such that $u_h > 0$, the default output is assigned to $\mathbf{z}$.

The device implementing the function $\hat{g}(\mathbf{x}) = \sigma(\mathbf{F}(\varphi(\mathbf{x})))$ is shown in Fig. 2. It can be considered a three layer feedforward neural network. The first layer is responsible for the latticization mapping φ; the second realizes the function $\mathbf{F}$ assigning a weight to each implicant. Finally, the third layer uses the weight vector $\mathbf{u} = \mathbf{F}(\mathbf{z})$ to decide the output value for the pattern $\mathbf{x}$.

Every AND port in the second layer is connected only to some of the outputs leaving the latticizers; they correspond to values 1 in the associated implicant. The choice of such values is performed by a *switch* port. For this reason the connectionist model shown in Fig. 2 is called *Switching Neural Network*.

Notice that the device can be subdivided into two parts: the left part includes the t_0 neurons characterizing the examples having output $y = 0$, whereas the right part involves the $t_1 = t - t_0$ implicants relative to the output $y = 1$. For this reason, the generalization of an SNN to a multiclass problem (where $y \in \{1,\ldots,c\}$, $c > 2$) is immediate: it is sufficient to create a set of implicants for each output value. In addition, the default decision has to be transformed into a *default decision list*, so that if two (or more) output values score the same level of confidence (according to the criterion fixed by σ), the device selects the one with a higher rank in the list.

It is interesting to observe that, unlike standard neural networks, SNNs do not involve floating point operations. In fact the weights $\mathbf{w}$ are provided by the training process and can be chosen from a set of integer values. Moreover, the antichain A can be converted into a set of intelligible rules in the form

if < *premise* > **then** < *consequence* >

through the application of a smart inverse operator [12] of φ to the elements of A.

4.1 Discretization

Since the exact behavior of the function g is not known, the approximating function $\hat{g}$ and the partition $\mathscr{B}$ have to be inferred from the samples $(\mathbf{x}_k, y_k) \in S$. It follows that at the end of the discretization task every set $B_i \in \mathscr{B}_i$ must be large enough to include the component x_{ki} of some point $\mathbf{x}_k$ in the training set. Nevertheless, the resulting partition $\mathscr{B}$ must be fine enough to capture the actual complexity of the function g.

Several different discretization methods for binary classification problems have been proposed in the literature [1, 5, 6, 7]. Usually, for each ordered input x_i a set of $m_i - 1$ consecutive values $r_{i1} < r_{i2} < \cdots < r_{i,m_i-1}$ is generated and the partition $\mathscr{B}_i$ is formed by the m_i sets $X_i \cap R_{ij}$, where $R_{i1} = (-\infty, r_{i1})$, $R_{i2} = (r_{i1}, r_{i2})$, $\ldots$, $R_{i,m_i-1} = (r_{i,m_i-2}, r_{i,m_i-1})$, $R_{im_i} = (r_{i,m_i-1}, +\infty)$. Excellent results have been obtained with the algorithms ChiMerge and Chi2 [5, 7], which employ the χ^2 statistic to decide the position of the points r_{ij}, $k = 1,\ldots,m_i - 1$, and with the technique Ent-MDL [6], which adopts entropy estimates to achieve the same goal. An alternative and promising approach is offered by the method used in the LAD system [1]: in this case an integer programming problem is solved to obtain optimal values for the cutoffs r_{ij}.

By applying a procedure of this kind, the discretization task defines for each ordered input x_i a mapping $\psi_i : X_i \to I_{m_i}$, where $\psi_i(z) = j$ if and only if $z \in R_{ij}$. If we assume that ψ_i is the identity function with $m_i = |X_i|$ when x_i is a nominal variable, the approximating function $\hat{g}$ is uniquely determined by a discrete function $h : I \to \{0,1\}$, defined by $h(\psi(\mathbf{x})) = \hat{g}(\mathbf{x})$, where $I = \prod_{i=1}^{d} I_{m_i}$ and $\psi(\mathbf{x})$ is the mapping from X to I, whose ith component is given by $\psi_i(x_i)$.

By definition, the usual ordering relation is induced by ψ_i on I_{m_i} when x_i is an ordered variable. On the other hand, since in general ψ_i is not 1-1, different choices for the metric on I_{m_i} are possible. For example, if the actual distances on X_i must be taken into account, the metric $d_i(j,k) = |r_{ij} - r_{ik}|$ can be adopted for any $j,k \in I_{m_i}$, having set $r_{i,m_i} = 2r_{i,m_i} - r_{i,m_i-1}$. Alternative definitions employ the mean points of the intervals R_{ik} or their lower boundaries.

According to statistical non parametric inference methods a valid choice can also be to use the absolute metric d_a on I_{m_i}, without caring about the actual value of the distances on X_i. This choice assumes that the discretization method has selected correctly the cutoffs r_{ij}, sampling with greater density the regions of X_i where the unknown function g changes more rapidly. In this way the metric d on $I = \prod_{i=1}^{d} I_{m_i}$ is given by

$$d(\mathbf{v}, \mathbf{v}') = \sum_{i=1}^{d} d_i(u_i, v_i)$$

where d_i is the absolute metric d_a (resp. the flat metric d_f) if x_i is an ordered (resp. nominal) input.

4.2 *Latticization*

It can be easily observed that the function ψ provides a mapping from the domain X onto the set $I = \prod_{i=1}^{d} I_{m_i}$, such that $\psi(\mathbf{x}) = \psi(\mathbf{x}')$ if $\mathbf{x}$ and $\mathbf{x}'$ belong to the same set $B \in \mathscr{B}$, whereas $\psi(\mathbf{x}) \neq \psi(\mathbf{x}')$ if $\mathbf{x} \in B$ and $\mathbf{x}' \in B'$, being B and B' two different sets in $\mathscr{B}$. Consequently, the 1-1 function φ from X to Q_n^l, required in the latticization step, can be simply determined by defining a proper 1-1 function β that maps the elements of I into the binary strings of Q_n^l. In this way, $\varphi(\mathbf{x}) = \beta(\psi(\mathbf{x}))$ for every $\mathbf{x} \in X$.

A possible way of constructing the function β is to define properly d mappings $\beta_i : I_{m_i} \rightarrow Q_{n_i}^{l_i}$; then, the binary string $\beta(\mathbf{v})$ for an integer vector $\mathbf{v} \in I$ is obtained by concatenating the strings $\beta_i(v_i)$ for $i = 1, \dots, d$. With this approach, $\beta(\mathbf{v})$ always produces a binary string with length $n = \sum_{i=1}^{d} n_i$ having $l = \sum_{i=1}^{d} l_i$ values 1 inside it.

The mappings β_i can be built in a variety of different ways; however, it is important that they fulfill the following two basic constraints in order to simplify the generation of an approximating function $\hat{g}$ that generalizes well:

1. β_i must be an *isometry*, i.e. $D_i(\beta_i(v_i), \beta_i(v_i')) = d_i(v_i, v_i')$, where $D_i(\cdot,\cdot)$ is the metric adopted on $Q_{n_i}^{l_i}$ and $d_i(\cdot,\cdot)$ is the distance on I_{m_i} (the absolute or the flat metric depending on the type of the variable x_i),
2. if x_i is an ordered input, β_i must be *full order-preserving*, i.e. $\beta_i(v_i) \preceq \beta_i(v_i')$ if and only if $v_i \leq v_i'$, where $\prec$ is a (partial or total) ordering on $Q_{n_i}^{l_i}$.

A valid choice for the definition of $\prec$ consists of adopting the *lexicographic ordering* on $Q_{n_i}^{l_i}$, which amounts to asserting that $\mathbf{z} \prec \mathbf{z}'$ if and only if $z_k < z_k'$ for some $k = 1, \dots, n_i$ and $z_i = z_i'$ for every $i = 1, \dots, k-1$. In this way $\prec$ is a total ordering on $Q_{n_i}^{l_i}$ and it can be easily seen that $Q_{n_i}^{l_i}$ becomes isomorphic to I_m with $m = \binom{n_i}{l_i}$. As a consequence the counter metric d_c can be induced on $Q_{n_i}^{l_i}$; this will be the definition for the distance D_i when x_i is an ordered input.

Note that if $l_i = n_i - 1$, binary strings in $Q_{n_i}^{l_i}$ contain a single value 0. Let us suppose that two elements $\mathbf{z}$ and $\mathbf{z}' \in Q_{n_i}^{n_i-1}$ have the value 0 at the jth and at the j' position, respectively; then, we have $\mathbf{z} \prec \mathbf{z}'$ if and only if $j < j'$. Moreover, the distance $D_i(\mathbf{z},\mathbf{z}')$ between $\mathbf{z}$ and $\mathbf{z}'$ is simply given by the absolute difference $|k - k'|$. As an example, consider for $n_i = 6$ the strings $\mathbf{z} = 101111$ and $\mathbf{z}' = 111101$, belonging to Q_6^5. The application of the above definitions gives $\mathbf{z} \prec \mathbf{z}'$ and $D_i(\mathbf{z},\mathbf{z}') = 3$, since the value 0 is at the 2nd place in $\mathbf{z}$ and at the 5th place in $\mathbf{z}'$.

If x_i is a nominal variable, the flat metric can also be adopted for the elements of $Q_{n_i}^{l_i}$, thus obtaining

$$D_i(\mathbf{z},\mathbf{z}') = \begin{cases} 0 \text{ if } \mathbf{z} = \mathbf{z}' \\ 1 \text{ if } \mathbf{z} \neq \mathbf{z}' \end{cases}, \quad \text{for every } \mathbf{z},\mathbf{z}' \in Q_{n_i}^{l_i}$$

With these definitions, a mapping β_i that satisfies the two above properties (isometry and full order-preserving) is the *inverse only-one code*, which maps an integer $v_i \in I_{m_i}$ into the binary string $\mathbf{z}_i \in Q_{m_i}^{m_i-1}$ having length m_i and jth component z_{ij} given by

$$z_{ij} = \begin{cases} 0 \text{ if } v_i = j \\ 1 \text{ otherwise} \end{cases}, \quad \text{for every } j = 1,\ldots,m_i$$

For example, if $m_i = 6$ we have $\beta_i(2) = 101111$ and $\beta_i(5) = 111101$.

It can be easily seen that the function β, obtained by concatenating the d binary strings produced by the components β_i, maps the integer vectors of I into the set Q_m^{m-d}, being $m = \sum_{i=1}^d m_i$. If the metric $D(\mathbf{z},\mathbf{z}') = \sum_{i=1}^d D_i(\mathbf{z}_i,\mathbf{z}_i')$ is employed on Q_m^{m-d}, where $\mathbf{z}_i$ is the binary string formed by the m_i bits of $\mathbf{z}$ determined by I_{m_i} through β_i, we obtain that the 1-1 mapping β is an isometry.

The behavior of the mapping β allows us to retrieve a convenient form for the 1-1 function φ from X to Q_n^l, to be introduced in the latticization step, if the discretization task has produced for each ordered input x_i a set of $m_i - 1$ cutoffs r_{ij}, as described in the previous subsection. Again, let $\varphi(\mathbf{x})$ be obtained by the concatenation of d binary strings $\varphi_i(x_i)$ in $Q_{m_i}^{m_i-1}$. To ensure that $\varphi(\mathbf{x}) = \beta(\psi(\mathbf{x}))$, it is sufficient to define the jth bit z_{ij} of $\mathbf{z}_i = \varphi_i(x_i)$ as

$$z_{ij} = \begin{cases} 0 \text{ if } x_i \in R_{ij} \\ 1 \text{ otherwise} \end{cases}, \quad \text{for every } j = 1,\ldots,m_i \tag{2}$$

if x_i is an ordered variable and as

$$z_{ij} = \begin{cases} 0 \text{ if } x_i = j \\ 1 \text{ otherwise} \end{cases}, \quad \text{for every } j = 1,\ldots,m_i \tag{3}$$

if x_i is a nominal input. Note that $x_i \in R_{ij}$ if and only if x_i exceeds the cutoff $r_{i,j-1}$ (if $j > 1$) and is lower than the subsequent cutoff r_{ij} (if $j < m_i$).

Consequently, the mapping φ_i can be implemented by a simple device that receives in input the value x_i and compares it with a sequence of integers or real numbers, according to definitions (2) or (3), depending on whether x_i is an ordered or a nominal input. This device will be called *latticizer*; it produces m_i binary outputs, but only one of them can assume the value 0. The whole mapping φ is realized by a parallel of d latticizer, each of which is associated with a different input x_i.

5 Training Algorithm

Many algorithms are present in literature to reconstruct a Boolean function starting from a portion of its truth table. However two drawbacks prevent the use of such techniques for the current purpose: these methods usually deal with general Boolean functions and not with positive ones and they lack generalization ability. In fact, the aim of most of these algorithms is to find a minimal set of implicants which satisfies all the known input-output relations in the truth table. However, for classification purposes, it is important to take into account the behavior of the generated function on examples not belonging to the training set. For this reason some techniques [1, 4, 13] have been developed in order to maximize the generalization ability of the resulting standard Boolean function. On the other hand, only one method, named *Shadow Clustering* [14], is expressly devoted to the reconstruction of positive Boolean functions.

In this section a novel constructive algorithm for building a single f_y (denoted only by f for simplicity) will be described. The procedure must be repeated for each value of the output y in order to find an optimal classifier for the problem at hand. In particular, if the function f_y is built, the Boolean output 1 will be assigned to the examples belonging to the class y, whereas the Boolean output 0 will be assigned to all the remaining examples.

The architecture of the SNN has to be constructed starting from the converted training set S', containing s_1 positive examples and s_0 negative examples. Let us suppose, without loss of generality, that the set S' is ordered so that the first s_1 examples are positive. Since the training algorithm sets up, for each output value, the switches in the second layer of the SNN, the constructive procedure of adding neurons step by step will be called *Switch Programming (SP)*.

5.1 Implicant Generation

When a Boolean string $\mathbf{z}$ is presented as input, the output of the logical product $\bigwedge_{j\in P(\mathbf{a})} z_j$ at a neuron is positive if and only if $\mathbf{a} \leq \mathbf{z}$ according to the standard ordering in the Boolean lattice. In this case $\mathbf{a}$ will be said to *cover* $\mathbf{z}$.

The aim of a training algorithm for an SNN is to find the simplest antichain A covering all the positive examples and no negative examples in the training set. This target will be reached in two steps: first an antichain A' is generated, then redundant elements of A' are eliminated thus obtaining the final antichain A. A constructive approach for constructing A' consists of generating implicants one at a time according to a smart criterion of choice determined by an objective function $\Phi(\mathbf{a})$.

In particular $\Phi(\mathbf{a})$ must take into account the number of examples in S' covered by $\mathbf{a}$ and the degree of complexity of $\mathbf{a}$, usually defined as the number of elements in $P(\mathbf{a})$ or, equivalently, as the sum $\sum_{i=1}^{m} a_i$. These parameters will be balanced in the objective function through the definition of two weights λ and μ.

In order to define the constraints to the problem, define, for each example $\mathbf{z}_k$, the number ξ_k of indexes i for which $a_i = 1$ and $z_{ki} = 0$. It is easy to show that $\mathbf{a}$ covers $\mathbf{z}_k$ if and only if $\xi_k = 0$. Then, the quantity

$$\sum_{k=1}^{s_1} \theta(\xi_k)$$

where θ represents the usual Heaviside function (defined by $\theta(u) = 1$ if $u > 0$, $\theta(u) = 0$ otherwise), is the number of positive patterns not covered by $\mathbf{a}$. However, it is necessary that $\xi_k > 0$ for each $k = s_1 + 1, \ldots, s$, so that any negative pattern is not covered by $\mathbf{a}$.

Starting from these considerations, the best implicant can be retrieved by solving the following optimization problem:

$$\begin{aligned}
&\min_{\xi,\mathbf{a}} \ \frac{\lambda}{s_1} \sum_{k=1}^{s_1} \xi_k + \frac{\mu}{m} \sum_{i=1}^{m} a_i \\
\text{subj to } &\sum_{i=1}^{m} a_i(a_i - z_{ki}) = \xi_k \quad \text{for } k = 1, \ldots, s_1 \\
&\sum_{i=1}^{m} a_i(a_i - z_{ki}) \geq 1 \quad \text{for } k = s_1 + 1, \ldots, s \\
&\xi_k \geq 0 \quad \text{for } k = 1, \ldots, s_1 \\
&a_i \in \{0, 1\} \quad \text{for } i = 1, \ldots, d
\end{aligned} \tag{4}$$

where the Heaviside function has been substituted by its argument in order to avoid nonlinearity in the cost function. Notice that the terms in the objective function are normalized in order to be independent of the complexity of the problem at hand.

Since the determination of a sufficiently great collection of implicants, from which the antichain A is selected, requires the repeated solution of problem (4), the generation of an already found implicant must be avoided at any extraction. This can be obtained by adding the following constraint

$$\sum_{i=1}^{m} a_{ji}(1 - a_i) \geq 1 \quad \text{for } j = 1, \ldots, q - 1 \tag{5}$$

where $\mathbf{a}$ is the implicant to be constructed and $\mathbf{a}_1, \ldots, \mathbf{a}_{q-1}$ are the already found $q - 1$ implicants.

Additional requirements can be added to problem (4) in order to improve the quality of the implicant $\mathbf{a}$ and the convergence speed. For example, in order to better differentiate implicants and to cover all the patterns in fewer steps, the set S_1'' of positive patterns not yet covered can be considered separately and weighted by a different factor $\nu \neq \lambda$.

Moreover, in the presence of noise it would be useful to avoid excessive adherence of $\mathbf{a}$ with the training set by accepting a small fraction ε of errors.

In this case a further term is added to the objective function, measuring the level of misclassification, and constraints in (4) have to be modified in order to allow at most εs_0 patterns to be misclassified by the implicant $\mathbf{a}$. In particular, slack variables ξ_k, $k = s_1 + 1, \dots, s$ are introduced such that $\xi_k = 1$ corresponds to a violated constraints (i.e. to a negative pattern covered by the implicant). For this reason the sum $\sum_{k=s_1+1}^{s} \xi$, which is just the number of misclassified patterns, must be less than $\varepsilon_0 s_0$. If the training set is noisy, the optimal implicant can be found by solving the following LP problem, where it is supposed that the first s'_1 positive patterns are not yet covered:

$$\begin{aligned}
\min_{\xi,\mathbf{a}} \quad & \frac{\nu}{s'_1} \sum_{k=1}^{s'_1} \xi_k + \frac{\lambda}{s_1 - s'_1} \sum_{k=s'_1+1}^{s_1} \xi_k + \frac{\mu}{m} \sum_{i=1}^{m} a_i + \frac{\omega}{s_0} \sum_{k=s_1+1}^{s} \xi_k \\
\text{subj to} \quad & \sum_{i=1}^{m} a_i(1 - z_{ki}) = \xi_k \quad \text{for } k = 1, \dots, s_1 \\
& \sum_{i=1}^{m} a_i(1 - z_{ki}) \geq 1 - \xi_k \quad \text{for } k = s_1 + 1, \dots, s \\
& \sum_{k=s_1+1}^{s} \xi_k \leq \varepsilon_0 s_0 \,, \quad a_i \in \{0,1\} \quad \text{for } i = 1, \dots, m \\
& \xi_k \geq 0 \quad \text{for } k = 1, \dots, s_1 \,, \quad \xi_k \in \{0,1\} \quad \text{for } k = s_1 + 1, \dots, s
\end{aligned} \tag{6}$$

If desired, only the implicants covering at least a fraction η_1 of positive examples may be generated. To this aim it is sufficient to add the following constraint

$$\sum_{k=1}^{s_1} \xi_k \leq (1 - \eta_1) s_1$$

Notice that further requirements have to be imposed when dealing with real-world problems. In fact, due to the coding (2) or (3) adopted in the latticization phase, only some implicants correspond to a condition consistent with the original inputs. In particular at least one zero must be present in the substring relative to each input variable.

5.2 *Implicant Selection*

Once the antichain A' has been generated, it will be useful to look for a subset A of A' which is able to describe the data in the training set with sufficient accuracy. To this aim both the number of implicants included in A and the number N_k of nonnull components in each element $\mathbf{a}_k \in A$ must be minimized.

Denote with $\mathbf{a}_1, \mathbf{a}_2, \ldots, \mathbf{a}_q$ the q implicants obtained in the generation step and with c_{kj} a binary variable asserting if the input vector $\mathbf{z}_k$, $k = 1, \ldots, s_1$, is covered by $\mathbf{a}_j$:

$$c_{kj} = \begin{cases} 1 & \text{if } \mathbf{z}_k \text{ is covered by } \mathbf{a}_j \\ 0 & \text{otherwise} \end{cases}$$

In addition, consider the binary vector ζ having as jth component the value $\zeta_j = 1$ if the corresponding implicant $\mathbf{a}_j$ is included in the final collection A. Then, an optimal subset $A \subset A'$ can be found by solving the following constrained optimization problem:

$$\begin{aligned} \min_{\zeta} \quad & \sum_{j=1}^{q} \zeta_j(\alpha + \beta N_j)) \\ \text{subj to} \quad & \sum_{j=1}^{q} c_{kj}\zeta_j \geq 1 \quad \text{for } k = 1, \ldots, s_1 \\ & \zeta_j \in \{0,1\} \quad \text{for } j = 1, \ldots, q \end{aligned} \tag{7}$$

where α and β are constants.

Additional requirements can be added to the problem (7) in order to improve the generalization ability of A. For example, if the presence of noise has to be taken into account, the antichain A can be allowed not to cover a small fraction of positive examples. In this case, the problem (7) becomes

$$\begin{aligned} \min_{\zeta} \quad & \sum_{j=1}^{q} \zeta_j(\alpha + \beta N_j)) \\ \text{subj to} \quad & \sum_{j=1}^{q} c_{kj}\zeta_j \geq 1 - \xi_k \quad \text{for } k = 1, \ldots, s_1 \\ & \sum_{j=1}^{q} c_{kj}\zeta_j \leq \xi_k \quad \text{for } k = s_1, \ldots, s \\ & \sum_{k=1}^{s_1} \xi_k \leq \varepsilon s_1 \qquad \sum_{k=s_1+1}^{s} \xi_k \leq \varepsilon s_0 \qquad \zeta_j \in \{0,1\} \quad \text{for } j = 1, \ldots, q \\ & \xi_j \in \{0,1\} \quad \text{for } j = 1, \ldots, s_1 \qquad \xi_j \geq 0 \quad \text{for } j = s_1, \ldots, s \end{aligned} \tag{8}$$

5.3 An Approximate Method for Solving the LP Problem

The solution of the problems (4) (or (6)) and (7) (or (8)) allows the generation of a minimal set of implicants for the problem at hand. Nevertheless, in the presence of a large amount of data, the number of variables and constraints for the LP problems increases considerably, thus making the SP algorithm very slow.

Conversion of the continuous solution into a binary vector

1 Set $a_i = 0$ for each $i = 1, \dots, m$.

2 While the constraints in (4) are violated

a. Select $\hat{\imath} = \arg\max_i a_i$.
b. Set $a_{\hat{\imath}} = 1$.

3. While the constraints in (4) are satisfied

a. Pick an index $\hat{\imath}$.
b. Set $a_{\hat{\imath}} = 0$.
c. If a constraints in (4) is violated, set $a_{\hat{\imath}} = 1$.

Fig. 3 A greedy method for converting a continuous solution of problem (4) into a binary vector.

In this subsection, an approximate algorithm able to reduce the execution time of the training algorithm for huge datasets will be introduced. The method will be described for the minimization problem (4), as its generalization to the case of (6), (7) or (8) is straightforward.

Most of the LP methods perform the minimization of a function $\Phi(\mathbf{a})$ through the following phases:

1. A continuous solution $\mathbf{a}_c$ is retrieved by suppressing the constraints $a_i \in \{0,1\}$.
2. Starting from $\mathbf{a}_c$ the optimal binary vector $\mathbf{a}$ is obtained through a branch and bound approach.

In particular, during phase 2, the algorithm must ensure that all the constraints of the original LP problem (4) are still satisfied by the binary solution. The search for an optimal integer solution can thus require the exploration of many combinations of input values. Preliminary tests have shown that, when the number of integer input variables in an LP problem increases, the time employed by Phase 2 may be much longer than that needed by Phase 1.

For this reason, an approximate algorithm will be proposed to reduce the number of combinations explored in the conversion of the continuous solution to the binary one.

Of course, the higher the number of 1s in a vector $\mathbf{a}$, the higher is the probability that it satisfies all the constraints in (4), since the number of covered patterns is usually smaller. Therefore, the vector $\mathbf{a}_c$ can be employed in order to retrieve a minimal subset of indexes to be set to one, starting from the assumption that a higher value of $(a_c)_i$ corresponds to a higher probability that a_i has to be set to 1 (and vice versa).

The algorithm for the conversion of the continuous solution to a binary one is shown in Fig. 3. The method starts by setting $a_i = 0$ for each i; then the a_i corresponding to the highest value of $(a_c)_i$ is set to 1. The procedure is repeated controlling at each iteration if the constraints (4) are satisfied. When no constraint is violated the procedure is stopped; smart lattice descent techniques [14] may be adopted to further reduce the number of active bits in the implicant.

These methods are based on the definition of proper criteria in the choice of the bit to be set to zero. Of course, when the implicant does not satisfy all the constraints, the bit is set to one again, the algorithm is stopped and the resulting implicant is added to the antichain A. The same approach may be employed in the the pruning phase, too.

The approximate version of the SP algorithm, obtained by employing the greedy procedure for transforming the continuous solution of each LP problem involved in the method into a binary one, is named *Approximate Switch Programming* (*ASP*).

6 Transforming the Implicants into a Set of Rules

If the discretization task described in Subsection 4.1 is employed to construct the latticizers, every implicant $\mathbf{a} \in \{0,1\}^m$ generated by SC can be translated into an intelligible rule underlying the classification at hand. This assertion can be verified by considering the substrings $\mathbf{a}_i$ of $\mathbf{a}$ that are associated with the ith input x_i to the network. The logical product $\bigwedge_{j \in P(\mathbf{a})} z_j$, performed by the AND port corresponding to $\mathbf{a}$, gives output 1 only if the binary string $\mathbf{z} = \varphi(\mathbf{x})$ presents a value 1 in all the positions where $\mathbf{a}_i$ has value 1.

If x_i is an ordered variable, this observation gives rise to the condition $x_i \in \bigcup_{j \in I_{m_i} \setminus P(\mathbf{a}_i)} R_{ij}$. However, in the analysis of real-world problems, the execution of SP and ASP is constrained to generate only binary strings $\mathbf{a}_i$ (for ordered variables) having a single sequence of consecutive values 0, often called a *run* of 0. In this case the above condition can simply be written in one of the following three ways:

- $x_i \le r_{ij}$, if the run of 0 begins at the first position and finishes at the jth bit of $\mathbf{a}_i$,
- $r_{ij} < x_i \le r_{ik}$, if the run of 0 begins at the $(j+1)$th position and finishes at the kth bit of $\mathbf{a}_i$,
- $x_i > r_{ij}$, if the run of 0 begins at the $(j+1)$th position and finishes at the last (m_ith) bit of $\mathbf{a}_i$.

As an example, suppose that an ordered variable x_i has been discretized by using the four cutoffs $0.1, 0.25, 0.3, 0.5$. If the implicant $\mathbf{a}$ with $\mathbf{a}_i = 10011$ has been produced by SC, the condition $0.1 < x_i \le 0.3$ has to be included in the **if** part of the **if-then** rule associated with $\mathbf{a}$.

On the other hand, if x_i is a nominal variable the portion $\mathbf{a}_i$ of an implicant $\mathbf{a}$ gives rise to the condition $x_i \in \bigcup_{j \in I_{m_i} \setminus P(\mathbf{a}_i)} \{j\}$. Again, if the implicant $\mathbf{a}$ with $\mathbf{a}_i = 01101$ has been produced by SC, the condition $x_i \in \{1,4\}$ has to be included in the **if** part of the **if-then** rule associated with $\mathbf{a}$. In any case, if the binary string $\mathbf{a}_i$ contains only values 0, the input x_i will not be considered in the rule for $\mathbf{a}$.

Thus, it follows that every implicant **a** gives rise to an **if-then** rule, having in its **if** part a conjunction of the conditions obtained from the substrings $\mathbf{a}_i$ associated with the d inputs x_i. If all these conditions are verified, to the output $y = \hat{g}(\mathbf{x})$ will be assigned the value 1.

Due to this property, SP and ASP (with the addition of discretization and latticization) become rule generation methods, being capable of retrieving from the training set some kind of intelligible information about the physical system underlying the binary classification problem at hand.

7 An Example of SNN Training

A simple example will be presented in this section in order to make the training of an SNN clearer. Consider the problem of forecasting the quality of the layer produced by a rolling mill starting from the knowledge of two continuous values: Pressure (x_1) and rolling Speed (x_2). The behavior of the rolling mill can be described by a function $g : \mathbb{R}^2 \to \{0,1\}$, whose output y may be either 0 (Bad layer) or 1 (Good layer). The aim of the classification task is therefore to realize a function $\hat{g}$ which constitutes a valid approximation for g starting from the training set S shown in Tab. 1.

As Tab. 1 shows, S is composed of 20 examples: 10 of those are Good and 10 are Bad. Suppose that the discretization process has subdivided the values of Pressure into three intervals $(-\infty, 1.63)$, $(1.63, 1.56)$, $(2.56, \infty)$, whereas the domain for Speed has been partitioned into 4 intervals $(-\infty, 3.27)$, $(3.27, 4.9)$, $(4.9, 6.05)$, $(6.05, \infty)$.

Through the discretization phase, it is possible to define a mapping $\psi : \mathbb{R}^2 \to I_3 \times I_4$, which associates two integer values v_1 and v_2 with each input pattern $\mathbf{x}$. Starting from the vector $\mathbf{v}$ obtained at the discretization step, the latticizer associates a binary string $\mathbf{z} = \varphi(\mathbf{x})$ with each pattern $\mathbf{x}$. Since the number of intervals for x_1 and x_2 is respectively 3 and 4, the latticizer produces a 7-dimensional binary string, obtained by concatenating the substrings relative to each input variable.

For example, the input pattern $\mathbf{x} = (3.12, 3.90)$ belongs to the interval $(2.56, \infty)$ for x_1 and to the interval $(3.27, 4.90)$ for x_2. Therefore the integer vector $\mathbf{v} = (3, 2)$ and the binary string 1101011, obtained through the inverse-only one coding, are associated with $\mathbf{x}$. Starting from the original data, it is possible to obtain a binary training set S', which is used to perform the classification. In fact the function $\hat{g}$ can be retrieved as $\hat{g}(\mathbf{x}) = \sigma(F(\varphi(\mathbf{x})))$.

Since the expected noise in the training data is negligible, the problem (6) can be solved by setting $\varepsilon = \omega = 0$, whereas a standard choice for the other coefficients in the cost function is given by $\lambda = \mu = \nu = 1$.

Suppose that the neurons for the output class 1 are generated initially. A first execution of the SP algorithm on the rolling mill problem produces the implicant 1000001. In fact it can be easily verified that this binary string satisfies all the constraints in (6) since it does not cover any example labelled by 0. Moreover, it covers 7 positive examples (all those after the third one) and has only two active bits, thus

Table 1 The original and transformed dataset for the problem of controlling the quality of a layer produced by a rolling mill. x_1 and x_2 are the original values for Pressure and Speed, v_1 and v_2 are the discretized values, whereas $\mathbf{z}$ is the binary string obtained through the latticizer. The quality of the resulting layer is specified by the value of the Boolean variable y.

x_1	x_2	v_1	v_2	$\mathbf{z}$	y
0.62	0.65	1	1	0110111	1
1.00	1.95	1	1	0110111	1
1.31	2.47	1	1	0110111	1
1.75	1.82	2	1	1010111	1
2.06	3.90	2	2	1011011	1
2.50	4.94	2	3	1011101	1
2.62	2.34	3	1	1100111	1
2.75	1.04	3	1	1100111	1
3.12	3.90	3	2	1101011	1
3.50	4.94	3	3	1011110	1
0.25	5.20	1	3	0111101	0
0.87	6.01	1	3	0111101	0
0.94	4.87	1	2	0111011	0
1.87	4.06	1	2	0111011	0
1.25	8.12	1	4	0111110	0
1.56	6.82	1	4	0111110	0
1.87	8.75	2	4	1011110	0
2.25	8.12	2	4	1011110	0
2.50	7.15	2	4	1011110	0
2.81	9.42	3	4	1101110	0

scoring a very low value of the objective function. Nevertheless, the first three examples are still to be covered, so the SP algorithm must be iterated.

The constraint (5) has to be added

$$(1 - a_1) + (1 - a_7) \geq 1$$

and the first three examples constituting the set S'' must be considered separately in the cost function (6).

A second execution of the SP algorithm generates the implicant 0000111, which covers 6 examples among which are the ones not yet covered. Therefore the antichain $A = \{1000001, 0000111\}$, corresponding to the PDNF $f(\mathbf{z}) = z_1 z_7 + z_5 z_6 z_7$, correctly describes all the positive examples. It is also minimal since the pruning phase cannot eliminate any implicant.

In a similar way an antichain is generated for the output class labelled by 0, thus producing the second layer of the SNN.

A possible choice for the weight u_h to be associated with the h-th neuron is given by its *covering*, i.e. the fraction of examples covered by it. For example, the weights associated with the neurons for the class 1 may be $u_1 = 0.7$ for 1000001 and $u_2 = 0.6$ for 0000111.

In addition the retrieved implicants can be transformed in intelligible rules involving the input variables. For example the implicant 1000001, associated with the output class 1, corresponds to the rule:

$$\textbf{if} \quad x_1 < 1.63 \text{ AND } x_2 > 6.05 \quad \textbf{then} \quad y = 1$$

8 Simulation Results

To obtain a preliminary evaluation of the performances achieved by SNNs trained with SP or ASP, the classification problems included in the well-known StatLog benchmark [9] have been considered. In this way the generalization ability and the complexity of resulting SNNs can be compared with those of other machine learning methods, among which are the backpropagation algorithm (BP) and rule generation techniques based on decision trees, such as C4.5 [15].

All the experiments have been carried out on a personal computer with an Intel Core Quad Q6600 (CPU 2.40 GHz, RAM 3 GB) running under the Windows XP operative system.

The tests contained in the Statlog benchmark presents different characteristics which allow the evaluation of different peculiarities of the proposed methods. In particular, four problems (Heart, Australian, Diabetes, German) have a binary output; two of them (Heart and German) are clinical datasets presenting a specific weight matrix which aims to reduce the number of misclassifications on ill patients. The remaining datasets present 3 (Dna), 4 (Vehicle) or 7 (Segment, Satimage, Shuttle) output classes. In some experiments, the results are obtained through a cross-validation test; however, in the presence of large amount of data, a single trial is performed since the time for many executions may be excessive.

The generalization ability of each technique is evaluated through the level of misclassification on a set of examples not belonging to the training set; on the other hand, the complexity of an SNN is measured using the number of AND ports in the second layer (corresponding to the number of intelligible rules) and the average number of conditions in the **if** part of a rule. Tab. 2 presents the results obtained on the datasets, reported in increasing order of complexity. Accuracy and complexity of resulting SNNs are compared to those of rulesets produced by C4.5. In the same table is also shown the best generalization error included in the StatLog report [9] for each problem, together with the rank scored by SNN when its generalization error is inserted into the list of available results.

The performances of the different techniques for training an SNN depend on the characteristics of the different problems. In particular the SP algorithm scores a better level of accuracy with respect to ASP in the datasets Heart and Australian. In fact, these problems are characterized by a small amount of data so that the execution of the optimal minimization algorithm may obtain a good set of rules within a reasonable execution time.

On the other hand, the misclassification of ASP is lower than that of SP in all the other problems (except for Shuttle), which are composed of a greater amount of

Table 2 Generalization error of SNN, compared with C4.5, BP and other methods, on the StatLog benchmark.

Test Problem	Generalization error					Rank	
	SP	ASP	C4.5	BP	Best	SP	ASP
HEART	0.439	0.462	0.781	0.574	0.374	6	9
AUSTRALIAN	0.138	0.141	0.155	0.154	0.131	3	3
DIABETES	0.246	0.241	0.270	0.248	0.223	7	5
VEHICLE	0.299	0.291	0.266	0.207	0.150	18	15
GERMAN	0.696	0.568	0.985	0.772	0.535	10	3
SEGMENT	0.0424	0.042	0.040	0.054	0.030	8	8
DNA	0.0658	0.056	0.076	0.088	0.041	7	3
SATIMAGE	0.168	0.149	0.150	0.139	0.094	19	10
SHUTTLE	0.0001	0.0001	0.001	0.43	0.0001	1	1

data. The decrease of the performances of SP in the presence of huge datasets is due to the fact that simplifications in the LP problem are necessary in order to make it solvable within a reasonable period of time. For example, some problems may be solved by setting $\varepsilon = \omega = 0$, since taking into account the possible presence of noise gives rise to an excessive number of constraints in (6).

Notice that in one case (Shuttle), SP and ASP achieve the best results among the methods in the StatLog archive, whereas in four other problems ASP achieves one of the first five positions. However, ASP is in the first ten positions in all the problems except for Vehicle.

Moreover, a comparison of the other methods reported in Tab. 2 with the best version of SNN for each problem illustrates that:

- Only in one case (Vehicle) the classification accuracy achieved by C4.5 is higher than that of SNN; in two problems (Satimage and Segment) the performances are similar, whereas in all the other datasets SNN scores significantly better results.
- In two cases (Vehicle and Satimage), BP achieves better results with respect to SNN; in all the other problems the performances of SNN are significantly better than those of BP.

These considerations highlight the good quality of the solutions offered by the SNNs, trained by the SP or ASP algorithm.

Nevertheless, the performances obtained by SP are conditioned by the number of examples s in the training set and by the number of input variables d. Since the number of constrains in (4) or (6) depends linearly on s, SP becomes slower and less efficient when dealing with complex training sets. In particular, the number of implicants generated by SP in many cases is higher than that of the rules obtained by C4.5, causing an increase in the training time.

However a smart combination of the standard optimization techniques with the greedy algorithm in Sec. 5.3 may allow complex datasets to be handled very efficiently. In fact the execution of ASP requires at most three minutes for each execution of the first six problems, about twenty minutes for the Dna dataset and about two hours for Satimage and Shuttle.

Notice that the minimization of (4) or (6) is obtained using the package *Gnu Linear Programming Kit (GLPK)* [8], a free library for the solution of linear programming problems. It is thus possible to improve the above results by adopting more efficient tools to solve the LP problem for the generation of implicants.

Concluding Remarks

In this paper a general schema for constructive methods has been presented and employed to train a Switching Neural Network (SNN), a novel connectionist model for the solution of classification problems. According to the SNN approach, the input-output pairs included in the training set are mapped to Boolean strings according to a proper transformation which preserves ordering and distance. These new binary examples can be viewed as a portion of the truth table of a positive Boolean function f, which can be reconstructed using a suitable algorithm for logic synthesis.

To this aim a specific method, named Switch Programming (SP), for reconstructing positive Boolean functions from examples has been presented. SP is based on the definition of a proper integer linear programming problem, which can be solved with standard optimization techniques. However, since the treatment of complex training sets with SP may require an excessive computational cost, a greedy version, named Approximate Swith Programming (ASP), has been proposed to reduce the execution time needed for training SNN.

The algorithms SP and ASP have been tested by analyzing the quality of the SNNs produced when solving the classification problems included in the Statlog archive. The results obtained show the good accuracy of classifiers trained with SP and ASP. In particular, ASP turns out to be very convenient from a computational point of view.

Acknowledgement. This work was partially supported by the Italian MIUR project "Laboratory of Interdisciplinary Technologies in Bioinformatics (LITBIO)".

References

1. Boros, E., Hammer, P.L., Ibaraki, T., Kogan, A., Mayoraz, E., Muchnik, I.: An implementation of Logical Analysis of Data. IEEE Trans. Knowledge and Data Eng. 9, 292–306 (2004)
2. Devroye, L., Györfi, L., Lugosi, G.: A probabilistic theory of Pattern Recognition. Springer, Heidelberg
3. Ferrari, E., Muselli, M.: A constructive technique based on linear programming for training switching neural networks. In: Kůrková, V., Neruda, R., Koutník, J. (eds.) ICANN 2008, Part II. LNCS, vol. 5164, pp. 744–753. Springer, Heidelberg (2008)
4. Hong, S.J.: R-MINI: An iterative approach for generating minimal rules from examples. IEEE Trans. Knowledge and Data Eng. 9, 709–717 (1997)

5. Kerber, R.: ChiMerge: Discretization of numeric attributes. In: Proc. 9th Int'l. Conf. on Art. Intell., pp. 123–128 (1992)
6. Kohavi, R., Sahami, M.: Error-based and Entropy based discretization of continuous features. In: Proc. 2nd Int. Conf. Knowledge Discovery and Data Mining, pp. 114–119 (1996)
7. Liu, H., Setiono, R.: Feature selection via discretization. IEEE Trans. on Knowledge and Data Eng. 9, 642–645 (1997)
8. Makhorin, A.: GNU Linear Programming Kit - Reference Manual (2008), `http://www.gnu.org/software/glpk/`
9. Michie, D., Spiegelhalter, D., Taylor, C.: Machine Learning, Neural and Statistical Classification. Ellis-Horwood, London (1994)
10. Muselli, M.: Sequential Constructive Techniques. In: Leondes, C. (ed.) Optimization Techniques, Neural Netw. Systems Techniques and Applications, pp. 81–144. Academic Press, San Diego (1998)
11. Muselli, M.: Approximation properties of positive boolean functions. In: Apolloni, B., Marinaro, M., Nicosia, G., Tagliaferri, R. (eds.) WIRN 2005 and NAIS 2005. LNCS, vol. 3931, pp. 18–22. Springer, Heidelberg (2006)
12. Muselli, M.: Switching neural networks: A new connectionist model for classification. In: Apolloni, B., Marinaro, M., Nicosia, G., Tagliaferri, R. (eds.) WIRN/NAIS 2005. LNCS, vol. 3931, pp. 23–30. Springer, Heidelberg (2006)
13. Muselli, M., Liberati, D.: Binary rule generation via Hamming Clustering. IEEE Trans. Knowledge and Data Eng. 14, 1258–1268 (2002)
14. Muselli, M., Quarati, A.: Reconstructing positive Boolean functions with Shadow Clustering. In: Proc. 17th European Conf. Circuit Theory and Design, pp. 377–380 (2005)
15. Quinlan, J.R.: C4.5: Programs for Machine Learning. Morgan Kaufmann, San Francisco (1994)
16. Reed, R.: Pruning Algorithm–A Survey. IEEE Trans. on Neural Netw. 4, 740–747 (1993)
17. Rumelhart, D.E., Hinton, G.E., Williams, R.J.: Learning representations by backpropagating errors. Nature 323, 533–536 (1988)
18. Vapnik, V.N.: Statistical learning theory. John Wiley & Sons, New York (1998)

Constructive Neural Network Algorithms That Solve Highly Non-separable Problems

Marek Grochowski and Włodzisław Duch

Abstract. Learning from data with complex non-local relations and multimodal class distribution is still very hard for standard classification algorithms. Even if an accurate solution is found the resulting model may be too complex for a given data and will not generalize well. New types of learning algorithms are needed to extend capabilities of machine learning systems to handle such data. Projection pursuit methods can avoid "curse of dimensionality" by discovering interesting structures in low-dimensional subspace. This paper introduces constructive neural architectures based on projection pursuit techniques that are able to discover simplest models of data with inherent highly complex logical structures. The key principle is to look for transformations that discover interesting structures, going beyond error functions and separability.

Keywords: Constructive neural networks, projection pursuit, non-separable problems, Boolean functions.

1 Introduction

Popular statistical and machine learning methods that rely solely on the assumption of local similarity between instances (equivalent to a smoothness prior) suffer from the curse of dimensionality [2]. When high-dimensional functions are not sufficiently smooth learning becomes very hard, unless extremly large number of training samples is provided. That leads to a dramatic increase in cost of computations and creates complex models which are hard to interpret. Many data mining problems in bioinformatics, text analysis and other areas, have inherent complex logic. Searching for the simplest possible model capable of representing that kind of data is still a great challenge that has not been fully addressed.

Marek Grochowski and Włodzisław Duch
Department of Informatics, Nicolaus Copernicus University, Toruń, Poland
e-mail: grochu@is.umk.pl, Google:WDuch

L. Franco et al. (Eds.): Constructive Neural Networks, SCI 258, pp. 49–70.
springerlink.com

One of the simplest examples of such hard problems is the n-bit parity problem. It has a very simple solution (one neuron with all weights $w_i = 1$ implementing a periodic transfer function with a single parameter [3]), but popular kernel methods and algorithms that depend only on similarity relations, or only on discrimination, have strong difficulties in learning this function. Linear methods fail completely, because this problem is highly non-separable. Gaussian-based kernels in SVMs use all training vectors as support vectors, because in case of parity function all points have closest neighbors from the opposite class. The nearest neighbor algorithms (with the number of neighbors smaller than $2n$) and the RBF networks have the same problem. For multilayer perceptrons convergence is almost impossible to achieve and requires many initiations to find accurate solution. Special feedforward neural network architectures have been proposed to handle parity problems [16, 30, 31, 28, 21] but they are designed only for this function and cannot be used for other Boolean functions, even very similar to parity.

Learning systems are frequently tested on benchmark datasets that are almost linearly separable and relatively simple to handle, but without a strong prior knowledge it is very hard to find satisfactory solution for really complex problems. One can estimate how complex a given data is using the k-separability index introduced in [3]. Consider a dataset $\mathscr{X} = \{\mathbf{x}_1, \ldots, \mathbf{x}_n\} \subset \mathscr{R}^d$, where each vector $\mathbf{x}_i$ belongs to one of the two classes.

Definition 1. Dataset $\mathscr{X}$ is called k-separable if a direction $\mathbf{w}$ exist such that all vectors projected on this direction $y_i = \mathbf{w}^T\mathbf{x}_i$ are clustered in k separated intervals, each containing instances from a single class only.

For example, datasets with two classes that can be separated by a single hyperplane have $k = 2$ and are thus 2-separable. XOR problem belongs to the 3-separable category, as projections have at least three clusters that contain even, odd and even instances (or odd, even and odd instances). n-bit parity problems are $n+1$-separable, because linear projection of binary strings exists that forms at least $n+1$ separated alternating clusters of vectors for odd and even cases. Please note that this is equivalent to a linear model with n parallel hyperplanes, or a nearest-prototype model in one dimension (along the line) with $n+1$ prototypes. This may be implemented as a Learning Vector Quantization (LVQ) model [19] with strong regularization. In both cases n linear parameters define direction $\mathbf{w}$, and n parameters define thresholds placed on the y line (in case of prototypes there are placed between thresholds, except for those on extreme left and extreme right, placed on the other side of the threshold in the same distance as the last prototype), so the whole model has $2n$ parameters.

It is obvious that complexity of data classification is proportional to the k-separability index, although for some datasets additional non-linear transformations are needed to avoid overlaps of projected clusters. For high values of k learning becomes very difficult and most classifiers, based on the Multi-Layer Perceptron (MLPs), Radial Basis Function (RBF) network, or Support Vector Machine (SVM) data models, as well as almost all other systems, are not able to discover simple data models. Linear projections are the simplest transformations with an easy interpretation. For many complicated situations proper linear mapping can discover interesting structures. Local and non-local distributions of data points can be clustered and discriminated by hyperplanes. Often a simple linear mapping exists that leaves only trivial non-linearities that may be separated using neurons that implement a window-like transfer function:

$$\tilde{M}(\mathbf{x};\mathbf{w},a,b) = \begin{cases} 1 \text{ if } \mathbf{wx} \in [a,b] \\ 0 \text{ if } \mathbf{wx} \notin [a,b] \end{cases} \quad (1)$$

This function is suitable for learning all 3-separable data (including XOR). The number of such Boolean functions for 3 or more bits is much greater than of the linearly separable functions [3]. For data which is more than $k = 3$ separable this will not give an optimal solution, but it will still be simpler than the solution constructed using hyperplanes. There are many advantages of using window-type functions in neural networks, especially in difficult, highly non-separable classification problems [7]. One of the most interesting learning algorithms in the field of learning Boolean functions is the constructive neural network with Sequential Window Learning (SWL), an algorithm described by Muselli [26]. This network also uses window-like transfer function, and in comparison with other constructive methods [26] outperforms similar methods with threshold neurons, leading to models with lower complexity, higher speed and better generalization [12]. SWL works only for binary data and therefore some pre-processing is needed to use it for different kind of problems.

The k-separability idea is a good guiding principle that facilitates searching for transformations that can create non-separable data distributions that will be easy to handle. In the next section constructive network is presented that uses window-like transfer functions to distinguish clusters created by linear projections. A lot of methods that search for optimal and most informative linear transformations have been developed. Projection pursuit is a branch of statistical methods that search for interesting data transformations by maximizing some "index of interest" [18, 10]. First a *c3sep* network that has nodes designed to discover 3-separable structures is presented and tested on learning Boolean functions and some benchmark classification problems. Second, a new "Quality of Projected Clusters" index designed to discover k-separable structures is introduced, and applied to visualization of data in low-dimensional spaces. Constructive networks described in section 3 use this index with the projection pursuit methodology for construction of an accurate neural QPCNN architecture for solving complex problems. The paper ends with a discussion and conclusion.

2 Constructive 3-Separability Model (*c3sep*)

With growing k-separability index problems quickly become intractable for general classification algorithms. Although some problems with high k may also be solved using complex models it is rather obvious that simplest linear solutions, or solutions involving smooth minimal-complexity non-linear mappings combined with interval non-linearities, should show better generalization and such solutions should be easier to comprehend. In feedforward multilayer neural networks hidden layers represents some data transformation which should lead to linear separation of samples in the output layer. For highly-nonseparable data this transformation in very hard to find.

In case of backpropagation procedure, when all network weights are adjusted in each iteration, convergence to the optimal solution is almost impossible. The final MLP model gives no information about the structure of the problem, representing data structures in completely distributed way. Using constructive methods, a single node can be trained separately, providing a partial solution. As a result each network node represents a chunk of knowledge about the whole problem, focusing on different subsets of data and facilitating interpretation of the data structure.

Window-Type Transfer Functions

In the brain neurons are organized in cortical column microcircuits [22] that resonate with certain frequencies whenever they receive specific signals. Threshold neurons split input space in two disjoint regions, with hyperplane defined by the $\mathbf{w}$ direction. For highly non-separable data searching for linear separation is useless, while finding interesting clusters for projected data, corresponding to an active microcircuit, is more likely. Therefore network nodes should implement a window-like transfer functions (Eq. 1) that solve 3-separable problems by combination of projection and clustering, separating some (preferably large) number of instances from a single class in the $[a,b]$ interval. This simple transformation may handle not only local neighborhoods, as Gaussian functions in SVM kernels or RBF networks do, but also non-local distributions of data that typically appear in the Boolean problems. For example, in the n-bit parity problem projection on the $[1,1..1]$ direction creates several large clusters with vectors that contain fixed number of 1 bits.

Optimization methods based on gradient descent used in the error backpropagation algorithm require continuous and smooth functions, therefore soft windowed-type functions should be used in the training phase. Good candidate functions include a combination of two sigmoidal functions:

$$M(\mathbf{x};\mathbf{w},a,b,\beta) = \sigma(\beta(\mathbf{wx}-a)) - \sigma(\beta(\mathbf{wx}-b)) \quad . \tag{2}$$

For $a > b$ an equivalent product form, called bicentral function [5], is:

$$M(\mathbf{x};\mathbf{w},a,b,\beta) = \sigma(\beta(\mathbf{wx}-a))(1-\sigma(\beta(\mathbf{wx}-b)) \quad . \tag{3}$$

Parameter β controls slope of the sigmoid functions and can be adjusted during training together with weights $\mathbf{w}$ and biases a and b. Bicentral function (3) has values in the range $[0,1]$, while function (2) for $b < a$ may become negative, giving values in the $[-1,+1]$ range. This property may be useful in constructive networks for "unlearning" instances, misclassified by previous hidden nodes. Another interesting window-type function is:

$$M(\mathbf{x};\mathbf{w},a,b,\beta) = \frac{1}{2}\left(1 - \tanh(\beta(\mathbf{wx}-a))\tanh(\beta(\mathbf{wx}-b))\right) \ . \tag{4}$$

This function has one interesting feature: for points $\mathbf{wx} = a$ or $\mathbf{wx} = b$ and for any value of slope β it is equal to $1/2$. By setting large value of β hard-window type function (1) is obtained

$$M(\mathbf{x};\mathbf{w},a,b,\beta) \stackrel{\beta\to\infty}{\longrightarrow} \tilde{M}(\mathbf{x};\mathbf{w},a',b') \ , \tag{5}$$

where for function (4) boundaries of the $[a,b]$ interval do not change ($a = a'$ and $b = b'$), while for the bicentral function (3) value of β has influence on the interval boundaries, so for $\beta \to \infty$ they are different than $[a,b]$. Another way to achieve sharp decision boundaries is by introduction of an additional threshold function and parameter:

$$\tilde{M}(\mathbf{x};\mathbf{w},a',b') = \operatorname{sgn}\left(M(\mathbf{x};\mathbf{w},a,b,\beta) - \theta\right) \ . \tag{6}$$

Many other types of transfer functions can be used for practical realization of 3-separable models. For detailed taxonomy of neural transfer functions see [7, 8].

Modified Error Function

Consider a dataset $\mathscr{X} \subset \mathscr{R}^d$, where each vector $\mathbf{x} \in \mathscr{X}$ belongs to one of the two classes $c(\mathbf{x}) \in \{0,1\}$. To solve this classification problem neural network should minimize an error measure:

$$E(\mathscr{X};\Gamma) = E_{\mathbf{x}}||y(\mathbf{x};\Gamma) - c(\mathbf{x}))|| \ , \tag{7}$$

where $y(\mathbf{x};\Gamma)$ is the network output and Γ denotes a set of all parameters that need to be adjusted during training (weights, biases, etc.). The expectation value is calculated over all training vectors using the mean square error, or cross entropy, or other norms suitable for error measures. However, in constructive networks nodes may be trained separately, one at a time, and a partial solution in form of pure clusters for some range of $[a_i,b_i]$ output values created by each $M(\mathbf{x};\Gamma_i)$ node are used to improve the network function. To evaluate a usefulness of a new node $M(\mathbf{x};\Gamma_i)$ for the network $y(\mathbf{x};\Gamma)$, where Γ represents all parameters, including Γ_i, an extra term is added to the standard error measure:

$$E(\mathscr{X};\Gamma;a,b,\lambda) = E_{\mathbf{x}}||y(\mathbf{x};\Gamma) - c(\mathbf{x}))|| + \lambda_i E_{M\in[a_i,b_i]}||M(\mathbf{x};\Gamma_i) - c(\mathbf{x}))|| \ , \tag{8}$$

where λ_i controls the tradeoff between the covering and the quality of solution after the new $M(\mathbf{x};\Gamma_i)$ hidden node is added. For nodes implementing $\mathbf{wx}$ projections (or any other functions with outputs restricted to $[a_i, b_i]$ interval) largest pure cluster will give the lowest contribution to the error, lowering the first term while keeping the second one equal to zero. If such cluster is rather small it may be worthwhile to create a slightly bigger one, but not quite pure, to lower the first term at the expense of the second. Usually a single λ_i parameter is taken for all nodes, although each parameter could be individually optimized to reduce the number of misclassification.

The current version of the *c3sep* constructive network assumes binary $0/1$ class labels, and uses the standard mean square error (MSE) measure with two additional terms:

$$E(\mathbf{x};\Gamma,\lambda_1,\lambda_2) = \frac{1}{2}\sum_{\mathbf{x}}(y(\mathbf{x};\Gamma) - c(\mathbf{x}))^2 + \\ +\lambda_1 \sum_{\mathbf{x}}(1 - c(\mathbf{x}))y(\mathbf{x};\Gamma) - \lambda_2 \sum_{\mathbf{x}} c(\mathbf{x})y(\mathbf{x};\Gamma) \ . \tag{9}$$

The term scaled by λ_1 represents additional the penalty for "unclean" clusters, increasing the total error for vectors from class 0 that falls into at least one interval created by hidden nodes. The term scaled by λ_2 represents reward for large clusters, decreasing the value of total error for every vector that belongs to class 1 and was correctly placed inside created clusters.

The *c3sep* Architecture and Training

The output of the *c3sep* network is given by:

$$y(\mathbf{x};\Gamma) = \sigma\left(\sum_i M(\mathbf{x};\Gamma_i) - \theta\right) \ , \tag{10}$$

where $\Gamma_i = \{\mathbf{w}_i, a_i, b_i, \beta_i\} \subset \Gamma$ denote subset of parameters of i-th node. All connections between hidden and output layer are fixed with strength 1, although in the final step they could be used as a linear layer for additional improvement of the network. One can use linear node as an output, but in practice sigmoidal function provides better convergence. The architecture of this network is shown in Fig. 1. Each hidden node tries to separate a large group of vectors that belong to class $c = 1$. Learning procedure starts with an empty hidden layer. In every phase of the training one new window-type unit is added and trained by minimization of the error function (9). Weights of each node are initialized with small random values before training. Initial values for biases a and b can be set to

$$a = (\mathbf{wx})_{min} + \frac{1}{3}|(\mathbf{wx})_{max} - (\mathbf{wx})_{min}|$$

$$b = (\mathbf{wx})_{min} + \frac{2}{3}|(\mathbf{wx})_{max} - (\mathbf{wx})_{min}|$$

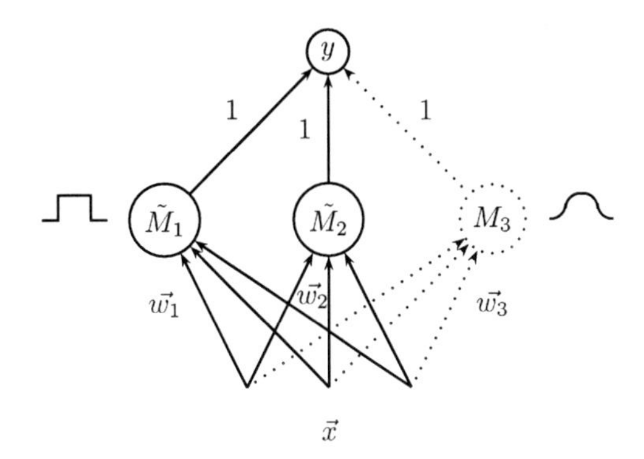

Fig. 1 Example of the *c3sep* network with three hidden neurons. Only parameters of the last node are adjusted during training (dotted line), the first and the second node have been frozen, with large value of β used to obtain sharp interval boundaries.

In most cases this should provide a good starting point for optimization with gradient based methods.

Network construction proceeds in a greedy manner. First node is trained to separate as large group of class 1 vectors as possible. After convergence is reached the slope of transfer function is set to a large value to obtain hard-windowed function, and the weights of this neuron are kept frozen during further learning. Samples from class 1 correctly handled by the network do not contribute to the error, and can be removed from the training data to further speed up learning (however, leaving them may stabilize learning, giving a chance to form more large clusters). After that, the next node is added and the training is repeated on the remaining data, until all vectors are correctly handled. To avoid overfitting, one may use pruning techniques, as it is done in the decision trees. The network construction should be stopped when the number of cases correctly classified by a new node becomes too small, or when the crossvalidation tests show that adding such node will decrease generality.

Experiments

2.1 Boolean Functions

The *c3sep* network has been tested on several types of problems. Figures 2 and 3 shows results of learning on Boolean problems with systematically increasing complexity. Results are compared with a few neural constructive algorithms designed to deal with Boolean functions. All these algorithms may be viewed as a realization of general sequential constructive method [26] (this method is briefly described in subsection 3), and differ by strategy of searching for the best hidden nodes. Irregular Partitioning algorithm [23] uses threshold perceptrons optimized with linear programing. Carve [32] is trained by the convex hull traversing algorithm. Oil Spot algorithm [24] searches for connected subgraphs and proper edge orientation in hypercube graph. Sequential Window Learning method [25] uses window transfer functions for which weights are obtained from solution of a system of algebraic equations. Target Switch algorithm [33] use traditional perceptron learning.

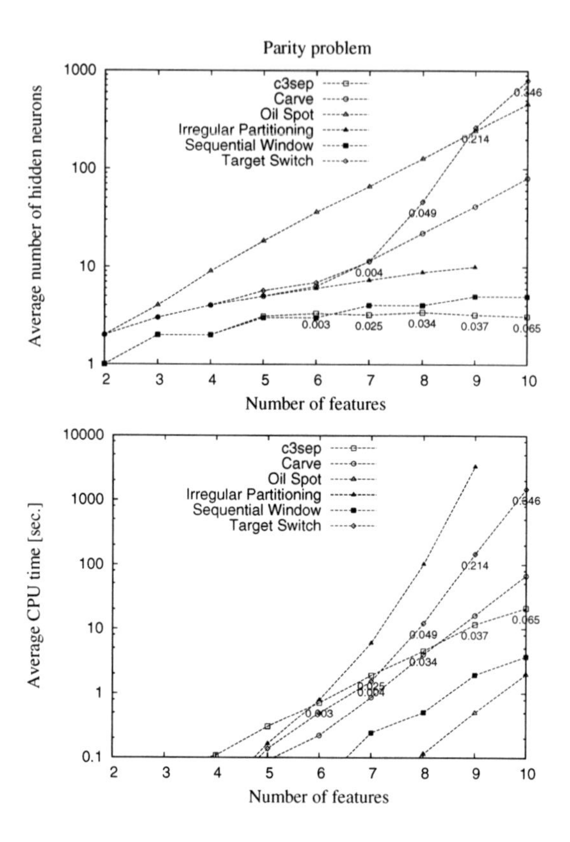

Fig. 2 Number of hidden units created, and time consumed during learning of parity problems. Each result is averaged over 10 trials.

Fig. 2 presents results of learning of constructive methods applied to the parity problems from 2 bits to 10 bits, and Fig. 3 shows the results of learning randomly selected Boolean functions, where labels for each string of bits have been drawn with equal probability $P(C(\mathbf{x}) = 1) = 0.5$. The same random function was used to train all algorithms. These figures show the size of networks constructed by a given algorithm, and the time needed for learning until a given problem has been solved without mistakes.

The *c3sep* network avoids small clusters increasing generalization, and uses stochastic gradient algorithm, that avoids local minima through multistarts, and thus leads to small errors in some runs. Values of the training error are placed in corresponding points of Fig. 2 and Fig. 3. The n-dimensional parity problem can be solved by a two-layer neural network with n threshold neurons or $(n+1)/2$ window-like neurons in the hidden layer[3]. Sequential window learning and irregular partitioning algorithms were able do obtain optimal solution for all dimensions.

Learning of random Boolean functions is much more difficult, and upper bound for the number of neurons needed for solving of this kind of functions is not known. This purpose of these tests is to check the ability of each algorithm to discover simple models of complex logical functions. Algorithms capable of exact learning of every example by creating separate node for single vectors are rarely useful as they

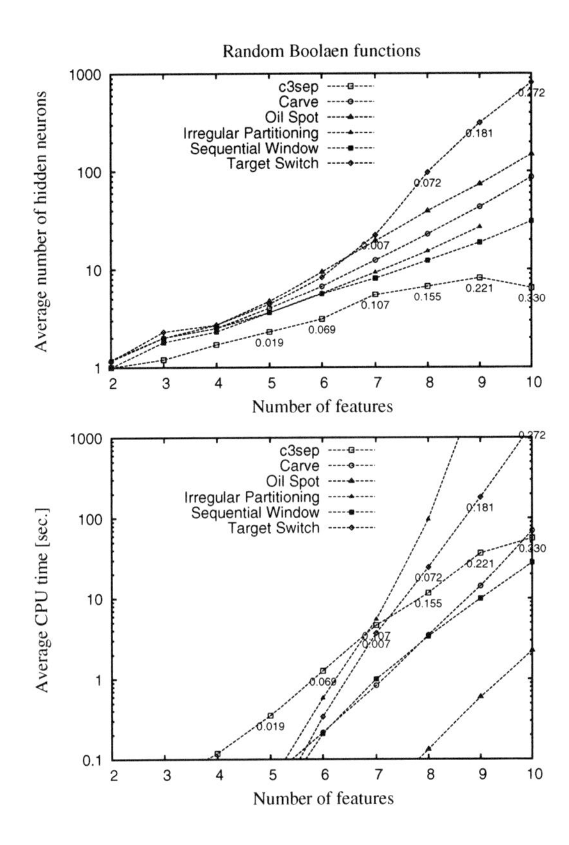

Fig. 3 Number of hidden units created and time consumed during learning of random Boolean functions. Each result is averaged over 10 trials.

will overfit the data. Therefore the ability to find simple, but approximate, solutions is very useful. One should expect that such approximate models should be more robust than perfect models if the training is carried on slightly different subset of examples for a given Boolean function.

Irregular partitioning produces small networks, but the training time is very high, while on the other hand the fastest methods (Oil Spot) needs many neurons. Sequential window learning gave solutions with a small number of neurons and rather low computational cost. The *c3sep* network was able to create smallest architectures, but the average times of computations are somehow longer than needed by most other algorithms. This network provides near optimal solution, as not all patterns were correctly classified.

2.2 *Real World Problems*

Tables 1 and 2 present results of a generalization tests for a few benchmark datasets from the UCI repository [1]. The Iris dataset is perhaps the most widely used simple problem, with 3 types of Iris flowers described by 4 real valued attributes. The Glass identification problem has 9 real valued features with patterns divided into

Table 1 30x3 CV test accuracy.

	Iris	Glass	Voting0	Voting1
Carve	90.18 ± 1.58	74.17 ± 3.28	93.24 ± 1.00	87.45 ± 1.37
Irregular Partitioning	90.98 ± 2.29	72.29 ± 4.39	93.41 ± 1.13	86.96 ± 1.43
Target Switch	65.45 ± 5.05	46.76 ± 0.91	94.64 ± 0.63	88.13 ± 1.47
c3sep	95.40 ± 1.30	70.68 ± 2.97	94.38 ± 0.72	90.42 ± 1.15
binary features				
Oil Spot	75.16 ± 2.86	66.05 ± 2.41	90.93 ± 0.90	86.68 ± 1.47
Carve	71.84 ± 3.46	62.08 ± 4.55	91.79 ± 1.22	86.77 ± 1.43
Irregular Partitioning	75.53 ± 3.20	62.38 ± 3.66	92.73 ± 1.19	86.79 ± 2.20
Target Switch	84.93 ± 3.28	71.69 ± 3.42	94.66 ± 0.69	88.36 ± 0.98
Sequential Window	77.36 ± 4.71	54.18 ± 3.50	91.40 ± 1.21	82.28 ± 1.82
c3sep	75.58 ± 6.15	60.92 ± 4.47	94.50 ± 0.89	89.78 ± 1.26

Table 2 Average number of hidden neurons generated during 30x3 CV test.

	Iris	Glass	Voting0	Voting1
Carve	5.72 ± 0.46	7.00 ± 0.50	4.99 ± 0.39	8.34 ± 0.45
Irregular Partitioning	5.49 ± 0.53	4.69 ± 0.26	2.04 ± 0.21	3.48 ± 0.30
Target Switch	22.76 ± 2.17	55.49 ± 2.38	3.69 ± 0.29	9.22 ± 0.85
c3sep	3.00 ± 0.00	1.14 ± 0.26	1.00 ± 0.00	1.02 ± 0.12
binary features				
Oil Spot	27.78 ± 1.41	21.54 ± 1.80	22.76 ± 1.39	37.32 ± 2.32
Carve	8.02 ± 0.52	6.79 ± 0.26	5.56 ± 0.32	8.59 ± 0.46
Irregular Partitioning	3.00 ± 0.00	1.00 ± 0.00	0.99 ± 0.06	2.50 ± 0.30
Target Switch	3.07 ± 0.14	1.72 ± 0.25	3.20 ± 0.26	7.46 ± 0.48
Sequential Window	9.90 ± 0.68	5.54 ± 0.40	5.46 ± 0.50	7.10 ± 0.61
c3sep	3.30 ± 0.35	1.03 ± 0.10	1.00 ± 0.00	1.00 ± 0.00

float-processed and non float-processed pieces of glass. United States congressional voting record database, denoted here as Voting0 dataset, contains 12 features that record decisions of congressmans who belong to a democratic or republican party. The Voting1 dataset has been obtained from the Voting0 by removing the most informative feature. Each input can assume 3 values: yes, no or missing.

Some algorithms used for comparison work only for binary features, therefore their application requires additional pre-processing of data vectors. Real valued features have been transformed to binary features by employing Gray coding [27]. Resulting Iris and Glass dataset in binary representation have 22 and 79 features, respectively. For the three-valued input of Voting dataset a separate binary feature has been associated with the presence of each symbolic value, resulting in 48 binary features for Voting0, and 45 for Voting1 datasets.

Algorithms that can handle real features have been applied to both original data and binarized data (Tab. 1). Although dimensionality of binary data is higher (and thus more adaptive parameters are used by standard MLP networks and other algorithms), results of most methods on binary data are significantly worse, particularly in the case of Iris and Glass where all features in the original data are real valued.

In all these tests *c3sep* network gave very good accuracy with low variance, better on statistically equivalent to the best solutions, with a very small number of neurons created in the hidden layer. The ability to solve complex problems in an approximate way is evidently helpful also for relatively simple data used here, showing the universality of constructive *c3sep* networks.

3 Projection Pursuit Constructive Neural Network

Projection pursuit (PP) is a generic name given to all algorithms that search for the most "interesting" linear projections of multidimensional data, maximizing (or minimizing) some objective functions or indices [11, 10]. Many projection pursuit indices may be defined to characterize different aspects or structures that the data may contain. Modern statistical dimensionality reduction approaches, such as the principal component analysis (PCA), Fisher's discriminant analysis (FDA) or independent component analysis (ICA) may be seen as special cases of projection pursuit approach. Additional directions may be generated in the space orthogonalized to the already found directions.

PP indices may be introduced both for unsupervised and for supervised learning. By working in a low-dimensional space based on linear projections projection pursuit methods are able to avoid the "curse of dimensionality" caused by the fact that high-dimensional space is mostly empty [15]. In this way noisy and non-informative variables may be ignored. In contrast to most similarity-based methods that optimize metric functions to capture local clusters, projection pursuit may discover also non-local structures. Not only global, but also local extrema of the PP index are of interest and may help to discover interesting data structures.

A large class of PP constructive networks may be defined, where each hidden node is trained by optimization of some projection pursuit index. In essence the hidden layer defines a transformation of data to low dimensional space based on sequence of projections. This transformation is then followed by linear discrimination in the output layer. PCA, FDA or ICA networks are equivalent to linear discrimination on pre-processed suitable components. In the next section more interesting index, in the spirit of *k*-separability, is defined.

The QPC Projection Index

Consider a dataset $\mathscr{X} = \{\mathbf{x}_1, \ldots, \mathbf{x}_n\} \subset \mathscr{R}^d$, where each vector $\mathbf{x}_i$ belongs to one of the k different classes. Let $\mathscr{C}_{\mathbf{x}}$ denote the set of all vectors that have the same label as $\mathbf{x}$. The following index achieves maximum value for projections on the direction $\mathbf{w}$ that groups all vectors from class $\mathscr{C}_{\mathbf{x}}$ into a compact cluster separated from vectors that belong to other classes:

$$Q(\mathbf{x};\mathbf{w}) = A^{+} \sum_{\mathbf{x}_k \in \mathscr{C}_{\mathbf{x}}} G\left(\mathbf{w}^T(\mathbf{x}-\mathbf{x}_k)\right) - A^{-} \sum_{\mathbf{x}_k \notin \mathscr{C}_{\mathbf{x}}} G\left(\mathbf{w}^T(\mathbf{x}-\mathbf{x}_k)\right) , \tag{11}$$

where $G(x)$ is a function with localized support and maximum in $x = 0$, for example a Gaussian function. The first term in $Q(\mathbf{x};\mathbf{w})$ function is large if all vectors from class $\mathscr{C}_{\mathbf{x}}$ are placed close to $\mathbf{x}$ after the projection on direction defined by $\mathbf{w}$, indicating how compact and how large is this cluster of vectors. The second term depends on distance beetwen $\mathbf{x}$ and all patterns that do not belong to class $\mathscr{C}_{\mathbf{x}}$, therefore it represents penalty for placing vector $\mathbf{x}$ too close to the vectors from opposite classes. The Quality of Projected Clusters (QPC) index is defined as an average of the $Q(\mathbf{x};\mathbf{w})$ for all vectors:

$$QPC(\mathbf{w}) = \frac{1}{n} \sum_{\mathbf{x} \in \mathscr{X}} Q(\mathbf{x};\mathbf{w}) , \tag{12}$$

providing a leave-one-out estimator that measures quality of clusters projected on $\mathbf{w}$ direction. This index achieves maximum value for projections $\mathbf{w}$ that create small number of pure, compact and well separated clusters. For linearly separable problems function $QPC(\mathbf{w})$ achieves maximum for projections $\mathbf{wx}$ that create two well-separated pure clusters of vectors. In case of k-separable dataset maximization of QPC index leads to a projection with k separated clusters. Thus optimization of QPC should discover k-separable solutions if they exist.

Parameters A^{+}, A^{-} control influence of each term in Eq. (11). If A^{-} is large strong separation between classes is enforced, while large A^{+} impacts mostly compactness and purity of clusters. For example, by setting $A^{+} = p(\mathscr{C}_{\mathbf{x}})$ and $A^{-} = 1 - p(\mathscr{C}_{\mathbf{x}})$ (where $p(\mathscr{C}_{\mathbf{x}})$ is the *a priori* class probability), projection index is balanced in respect to the size of classes. If in addition $G(x)$ is normalized, such that $G(0) = 1$, then the upper limit of QPC index is 1 and it occurs only when all vectors from the same class after projection are placed in a single very narrow cluster and the gap beetwen each cluster is greater than the range of $G(\mathbf{x})$ function. All bell-shaped functions that achieve maximum value for $x = 0$ and vanish for $x \to \pm\infty$ are suitable for $G(x)$, including Gaussian, bicentral functions Eq. (3) and Eq. (2), or an inverse quartic function:

$$G(x) = \frac{1}{1 + (bx)^4} , \tag{13}$$

where parameter b controls the width of $G(x)$.

These functions are continuous and thus may be used in gradient-based methods. Iterative gradient optimization procedures applied to functions with multiple local minima do no guarantee that an optimal solution will be found, and may converge slowly. Direct calculation of the QPC index (12) requires $O(n^2)$ operations (after projection distances beetwen all pairs of vectors are computed), as in the nearest neighbor methods. For large datasets this may be excessive.To overcome this problem various “editing techniques”, or instance selection algorithms developed for the nearest neighbor methods may be used [17, 13, 29]. By sorting projected vectors and restricting computations of the sum in Eq. (11) only to vectors $\mathbf{x}_i$ with $G(\mathbf{w}(\mathbf{x} - \mathbf{x}_i)) > \varepsilon$ computational time is easily decreased to $O(n \log n)$. Further improvements in speed may be achieved if the sum in Eq. (12) is restricted only to a few centers of projected clusters $\mathbf{t}_i$. This may be done after projection $\mathbf{w}$ stabilizes, as at the beginning of the training the number of clusters in the projection is not known without some prior knowledge about the problem. For k-separable datasets k centers are sufficient and the cost of computing QPC index drops to $O(kn)$. Gradient descent methods may be replaced by more sophisticated approaches [14, 20]), although in practice multistart gradient methods have been quite effective in searching for interesting projections. It is worth to notice that although global extrema of QPC index give most valuable projections, suboptimal solutions may also provide useful insight into the structure of data.

First QPC Direction

Figure 4 presents projections for 4 very different kinds of datasets: Wine, Monk1, 10-bit Parity and Concentric Rings. All projections were obtained taking quartic function (13) for $G(x)$, with $b = 3$, and using simple gradient descent maximization initialized 10 times, selecting after a few iterations the most promising solution that is trained until convergence. Values of weights and the value of $QPC(\mathbf{w})$ are shown in the corresponding figures. Positions of projected vectors on the line are shown for each class below the projection line. Smoothed histograms for these projections may be normalized and taken as estimations of class conditionals $p(x|C)$, from which posterior probabilities $p(C|x) = p(x|C)p(C)/p(x)$ are easily calculated.

The first two datasets are taken from the UCI repository [1]. The Wine data consist of 178 vectors, 13 continuous features and 3 classes. It can be classified using a single linear projection that gives 3 groups of vectors (one for each class). The weight for "flavanoids" feature dominates and is almost sufficient to separate all 3 classes. Monk 1 is an artificial datasets [1], with 6 symbolic features and two classes, defined by two simple rules: given object is "a monk" if "the head shape" (first attribute) = "body shape" (second attribute) or "jacket color" (fifth attribute) = red. Direction generated by maximization of the QPC index produces large cluster of vectors in the middle of the projection. First two coefficients are large and equal, others are essentially zero. This corresponds to the first rule, but the second rule cannot be captured by the same projection. To separate the remaining cases a second projection is needed (see below). These logical rules have also been extracted using a special version of MLP network [5]. The 10 bit parity is an example of a hard Boolean problem, where 1024 samples are divided into even and odd binary strings. This problem is 11-separable, with a maximum value of projection index obtained for diagonal direction in the 10 dimensional hypercube, therefore all weights have the same value. Although a perfect solution using a single projection has been found clusters at the extreme left and right of the projection are quite small, therefore finding another direction that puts these vectors in larger clusters may be useful. Convergence of MLP or RBF networks for such complex data is quite unlikely, but also standard SVM approaches fail completely if crossvalidation tests are performed. The final dataset (Concentric Rings) contains 800 samples distributed in

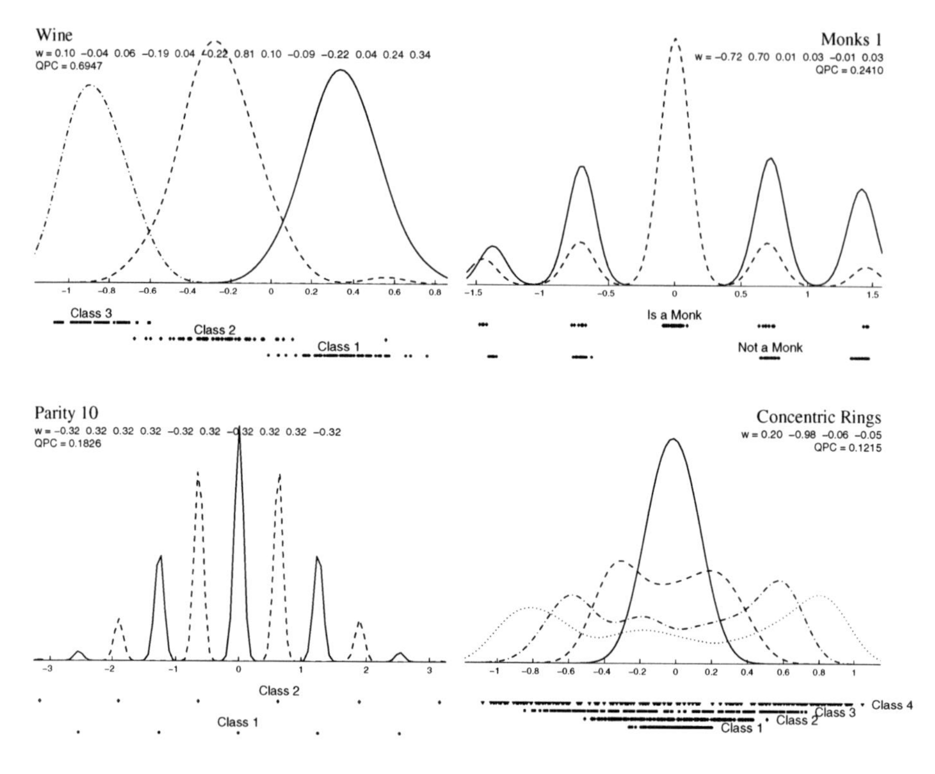

Fig. 4 Examples of four projections found by maximization of the QPC index using gradient descent for the Wine data (top-left), the Monk1 problem (top-right), the 10-bit Parity (bottom-left) and the noisy Concentric Rings (bottom-right).

4 classes, each with 200 samples defined by 4 continuous features. Only the first and the second feature is relevant, vectors belonging to the same class are located inside one of the 4 concentric rings. The last two noisy features are uniformly distributed random numbers. For this dataset the best projection that maximizes the QPC index reduces influence of noisy features, with weights for dimensions 3 and 4 close to zero. This shows that the QPC index may be used for feature selection, but also that linear projections have limited power: a complicated solution requiring many projections at different angles to delineate the rings is needed. Of course a much simpler network using localized functions will solve this problem more accurately. The need for networks with different types of transfer functions [7, 9] has been stressed some time ago, but still there are no programs capable of finding the simplest data models in all cases.

Second QPC Direction

For complex problems usually more than one projection is required. Using QPC index searching for additional interesting projections can be realized in several ways.

Sequence of unique directions may be generated applying repeatedly QPC optimization in the subspace orthogonal to all directions found earlier. Another possible approach is to focus on subsets of vectors with poor separation and search for another direction only for overlapping clusters until separation is attained. The third possibility is to search for the next linear projection with additional penalty term that will punish solutions similar to those found earlier:

$$QPC(\mathbf{w};\mathbf{w}_1) = QPC(\mathbf{w}) - \lambda f(\mathbf{w},\mathbf{w}_1) \tag{14}$$

The value of $f(\mathbf{w},\mathbf{w}_1)$ should be large if the current direction $\mathbf{w}$ is close to the previous direction $\mathbf{w}_1$. For example, some power of the scalar product between these directions may be used: $f(\mathbf{w},\mathbf{w}_1) = (\mathbf{w}_1{}^T \cdot \mathbf{w})^2$. Parameter λ scales the importance of enforcing this condition during the optimization process.

Scatterplots of data vectors projected on two directions may be used for visualization. Figure 5 presents such scatterplots for the four datasets used in the previous section. The second direction $\mathbf{w}$, found by gradient descent optimization of function (14) with $\lambda = 0.5$, is used for the horizontal axis. The final weights of the second

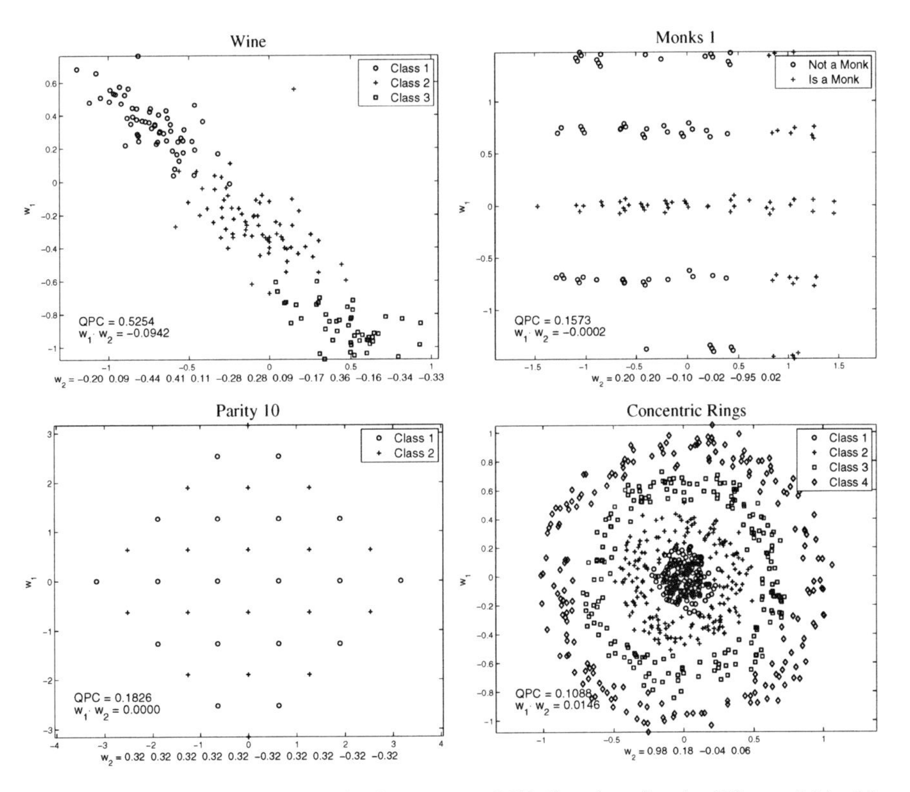

Fig. 5 Scatterplots created by projection on two QCP directions for the Wine and Monk1 data (top-left/right), 10-bit parity and the noisy Concentric Rings data (bottom-left/right).

direction, value of the projection index $QPC(\mathbf{w})$ and the inner product of $\mathbf{w}_1$ and $\mathbf{w}$ are shown in the corresponding figures. For the Wine problem first projection was able to separate almost perfectly all three classes. Second projection (Fig. 5) gives additional insight into the structure of this data, leading to a better separation of vectors placed near decision boundary. Two-dimensional projection of Monk1 data shows separate and compact clusters. The 5th feature (which forms the second rule describing this dataset) has significant value, and all unimportant features have weights equal almost zero, allowing for simple extraction of correct logical rules. In case of the 10-bit parity problem each diagonal direction of a hypercube representing Boolean function gives a good solution with large cluster in the center. Two such orthogonal directions have been found, projecting each data vector into large pure cluster, either in the first or in the second dimension. In particular small, one or two-vector clusters at the extreme ends of the first projection belong to the largest clusters in the second direction, ensuring good generalization in this two-dimensional space using naive Bayes estimation of classification probabilities. Results for the noisy Concentric Rings dataset show that maximization of the QPC index has caused vanishing of noisy and uninformative features, and has been able to discover two-dimensional relations hidden in this data. Although linear projections in two directions cannot separate this data, such dimensionality reduction is sufficient for any similarity-based method, for example the nearest neighbor method, to perfectly solve this problem.

A single projection allows for estimation and drawing class-conditional and posterior probabilities, but may be not sufficient for optimal classification. Projections on 2 or 3 dimensions allow for visualization of scatterograms, showing data structures hidden in the high-dimensional distributions, suggesting how to handle the problem in the simplest way: adding linear output layer (Wine), employing localized functions, decision trees or covering algorithms, using intervals (parity) or naive Bayes, or using the nearest neighbor rule (Concentric Rings). If this is not sufficient more projections should be used as a pre-processing for final classification, trying different approaches in a meta-learning scheme [6].

Coefficients of the projection vectors may be used directly for feature ranking/selection, because maximization of the QPC index gives negligible weights to noisy or insignificant features, while important attributes have distinctly larger values. This method might be used to improve learning for many machine learning models sensitive to feature weighting, such as all similarity-based methods. Interesting projections may also be used to initialize weights in various neural network architectures.

Constructive Neural Network Based on the QPC Index

Projection pursuit methods in a natural way may be used for constructive neural networks learning, where each hidden node coresponds to a linear mapping obtained by optimization of a projection index. To build a neural network architecture using

QPC index general sequential constructive method may be used [26]. For the two-class problems this method is described as follows:

1. start learning with an empty hidden layer;
2. if there are some misclassified vectors *do*:
3. add a new node;
4. train the node to obtain a partial classiffier;
5. remove all vectors for which the current node outputs $+1$;
6. *enddo*.

A partial classiffier is a node with output $+1$ for at least one vector from one of the classes, and -1 for all vectors from the opposite classes. After a finite number of iterations this procedure leads to a construction of neural network that classifies all training vectors (unless there are conflicting cases, i.e. identical vectors with different labels, that should be removed). Weights in the output layer do not take part in the learning phase and their values can be determined from a simple algebraic equation, assigning the largest weight to the node created first, and progressively smaller weights to subsequently created nodes, for example:

$$u_0 = \sum_{j=1}^{h} u_j + d_{h+1} \quad , \qquad u_j = d_j 2^{h-j} \quad \text{for} \quad j = 1, \ldots, h \tag{15}$$

where h is the number of hidden neurons, $d_i = \{-1, +1\}$ denotes label for which i-th hidden node gives output $+1$ and $d_{h+1} = d_h$.

The sequential constructive method critically depends on construction of good partial classifier. A method to create it is described below. Consider a node M implementing the following function:

$$M(\mathbf{x}) = \begin{cases} 1 & \text{if} \quad |G(\mathbf{w}(\mathbf{x}-\mathbf{t})) - \theta| \geq 0 \\ -1 & \text{otherwise} \end{cases} \tag{16}$$

where the weights $\mathbf{w}$ are obtained by maximization of the QPC index, θ is a threshold parameter which determines the window width, and $\mathbf{t}$ is the center of a cluster of vectors projected on $\mathbf{w}$, estimated by:

$$\mathbf{t} = \arg\max_{\mathbf{x} \in \mathscr{X}} Q(\mathbf{w}, \mathbf{x}) \tag{17}$$

If the direction $\mathbf{w}$ corresponds to the maximum of QPC index then $\mathbf{t}$ should be at the center of a large, clean and well separated cluster. Thus the node (16) splits input space into two disjoint subspaces, with output $+1$ for each vector that belongs to the cluster, and -1 for all other vectors. Large, clean and well-separated clusters may be achieved by maximization of the function $Q(\mathbf{t};\mathbf{w})$ with respect to weights $\mathbf{w}$ and cluster center $\mathbf{t}$, or by minimization of an error function:

$$E(\mathbf{x}) = E_{\mathbf{x}} ||G(\mathbf{w}(\mathbf{x}-\mathbf{t})) - \delta(c_{\mathbf{x}}, c_{\mathbf{t}})|| \tag{18}$$

where $\delta(c_{\mathbf{x}}, c_{\mathbf{t}})$ is equal to 1 when $\mathbf{x}$ belongs to the class associated with the cluster centered at $\mathbf{t}$, and 0 if it does not. This error function has twice as many parameters to optimize (both the weights and the center are adjusted), but computational cost of calculations here is linear in the number of vectors $O(n)$, and since only a few iterations are needed this part of learning is quite fast.

If all vectors for which the trained node gives output $+1$ have the same label then this node is a good partial classifier and sequential constructive method described above can be used directly for network construction. However, for some datasets linear projections cannot create pure clusters, as for example in the Concentric Rings case. Creation of a partial classifier may then be done by searching for additional directions by optimization of $Q(\mathbf{t};\mathbf{w})$ function (11) in respect to weights $\mathbf{w}$ and center $\mathbf{t}$ restricted to the subset of vectors that fall into the impure cluster. Resulting direction and center define the next network node according to Eq. 16. If this node is not pure, that is it provides $+1$ output for vectors from more than one class, then more nodes are required. This leads to the creation of a sequence of neurons $\{M_i\}_{i=1}^{K}$, where the last neuron M_K separates some subset of training vectors without mistakes. Then the following function:

$$\bar{M}(\mathbf{x}) = \begin{cases} +1 & \text{if } \frac{1}{K}\sum_{i=1}^{K} M_i(\mathbf{x}) - \frac{1}{2} > 0 \\ -1 & \text{otherwise} \end{cases} \tag{19}$$

is a partial classifier. In neural network function Eq. 19 is realized by group of neurons M_i placed in the first hidden layer and connected to a threshold node $\bar{M}$ in the second hidden layer with weight equal to $\frac{1}{K}$ and bias $\frac{1}{2}$. This approach has been implemented and the test results are reported below.

QPCNN Tests

Table 3 presents comparison of results of the nearest neighbor (1-NN), naive Bayes classifier, support vector machine (SVM) with Gaussian kernel, the *c3sep* network described in this article, and the constructive network based on the QPC index (QPCNN). 9 datasets from the UCI repository [1] have been used in 10-fold cross-validation to test generalization capabilities of these systems. For the SVM classifier parameters γ and C have always been optimized using an inner 10-fold crossvalidation procedure, and those that produced the lowest error have been used to learn the model on the whole training data.

Most of these datasets are relatively simple and require networks with only a few neurons in the hidden layer. Both the *c3sep* and the QPCNN networks achieve good accuracy, in most cases comparable with 1-NN, Naive Bayes and SVM algorithms. General constructive sequence learning in original formulation applied to QPCNN may lead to overfitting. This effect have occurred for Glass and Pima-diabetes where average size of created networks is higher than in the case of *c3sep* network, while the average accuracy is lower. To overcome this problem proper stop criterion for growing the network should be considered, e.g. by tracking test error changes estimated on a validation subset.

Table 3 Average classification accuracy for 10 fold crossvalidation test. Results are averaged over 10 trials. For SVM average number of support vectors (#SV) and for neural networks average number of neurons (#N) are reported.

dataset	**1-NN**	**N. Bayes**	**SVM**		**C3SEP**		**QPCNN**	
	acc.	acc.	acc.	# SV	acc.	# N	acc.	# N
Appendicitis	81.3 ± 1.5	85.3 ± 1.0	86.5 ± 0.3	32.1	85.3 ± 1.0	4.2	83.4 ± 1.0	4.0
Flag	50.1 ± 1.1	41.1 ± 1.1	51.1 ± 1.1	315.2	53.6 ± 1.8	26.7	52.9 ± 2.8	10.7
Glass	76.9 ± 2.0	61.5 ± 1.1	78.9 ± 2.0	80.3	70.5 ± 1.8	1.9	64.6 ± 2.7	12.1
Ionosphere	85.2 ± 1.2	82.2 ± 0.2	90.8 ± 1.1	63.9	85.1 ± 1.5	7.9	81.3 ± 1.5	6.3
Iris	95.9 ± 0.5	94.9 ± 0.2	95.5 ± 0.3	43.4	95.7 ± 1.0	5.0	95.3 ± 1.0	3.0
Pima-diabetes	70.5 ± 0.5	75.2 ± 0.5	70.3 ± 1.0	365.3	76.3 ± 0.4	9.1	65.2 ± 0.4	13.1
Promoters	78.5 ± 1.8	85.8 ± 1.3	73.1 ± 1.5	77.2	74.7 ± 5.6	3.7	78.8 ± 2.4	2.6
Sonar	86.8 ± 1.8	67.8 ± 1.2	84.2 ± 1.1	109.7	77.9 ± 2.4	8.5	80.2 ± 2.4	5.1
Wine	95.1 ± 0.8	98.1 ± 0.3	95.1 ± 0.2	63.3	97.1 ± 0.8	4.0	97.4 ± 0.8	3.0

4 Discussion and Conclusions

The big challenge facing computational intelligence is to discover correct bias for a given data, creating a simple but accurate models [4]. Many datasets, such as those arising from the natural language processing and problems in bioinformatics, may have an inherent complex logics that we are unable to decipher. This challenge has not yet been met by the existing systems and may require a tailor-made methods for a given data that may be created by meta-learning [6, 4]. Neural networks and kernel classifiers are universal approximators and thus they may learn any problem creating a highly complex solution. However, this leads to a poor generalization, because the correct underlying model that represents data cannot be discovered.

Each learning procedure is based on some guiding principles. Minimization of error rarely leads to the discovery of the simplest data models and thus cannot be the only basis for optimal learning systems. Linear separability is also not the best goal for learning. In many cases k-separable solutions are much simpler to achieve, leaving non-separable clusters that are easily handled. They may be treated as strongly regularized (all prototypes on a single line) nearest prototype models. The QPC index provides one way to find k-separable projections. It allows to solve problems that go well beyond capabilities of standard neural networks, such as the classification of Boolean functions in high-dimensional cases. It also enables visualization of data in one or more dimensions, allowing for estimation of reliability of classification for individual cases. It will also be useful for dimensionality reduction and feature selection.

The *c3sep* and QPCNN networks presented in this article are also designed to deal with complex data using a very simple model. The *c3sep* approach tries to find a simplification of the k-separable projection, with each node designed to discriminate a single large cluster. This is done using the error function with additional penalty and reward terms, showing many advantages when dealing with complex logical problems. This network is able to discover simple models for difficult Boolean functions and works also well for real benchmark problems.

Many other variants of the constructive networks based on the guiding principles that may be implemented using projection pursuit indices are possible. From Fig. 5 it is evident that an optimal model should use transformations that discover important features, followed in the reduced space by a specific approach, depending on the character of a given data. The class of PP networks is quite broad. One can implement various transformations in the hidden layer, explicitly creating hidden representations that are used as new inputs for further network layers, or used for initialization of standard networks. Brains are capable of deep learning, with many specific transformations that lead from simple contour detection to final invariant object recognition. Studying linear and non-linear projection pursuit networks will be most fruitful in combination with the meta-learning techniques, searching for the simplest data models in the low-dimensional spaces after initial PP transformation. This approach should bring us a bit closer to the powerful methods required for deep learning and for discovering hidden knowledge in complex data.

References

1. Asuncion, A., Newman, D.: UCI repository of machine learning databases (2007), `http://www.ics.uci.edu/~mlearn/MLRepository.html`
2. Bengio, Y., Delalleau, O., Roux, N.L.: The curse of dimensionality for local kernel machines. Technical Report 1258, Dṕartement d'informatique et recherche opérationnelle, Université de Montréal (2005)
3. Duch, W.: K-separability. In: Kollias, S.D., Stafylopatis, A., Duch, W., Oja, E. (eds.) ICANN 2006. LNCS, vol. 4131, pp. 188–197. Springer, Heidelberg (2006)
4. Duch, W.: Towards comprehensive foundations of computational intelligence. In: Duch, W., Mandziuk, J. (eds.) Challenges for Computational Intelligence, vol. 63, pp. 261–316. Springer, Heidelberg (2007)
5. Duch, W., Adamczak, R., Grąbczewski, K.: A new methodology of extraction, optimization and application of crisp and fuzzy logical rules. IEEE Transactions on Neural Networks 12, 277–306 (2001)
6. Duch, W., Grudziński, K.: Meta-learning via search combined with parameter optimization. In: Rutkowski, L., Kacprzyk, J. (eds.) Advances in Soft Computing, pp. 13–22. Physica Verlag, Springer, New York (2002)
7. Duch, W., Jankowski, N.: Survey of neural transfer functions. Neural Computing Surveys 2, 163–213 (1999)
8. Duch, W., Jankowski, N.: Taxonomy of neural transfer functions. In: International Joint Conference on Neural Networks, Como, Italy, vol. III, pp. 477–484. IEEE Press, Los Alamitos (2000)
9. Duch, W., Jankowski, N.: Transfer functions: hidden possibilities for better neural networks. In: 9th European Symposium on Artificial Neural Networks, Brusells, Belgium, pp. 81–94. De-facto publications (2001)

10. Friedman, J.H.: Exploratory projection pursuit. Journal of the American Statistical Association 82, 249–266 (1987)
11. Friedman, J.H., Tukey, J.W.: A projection pursuit algorithm for exploratory data analysis. IEEE Trans. Comput. 23(9), 881–890 (1974)
12. Grochowski, M., Duch, W.: A Comparison of Methods for Learning of Highly Non-separable Problems. In: Rutkowski, L., Tadeusiewicz, R., Zadeh, L.A., Zurada, J.M. (eds.) ICAISC 2008. LNCS (LNAI), vol. 5097, pp. 566–577. Springer, Heidelberg (2008)
13. Grochowski, M., Jankowski, N.: Comparison of instance selection algorithms II. Results and comments. In: Rutkowski, L., Siekmann, J.H., Tadeusiewicz, R., Zadeh, L.A. (eds.) ICAISC 2004. LNCS (LNAI), vol. 3070, pp. 580–585. Springer, Heidelberg (2004)
14. Haykin, S.: Neural Networks - A Comprehensive Foundation. Maxwell MacMillian Int., New York (1994)
15. Huber, P.J.: Projection pursuit. Annals of Statistics 13, 435–475 (1985), `http://www.stat.rutgers.edu/~rebecka/Stat687/huber.pdf`
16. Iyoda, E.M., Nobuhara, H., Hirota, K.: A solution for the n-bit parity problem using a single translated multiplicative neuron. Neural Processing Letters 18(3), 233–238 (2003)
17. Jankowski, N., Grochowski, M.: Comparison of instance selection algorithms. i. algorithms survey. In: Rutkowski, L., Siekmann, J.H., Tadeusiewicz, R., Zadeh, L.A. (eds.) ICAISC 2004. LNCS (LNAI), vol. 3070, pp. 598–603. Springer, Heidelberg (2004)
18. Jones, C., Sibson, R.: What is projection pursuit. Journal of the Royal Statistical Society A 150, 1–36 (1987)
19. Kohonen, T.: Self-organizing maps. Springer, Heidelberg (1995)
20. Kordos, M., Duch, W.: Variable Step Search MLP Training Method. International Journal of Information Technology and Intelligent Computing 1, 45–56 (2006)
21. Liu, D., Hohil, M., Smith, S.: N-bit parity neural networks: new solutions based on linear programming. Neurocomputing 48, 477–488 (2002)
22. Maass, W., Markram, H.: Theory of the computational function of microcircuit dynamics. In: Grillner, S., Graybiel, A.M. (eds.) Microcircuits,The Interface between Neurons and Global Brain Function, pp. 371–392. MIT Press, Cambridge (2006)
23. Marchand, M., Golea, M.: On learning simple neural concepts: from halfspace intersections to neural decision lists. Network: Computation in Neural Systems 4, 67–85 (1993)
24. Mascioli, F.M.F., Martinelli, G.: A constructive algorithm for binary neural networks: The oil-spot algorithm. IEEE Transactions on Neural Networks 6(3), 794–797 (1995)
25. Muselli, M.: On sequential construction of binary neural networks. IEEE Transactions on Neural Networks 6(3), 678–690 (1995)
26. Muselli, M.: Sequential constructive techniques. In: Leondes, C. (ed.) Optimization Techniques,of Neural Network Systems, Techniques and Applications, vol.2, pp. 81–144. Academic Press, San Diego (1998)
27. Press, W.H., Teukolsky, S.A., Vetterling, W.T., Flannery, B.P.: Numerical Recipes: The Art of Scientific Computing. Cambridge University Press, Cambridge (2007)
28. Sahami, M.: Learning non-linearly separable boolean functions with linear threshold unit trees and madaline-style networks. In: National Conference on Artificial Intelligence, pp. 335–341 (1993)
29. Shakhnarovish, G., Darrell, T., Indyk, P. (eds.): Nearest-Neighbor Methods in Learning and Vision. MIT Press, Cambridge (2005)
30. Stork, D.G., Allen, J.: How to solve the n-bit parity problem with two hidden units. Neural Networks 5, 923–926 (1992)

31. Wilamowski, B., Hunter, D.: Solving parity-n problems with feedforward neural network. In: Proc. of the Int. Joint Conf. on Neural Networks (IJCNN 2003), vol. I, pp. 2546–2551. IEEE Computer Society Press, Los Alamitos (2003)
32. Young, S., Downs, T.: Improvements and extensions to the constructive algorithm carve. In: Vorbrüggen, J.C., von Seelen, W., Sendhoff, B. (eds.) ICANN 1996. LNCS, vol. 1112, pp. 513–518. Springer, Heidelberg (1996)
33. Zollner, R., Schmitz, H.J., Wünsch, F., Krey, U.: Fast generating algorithm for a general three-layer perceptron. Neural Networks 5(5), 771–777 (1992)

On Constructing Threshold Networks for Pattern Classification

Martin Anthony

Abstract. This paper describes a method of constructing one-hidden layer feedforward linear threshold networks to represent Boolean functions (or partially-defined Boolean functions). The first step in the construction is sequential linear separation, a technique that has been used by a number of researchers [7, 11, 2]. Next, from a suitable sequence of linear separations, a threshold network is formed. The method described here results in a threshold network with one hidden layer. We compare this approach to the standard approach based on a Boolean function's disjunctive normal form and to other approaches based on sequential linear separation [7, 11].

1 Introduction

It is well known that any Boolean function can be represented by a feedforward linear threshold network with one hidden layer. The simplest way to see this is via the disjunctive normal form representation of the function (see later). Here, we discuss an alternative way of representing Boolean functions (or partially-defined Boolean functions, by which we mean restrictions Boolean functions to a specified domain). This alternative approach arises from considering a fairly natural way of classifying points by iterative or sequential linear separation.

The problem considered, to be more precise, is the following. Given disjoint subsets T and F of $\{0,1\}^n$, for some natural number n, we want to produce a feedforward linear threshold network whose output is 1 if its input is in T, and whose output is 0 if its output is in F. We refer to the pair (T,F) as a partially-defined Boolean function (pdBf), and if $T \cup F = \{0,1\}^n$, then the partially-defined Boolean function is simply a Boolean function (since its value is defined for all elements of $\{0,1\}^n$). The set T is called the set of true points, or those labelled 1; and the set

Martin Anthony
Department of Mathematics, London School of Economics and Political Science,
London WC2A 2AE, UK
e-mail: m.anthony@lse.ac.uk

L. Franco et al. (Eds.): Constructive Neural Networks, SCI 258, pp. 71–82.
springerlink.com

F is the set of false points, labelled with 0. We focus on two-class classification problems in the Boolean domain, but much of what we say can be generalised to deal with multi-class classification, or classification on more general domains (such as the whole of $\mathbb{R}^n$) [2].

We start by describing what we call the 'standard' approach, which is based on a disjunctive normal form representation of the Boolean function (or of a Boolean function that is consistent with a partially-defined Boolean function). Then we describe an approach in which we first find a *threshold decision list* that represents the pdBf (T,F) and, from this threshold decision list, produces a threshold network. We compare our method with other approaches. The threshold networks we produce have a single hidden layer of units, as do those resulting from the standard approach, but we show that there can be some advantages in the method we discuss here. Some previous work [7, 11, 12] also involved sequential linear separation, but resulted in networks with a different structure, in which there were many single-unit hidden layers, with connectivity between them.

2 Simple Threshold Networks Representing Boolean Functions: The Standard Approach

There is a very straightforward way in which to represent partially-defined Boolean functions by threshold networks having one hidden layer of units. This is based on the existence, for each Boolean function, of a *disjunctive normal form* for the function. We first briefly review key ideas on threshold networks and Boolean functions.

2.1 Threshold Functions and Threshold Networks

A function $t : \{0,1\}^n \to \{0,1\}$ is a *(Boolean) threshold function* if there are $w \in \mathbb{R}^n$ and $\theta \in \mathbb{R}$ such that

$$t(x) = \begin{cases} 1 \text{ if } \langle w,x \rangle \geq \theta \\ 0 \text{ if } \langle w,x \rangle < \theta, \end{cases}$$

where $\langle w,x \rangle$ is the standard inner product of w and x. Thus, $t(x) = \text{sgn}(\langle w,x \rangle - \theta)$, where $\text{sgn}(z) = 1$ if $z \geq 0$ and $\text{sgn}(z) = 0$ if $z < 0$. Given such w and θ, we say that t is represented by $[w, \theta]$ and we write $t \leftarrow [w, \theta]$. The vector w is known as the *weight vector*, and θ is known as the *threshold*.

A *threshold network* is formed when combine together *threshold units*, each of which computes a threshold function. More precisely, we have a directed graph, at each vertex of which is a 'unit', and with the arcs of the digraph representing the flows of signals between units. Some of the units are termed *input units*: these receive signals not from other units, but have their signals applied from outside. In our case, there will be n input units, each of which receives 0 or 1 as an input. In this situation, the set of all input patterns, or just 'inputs', is $\{0,1\}^n$. Units that do not transmit signals to other units are termed *output units*. We will be interested in networks with one output unit. The network is said to be a *feed-forward network* if the underlying directed graph is acyclic (that is, it has no directed cycles). This feed-forward condition means that the units (both the input units and the threshold units) can be labeled with integers in such a way that if there is a connection from the unit labeled i to the unit labeled j then $i < j$. In any feed-forward network, the units may be grouped into *layers*, labeled $0, 1, 2, \ldots, \ell$, in such a way that the input units form layer 0, these feed into the threshold units, and if there is a connection from a threshold unit in layer r to a threshold unit in layer s, then we must have $s > r$. Note, in particular, that there are no connections between any two units in a given layer. The 'top' layer consists of output units. The layers that are not inputs or outputs are called hidden layers.

We will be primarily interested in linear threshold networks having just one hidden layer, and it is useful to give an explicit description in this case of the functionality of the network. Such a network will consist of n inputs and some number, k, of threshold units in a single hidden layer, together with one output threshold unit. Each threshold unit computes a threshold function of the n inputs. The (binary-valued) outputs of these hidden nodes are then used as the inputs to the output node, which calculates a threshold function of these. Thus, the threshold network computes a threshold function of the outputs of the k threshold functions computed by the hidden nodes. If the threshold function computed by the output node is described by weight vector $\beta \in \mathbb{R}^k$ and threshold ϕ, and the threshold function computed by hidden node i is $f_i \leftarrow [w^{(i)}, \theta^{(i)}]$, then the threshold network as a whole computes the function $f : \{0,1\}^n \to \{0,1\}$ given by

$$f(y) = 1 \Longleftrightarrow \sum_{i=1}^{k} \beta_i f_i(y) \geq \phi;$$

that is,

$$f(y_1 y_2 \ldots y_n) = \operatorname{sgn}\left(\sum_{i=1}^{k} \beta_i \operatorname{sgn}\left(\sum_{j=1}^{n} w_j^{(i)} y_j - \theta^{(i)} \right) - \phi \right),$$

where $\operatorname{sgn}(x) = 1$ if $x \geq 0$ and $\operatorname{sgn}(x) = 0$ if $x < 0$. The *state* of the network is the (concatenated) vector

$$\omega = (w^{(1)}, \theta^{(1)}, w^{(2)}, \theta^{(2)}, \ldots, w^{(k)}, \theta^{(k)}, \beta, \phi) \in \mathbb{R}^{nk+2k+1}.$$

A fixed network architecture of this type (that is, fixing n and k), computes a parameterised set of functions $\{f_\omega : \omega \in \mathbb{R}^{nk+2k+1}\}$. In state ω, the network computes the function $f_\omega : \{0,1\}^n \to \{0,1\}$.

2.2 Boolean Functions and DNF Representations

Any Boolean function can be expressed by a *disjunctive normal formula* (or DNF), using *literals* $u_1, u_2, \ldots, u_n, \bar{u}_1, \ldots, \bar{u}_n$, where the $\bar{u}_i$ are known as *negated literals*. A disjunctive normal formula is one of the form

$$T_1 \vee T_2 \vee \cdots \vee T_k,$$

where each T_l is a *term* of the form

$$T_l = \left(\bigwedge_{i \in P} u_i \right) \bigwedge \left(\bigwedge_{j \in N} \bar{u}_j \right),$$

for some disjoint subsets P, N of $\{1, 2, \ldots, n\}$.

Given a disjunctive normal form for a Boolean function, there may be a number of ways of simplifying it. For two Boolean functions f and g, we write $f \leq g$ if $f(x) \leq g(x)$ for all x; that is, if $f(x) = 1$ implies $g(x) = 1$. Similarly, for two Boolean formulae ϕ, ψ, we shall write $\phi \leq \psi$ if, when f and g are the functions represented by ϕ and ψ, then $f \leq g$. A term T of a DNF is said to *absorb* another term T' if $T' \leq T$. For example, $T = \bar{u}_1 u_4$ absorbs the term $T' = \bar{u}_1 u_3 u_4$. That is, whenever T' is true, so is T. This means, for example, that the formula

$$\bar{u}_1 u_4 \vee u_1 u_2 \bar{u}_3 \vee \bar{u}_1 u_3 u_4$$

is equivalent to $\bar{u}_1 u_4 \vee u_1 u_2 \bar{u}_3$. A term T is an *implicant* of f if $T \leq f$; in other words, if T true implies f true. The terms in any DNF representation of a function f are implicants of f. The most important type of implicants are the *prime implicants*. These are implicants with the additional property that there is no other implicant of f absorbing T. Thus, a term is a prime implicant of f if it is an implicant, and the deletion of any literal from T results in a non-implicant T' of f (meaning that there is some x such that $T'(x) = 1$ but $f(x) = 0$). If we form the disjunction of all prime implicants of f, we have a particularly important DNF representation of f.

2.3 *From DNF to Threshold Network*

Suppose that (T, F) is a partially-defined Boolean function and that the Boolean function f is some 'extension' of (T, F), meaning that $f(x) = 1$ for $x \in T$ and $f(x) = 0$ for $x \in F$. Let ϕ be a DNF formula for f. Suppose $\phi = T_1 \vee T_2 \vee \cdots \vee T_k$, where each T_i is a term of the form $T_i = \left(\bigwedge_{j \in P_i} u_j \right) \bigwedge \left(\bigwedge_{j \in N_i} \bar{u}_j \right)$, for some disjoint subsets P_i, N_i of $\{1, 2, \ldots, n\}$. We form a network with k hidden units, one corresponding to each term of the DNF. Labelling these threshold units $1, 2, \ldots, k$, we set the weight vector $w^{(i)}$ from the inputs to hidden threshold unit i to correspond directly to T_i, in the sense that $w_j^{(i)} = 1$ if $j \in P_i$, $w_j^{(i)} = -1$ if $j \in N_i$, and $w_j^{(i)} = 0$ otherwise. We take the threshold $\theta^{(i)}$ on hidden unit i to be $|P_i|$. We set the weight on the connection between each hidden threshold unit and the output unit to be 1, and the threshold on the output unit to be $1/2$. That is, we set β to be the all-1 vector of dimension k, and set the threshold ϕ to be $1/2$. It is clear that hidden threshold unit i outputs 1 on input x precisely when x satisfies the term T_i, and that the output unit computes the 'or' of all the outputs of the hidden units. Thus, the output of the network is the disjunction of the terms T_i, and hence equals f.

Note that this does not describe a unique threshold network representing the pdBf (T,F), for there may be many choices of extension function f and, given f, there may be many possible choices of DNF for f. In the case in which $T \cup F = \{0,1\}^n$, so that the function is fully defined, we could, for the sake of definiteness, use the particular DNF formula described above, the disjunction of all 'prime implicants'.

In general, a simple counting argument establishes that, whatever method is being used to represent Boolean functions by threshold networks, for most Boolean functions a high number of units will be required in the resulting network. Explicitly, suppose we have an n-input threshold network with one output and one hidden layer comprising k threshold units. Then, since the number of threshold functions is at most 2^{n^2} (see [1, 3], for instance), the network computes no more than $(2^{n^2})^{k+1}$ different Boolean functions, this being an upper bound on the number of possible mappings from the input set $\{0,1\}^n$ to the vector of outputs of all the $k+1$ threshold units. This bound, $2^{n^2(k+1)}$ is, for any fixed k, a tiny proportion of all the 2^{2^n} Boolean functions and, to be comparable, we need $k = \Omega(2^n/n^2)$. (This is a very quick and easy observation. For more detailed bounds on the sizes of threshold networks required to compute general and specific Boolean functions, see [10], for instance.)

It is easy to give an explicit example of a function for which this standard method produces an exponentially large threshold network. The parity function f on $\{0,1\}^n$ is given by $f(x) = 1$ if and only if x has an odd number of ones. It is well known that any DNF formula ϕ for f must have 2^{n-1} terms. To see this, note first that each term of ϕ must have degree n. For, suppose some term T_i contained fewer than n literals, and that neither u_j nor $\bar{u}_j$ were present in T_i. Then there are $x, y \in \{0,1\}^n$ which are true points of T_i, but which differ only in position j. Then, since T_i is a term in the DNF representation of the parity function f, we would have $f(x) = f(y) = 1$. But this cannot be: one of x, y will have an odd number of entries equal to 1, and one will have an even number of such entries. It follows that each term must contain n literals, in which case each term has only one true point, and so we must have 2^{n-1} distinct terms, one for each true point. It follows that the resulting network has 2^{n-1} threshold units in the hidden layer.

3 Decision Lists and Threshold Decision Lists

We now present a different approach to the problem of finding a threshold network representation of a partially-defined Boolean function. To explain this, we first discuss decision lists and threshold decision lists.

3.1 Decision Lists

We start by describing *decision lists*, introduced by Rivest [9]. Suppose that G is any set of Boolean functions. A function $f : \{0,1\}^n \to \{0,1\}$ is said to be a *decision list* based on G if for some positive integer r, there are functions $f_1, f_2, \ldots, f_r \in G$ and bits $c_1, c_2, \ldots, c_r \in \{0,1\}$ such that f acts as follows. Given an example y, we first evaluate $f_1(y)$. If $f_1(y) = 1$, we assign the value c_1 to $f(y)$; if not, we evaluate $f_2(y)$, and if $f_2(y) = 1$ we set $f(y) = c_2$, otherwise we evaluate $f_3(y)$, and so on. If y fails to satisfy any f_i then $f(y)$ is given the default value 0. The evaluation of a decision list f can therefore be thought of as a sequence of `'if then else'` commands, as follows:

```
if f_1(y) = 1 then set f(y) = c_1
          else if f_2(y) = 1 then set f(y) = c_2
                          ...
                          ...
                          else if f_r(y) = 1 then set f(y) = c_r
                                  else set f(y) = 0.
```

We define $DL(G)$, the class of *decision lists based on* G, to be the set of finite sequences

$$f = (f_1, c_1), (f_2, c_2), \ldots, (f_r, c_r)$$

such that $f_i \in G$, $c_i \in \{0, 1\}$ for $1 \leq i \leq r$. The values of f are defined by $f(y) = c_j$ where $j = \min\{i : f_i(y) = 1\}$, or 0 if there are no j such that $f_j(y) = 1$. We call each f_j a *test* (or, following Krause [6], a *query*) and the pair (f_j, c_j) is called a *term* of the decision list.

3.2 Threshold Functions and Threshold Decision Lists

We now consider the class of decision lists in which the tests are threshold functions. We shall call such decision lists *threshold decision lists*, but they have also been called *neural* decision lists [7] and *linear* decision lists [13]. Formally, a threshold decision list

$$f = (f_1, c_1), (f_2, c_2), \ldots, (f_r, c_r)$$

has each $f_i : \mathbb{R}^n \to \{0, 1\}$ of the form $f_i(x) = \text{sgn}(\langle w_i, x\rangle - \theta_i)$ for some $w_i \in \mathbb{R}^n$ and $\theta_i \in \mathbb{R}$. The value of f on $y \in \mathbb{R}^n$ is $f(y) = c_j$ if $j = \min\{i : f_i(y) = 1\}$ exists, or 0 otherwise (that is, if there are no j such that $f_j(y) = 1$).

3.3 A Geometrical Interpretation: Iterative Linear Separation

Threshold decision lists are, in fact, quite a natural way in which to classify points, and a useful geometrical motivation can be given. Suppose we are given a partially-defined Boolean function (T, F). We can use a hyperplane to separate off a set of points all having the same classification label (that is, all of which are from T, or all of which are from F). At least one point can always be separated off in this way. For, given any $x \in \{0, 1\}^n$, x and $\{0, 1\}^n \setminus \{x\}$ are linearly separable. To see this, we can suppose, without any loss of generality, that x is the origin. Then the hyperplane with equation $\sum_{i=1}^{n} x_i = 1/2$ achieves the required separation. (Note that this argument is contingent on the geometry of $\{0, 1\}^n$. For more general subsets of $\mathbb{R}^n$, some additional properties, such as general position, would need to hold to make the argument work.) The points that have been 'chopped off' can then be removed from consideration and the procedure iterated until no points remain. In general, we would hope to be able to separate off more than one point at each stage, but the argument given above establishes that, at each stage, at least one point can indeed be 'chopped off', so since the set of points is finite, the procedure does indeed terminate.

We may regard the chopping procedure as a means of constructing a threshold decision list consistent with the data set. If, at stage i of the procedure, the hyperplane with equation $\sum_{i=1}^{n} \alpha_i y_i = \theta$ chops off points all having label j, with these points in the half-space with equation $\sum_{i=1}^{n} \alpha_i y_i \geq \theta$, then we take as the ith term of the threshold decision list the pair (f_i, j), where $f_i \leftarrow [\alpha, \theta]$. Therefore, given any partially-defined Boolean function (T, F), there will always be some threshold decision list representing the pdBf.

3.4 A Related Approach

This sequential linear separation, or 'chopping', procedure is similar to one employed by Jeroslow [5], but at each stage in his procedure, only examples from T may be 'chopped off' (and one cannot choose instead to chop off a subset of points from F).

Note that if the 'chopping' method of constructing a threshold decision list is applied to the sequence of hyperplanes resulting from the Jeroslow method, a restricted form of decision list results, namely one in which all terms are of the form $(f_i, 1)$. But such a decision list is quite simply the disjunction $f_1 \vee f_2 \vee \cdots$. For Boolean functions, the problem of decomposing a function into the disjunction of threshold functions has been given substantial consideration by Hammer *et al.* [4] and Zuev and Lipkin [14]. Hammer *et al.* defined the *threshold number* of a Boolean function to be the minimum s such that f is a disjunction of s threshold functions, and they showed that there is an increasing function with threshold number $\binom{n}{n/2}/n$. (A function is increasing if, when $f(x) = 1$ and $x_i = 0$, then $f(x + e_i) = 1$ too, where e_i is the unit basis vector with ith entry equal to 1 and all other entries equal to 0.) Zuev and Lipkin showed that almost all increasing functions have this order of threshold number, and that almost all Boolean functions have a threshold number that is $\Omega(2^n/2)$ and $O(2^n \ln n/n)$.

We give an example for illustration, which demonstrates the advantages to be gained by the threshold decision list approach over the Jeroslow approach.

Example: Consider again the parity function f on $\{0,1\}^n$, given by $f(x) = 1$ if and only if x has an odd number of ones. We first find a hyperplane such that all points on one side of the plane are either positive or negative. It is clear that all we can do at this first stage is chop off one of the points since the nearest neighbours of any given point have the opposite classification. Let us suppose that we decide to chop off the origin. We may take as the first hyperplane the plane with equation $y_1 + y_2 + \cdots + y_n = 1/2$. We then ignore the origin and consider the remaining points. We can next chop off all neighbours of the origin, all the points which have precisely one entry equal to 1. All of these are positive points and the hyperplane $y_1 + y_2 + \cdots + y_n = 3/2$ will separate them from the other points. These points are then deleted from consideration. We can continue in this way. The procedure iterates n times, and at stage i in the procedure we 'chop off' all data points having precisely $(i-1)$ ones, by using the hyperplane $y_1 + y_2 + \cdots + y_n = i - 1/2$, for example. (These hyperplanes are in fact all parallel, but this is not in general possible.) So we can represent the parity function by a threshold decision list with n terms. By contrast, Jeroslow's method requires 2^{n-1} iterations, since at each stage it can only 'chop off' one positive point: that is, it produces a disjunction of threshold functions (or a special type of threshold decision list) with an exponential number of terms.

3.5 Algorithmics

The chopping procedure as we have described it is in some ways merely a device to help us see that threshold decision lists have a fairly natural geometric interpretation. But the algorithmic practicalities have been investigated by Marchand *et al.* [7, 8] and Tajine and Elizondo [11]. Marchand *et al.* derive a greedy heuristic for constructing a sequence of 'chops', where the aim is to separate as large a set (all of the same class) as possible at each stage. This relies on an incremental heuristic for the NP-hard problem of finding at each stage a hyperplane that chops off as many remaining points as possible. Tajine and Elizondo consider batch and incremental and modular algorithms and also focus on greedy strategies.

4 Threshold Networks from Threshold Decision Lists

4.1 From a Threshold Decision List to a Threshold Network with One Hidden Layer

We now show how we can make use of the chopping procedure to find a threshold network representing a given Boolean function by giving an explicit way in which a threshold decision list can be represented by a threshold network with one hidden layer.

Theorem 1. *Suppose we have a threshold decision list*

$$f = (f_1, c_1), (f_2, c_2), \ldots, (f_k, c_k)$$

in which f_i is represented by weight vector $w^{(i)}$ and threshold $\theta^{(i)}$, so that $f_i \leftarrow [w^{(i)}, \theta^{(i)}]$. Consider a threshold network architecture having n inputs, k threshold units in a single hidden layer, and one output. Let ω be the state given as follows:

$$\omega = (w^{(1)}, \theta^{(1)}, w^{(2)}, \theta^{(2)}, \ldots, w^{(k)}, \theta^{(k)}, \beta, 1),$$

where

$$\beta = (2^{k-1}(2c_1-1), 2^{k-2}(2c_2-1), \ldots, 2(2c_{k-1}-1), (2c_k-1));$$

that is, $\beta_i = 2^{k-i}(2c_i - 1)$. *Then* f_ω, *the function computed by the network in state* ω, *equals* f.

Proof: We prove the result by induction on k, the length of the decision list (and number of hidden threshold units in the network).

The base case is $k = 1$. Since the default output of any decision list is 0, we may assume that f takes the form $f = (f_1, 1)$ where $f_1 \leftarrow [w, \theta]$ for some $w \in \mathbb{R}^n$ and $\theta \in \mathbb{R}$. Then, β is the single number $2^{1-1}(2c_1 - 1) = 1$ and $\phi = 1$. So

$$f_\omega(y_1y_2\ldots y_n) = \text{sgn}\left(\text{sgn}\left(\sum_{j=1}^{n} w_j^{(i)} y_j - \theta^{(i)}\right) - 1\right) = \text{sgn}\left(\sum_{j=1}^{n} w_j^{(i)} y_j - \theta^{(i)}\right) = f_1(y_1y_2\ldots y_n),$$

so $f_\omega = f_1 = f$.

Now suppose that the result is true for threshold decision lists of length k, where $k \geq 1$. Consider a threshold decision list

$$f = (f_1, c_1), (f_2, c_2), \ldots, (f_k, c_k), (f_{k+1}, c_{k+1}).$$

Let g denote the threshold decision list

$$g = (f_2, c_2), \ldots, (f_k, c_k), (f_{k+1}, c_{k+1}).$$

Then, the inductive assumption implies that, for all y,

$$g(y) = \text{sgn}\left(\sum_{i=1}^{k} 2^{k-i}(2c_{i+1} - 1) f_{i+1}(y) - 1\right) = \text{sgn}(G(y))),$$

say. What we need to prove is that for all y,

$$f(y) = \text{sgn}(F(y)),$$

where

$$F(y) = \sum_{i=1}^{k+1} 2^{k+1-i}(2c_i - 1) f_i(y) - 1.$$

Now,

$$\begin{aligned}
F(y) &= 2^k(2c_1 - 1) f_1(y) + \sum_{i=2}^{k+1} 2^{k+1-i}(2c_i - 1) f_i(y) - 1 \\
&= 2^k(2c_1 - 1) f_1(y) + \sum_{i=1}^{k} 2^{k-i}(2c_{i+1} - 1) f_{i+1}(y) - 1 \\
&= 2^k(2c_1 - 1) f_1(y) + G(y).
\end{aligned}$$

Now, suppose $f_1(y) = 0$. In this case, by the way in which decision lists are defined to operate, we should have $f(y) = g(y)$. This is indeed the case, since

$$\mathrm{sgn}(F(y)) = \mathrm{sgn}(2^k(2c_1 - 1)f_1(y) + G(y)) = \mathrm{sgn}(0 + G(y)) = \mathrm{sgn}(G(y)) = g(y).$$

Suppose now that $f_1(y) = 1$. In this case we have $f(y) = c_1$ and so we need to verify that $\mathrm{sgn}(F(y)) = c_1$. We have

$$\begin{aligned}(2c_1 - 1)F(y) &= 2^k(2c_1 - 1)^2 f_1(y) + (2c_1 - 1)G(y)\\ &= 2^k + (2c_1 - 1)\sum_{i=1}^{k} 2^{k-i}(2c_{i+1} - 1)f_{i+1}(y) - 1\\ &\geq 2^k - \sum_{i=1}^{k} 2^{k-i} - 1\\ &= 2^k - (2^k - 1) - 1\\ &= 0.\end{aligned}$$

That is, $(2c_1 - 1)F(y) \geq 0$, so $\mathrm{sgn}(F(y)) = \mathrm{sgn}(2c_1 - 1) = c_1$. This completes the proof.

4.2 *Using Other Types of Threshold Network*

Marchand *et al.* [7, 8] and Tajine and Elizondo [11] have also studied the construction of threshold networks through a consideration of how the points to be classified can be separated iteratively by hyperplanes. However, the threshold networks arising in [7] and (from the batch algorithm) in [11], have different architectures to those constructed above, in that there are connections between hidden units (making the networks have more than one layer). By contrast, like the standard representation based on DNF, our construction gives a network with only one hidden layer. A characteristic feature of decision lists which must be captured by the corresponding threshold networks is the 'if-then-else' nature of their definition: there is a *precedence* or hierarchy among the tests. The first test is conducted and, if passed, the output is determined. Only if the first test is failed, do we move on to the next test. In the construction of Theorem 1, the precedence structure is encoded into the network by the exponentially-decreasing weights in the β-vector: the output if the first hidden unit is weighted twice as much as that of the second, and so on. In [7, 11], the precedence structure is built in with lateral connections between hidden units. For instance, in [7], the network constructed has a 'cascade' structure: the hidden threshold units are labelled 1 to k and there are connections between unit i and unit j for all $j < i$. The weights on these connections are large enough to enable the output of unit i to dominate (or inhibit) that of unit j.

5 Comparison with an Approach Based on Disjunctive Normal Form

5.1 Comparing the DNF-Based Approach and the Threshold Decision List Approach

The parity function demonstrates that the representation arising from Theorem 1 can differ considerably from the one described earlier. For, we have seen that the parity function can be represented by a threshold decision list with n terms, and hence the network given by Thoerem 1 has only n hidden units. By contrast, as noted earlier, the standard DNF-based construction will, necessarily, have at least 2^{n-1} hidden units.

A useful observation in comparing the two approaches is the following: if T is any term of a DNF formula, then T can be represented by a threshold function. This is quite easy to see and, indeed, is implicit in our description of the standard construction of a network from a DNF. For, suppose that

$$T = \left(\bigwedge_{j \in P} u_j\right) \wedge \left(\bigwedge_{j \in N} \bar{u}_j\right),$$

where $P \cap N = \emptyset$. Then $T \leftarrow [w, |P|]$ where $w_j = 1$ if $j \in P$, $w_j = -1$ if $j \in N$, and $w_j = 0$ otherwise. So if $\phi = T_1 \vee T_2 \vee \cdots \vee T_k$ is a DNF representation of the function f, then f is also represented by the threshold decision list

$$(T_1, 1), (T_2, 1), \ldots, (T_k, 1).$$

Applying Theorem 1 now to this threshold decision list would give a threshold network representing f. That network would have exactly the same structure as the one obtained by using the standard DNF-based method, using DNF formula ϕ. (However, the weights from the hidden layer to the output would be different, with exponentially decreasing, rather than constant, values.) What this demonstrates is that, in particular, there is always a threshold decision list representation whose length is no more than that of any given DNF representation of the function. There may, as in the case of parity, be a significantly shorter threshold decision list. So the decision list approach (and application of Theorem 1) will, for any function (or partially-defined function), in the best case, give a network that is no larger than that obtained by the standard method.

6 Conclusions

We have shown that a natural approach to data classification by successive linear separation can be used to construct threshold networks of simple architecture to represent Boolean or partially-defined Boolean functions. Such an approach differs from previous constructions which have also been based on iterative linear separation, in that the networks constructed have only one hidden layer. Furthermore, it can always produce a network that is no larger than that which follows from the standard translation from a Boolean function's disjunctive normal form into a threshold network.

References

1. Anthony, M.: Discrete Mathematics of Neural Networks: Selected Topics. Society for Industrial and Applied Mathematics, Philadeplhia (2001)
2. Anthony, M.: On data classification by iterative linear partitioning. Discrete Applied Mathematics 144(1-2), 2–16 (2004)
3. Cover, T.M.: Geometrical and Statistical Properties of Systems of Linear Inequalities with Applications in Pattern Recognition. IEEE Trans. on Electronic Computers EC-14, 326–334 (1965)
4. Hammer, P.L., Ibaraki, T., Peled., U.N.: Threshold numbers and threshold completions. Annals of Discrete Mathematics 11, 125–145 (1981)
5. Jeroslow, R.G.: On defining sets of vertices of the hypercube by linear inequalities. Discrete Mathematics 11, 119–124 (1975)
6. Krause, M.: On the computational power of boolean decision lists. In: Alt, H., Ferreira, A. (eds.) STACS 2002. LNCS, vol. 2285, pp. 372–383. Springer, Heidelberg (2002)
7. Marchand, M., Golea, M.: On Learning Simple Neural Concepts: from Halfspace Intersections to Neural Decision Lists. Network: Computation in Neural Systems 4, 67–85 (1993)
8. Marchand, M., Golea, M., Ruján, P.: A convergence theorem for sequential learning in two-layer perceptrons. Europhys. Lett. 11, 487 (1990)
9. Rivest, R.R.: Learning Decision Lists. Machine Learning 2(3), 229–246 (1987)
10. Siu, K.Y., Rowchowdhury, V., Kalaith, T.: Discrete Neural Computation: A Theoretical Foundation. Prentice Hall, Englewood Cliffs (1995)
11. Tajine, M., Elizondo, D.: Growing methods for constructing Recursive Deterministic Perceptron neural networks and knowledge extraction. Artificial Intelligence 102, 295–322 (1998)
12. Tajine, M., Elizondo, D.: The recursive deterministic perceptron neural network. Neural Networks 11, 1571–1588 (1998)
13. Turán, G., Vatan, F.: Linear decision lists and partitioning algorithms for the construction of neural networks. In: Foundations of Computational Mathematics: selected papers of a conference, Rio de Janeiro, pp. 414–423. Springer, Heidelberg (1997)
14. Zuev, A., Lipkin, L.I.: Estimating the efficiency of threshold representations of Boolean functions. Cybernetics 24, 713–723 (1988); Translated from Kibernetika (Kiev) 6, 29–37 (1988)

Self-Optimizing Neural Network 3

Adrian Horzyk

Abstract. This paper describes an efficient construction of a partially-connected multilayer architecture and a computation of weight parameters of Self-Optimizing Neural Network 3 (SONN-3) that can be used as a universal classifier for various real, integer or binary input data, even for highly non-separable data. The SONN-3 consists of three types of neurons that play an important role in a process of extraction and transformation of important features of input data in order to achieve correct classification results. This method is able to collect and to appropriately reinforce values of the most important input features so that achieved generalization results can compete with results achieved by other existing classification methods. The most important aspect of this method is that it neither loses nor rounds off any important values of input features during this computation and propagation of partial results through a neural network, so the computed classification results are very exact and accurate. All the most important features and their most distinguishing ranges of values are effectively compressed and transformed into an appropriate network architecture with weight values. The automatic construction process of this method and all optimization algorithms are described here in detail. Classification and generalization results are compared by means of some examples.

Keywords: Lossy binarization, data conversion, discrimination, architecture construction and optimization, classification, compression of representation.

1 Introduction

Nowadays, various types of constructive neural networks and other incremental learning algorithms play an increasingly important role in neural computations. These algorithms usually provide an incremental method of building neural networks with reduced topologies for classification problems. Furthermore, this method

Adrian Horzyk
AGH University of Science and Technology, Department of Automatics
e-mail: horzyk@agh.edu.pl

L. Franco et al. (Eds.): Constructive Neural Networks, SCI 258, pp. 83–101.
springerlink.com

produces a multilayer network architecture, which together with the weights, is determined automatically by the introduced constructive algorithm. Another advantage of these algorithms is that convergence is guaranteed by the method [1], [3],[4], [6], [8], [13]. A growing amount of the current research in neural networks is oriented towards this important topic. Providing constructive methods for building neural networks can potentially create more compact models which can easily be implemented in hardware and used in various embedded systems.

There are many classification methods in the world today. Many of them need to set up some training parameters, initialize network parameters or build an initial network architecture before a training process can be initiated. Some of them suffer from limitations in the input value ranges or from the curse of the dimensionality problem [1], [8]. Some methods favour certain data values or treat data in a different way depending on the number of cases that represent classes [1], [3]. Not many methods can automatically manage, reduce or simplify an input data space and their training processes are sometimes unsuccessful when exposed to the influence of minor values of weakly differentiating input features. Moreover, the model size (e.g. the architecture size) and the time necessary for training and evaluating are also significant. Many neural methods compete for a better generalization using various training strategies and parameters, neuron functions, various quantities of layers, neurons and interconnections [1], [3], [8], [13]. This paper confirms that generalization potential is also hidden in suitable preprocessing of training data and an appropriate covering of an input data space.

This paper describes constructive Self-Optimizing Neural Network 3 [4] that is devoided of many of the limitations described above and thanks to the proposed formulas for estimating of discrimination of input features it can build a suitable neural network architecture and precisely compute weight values. This neural solution can automatically reduce and simplify an input data space, specifically converting real input vectors into bipolar binary vectors $\{-1;+1\}$, which can be used to aggregate equal binary values into a compact neural model of training data. A human brain works in a very similar way. It gathers data using sensors that are able to convert various physical data from the surrounding world into some frequencies of binary signals. Binary information is then used to perform the relevant computations until actuators are activated and the information is transformed into physical reactions [6], [10], [12]. It is very interesting that nature has also chosen binary signals to carry out complex computations in our brains instead of real signals. The SONN-3 converts all real vectors into binary ones using the specialist algorithm (ADLBCA), which cleverly transforms values of real input features into bipolar binary values so that discrimination properties of real input data are not lost. It also automatically rejects all useless values of real input features that have minor significance for classification. All complex computations for discrimination of classes are carried out on the bipolar binary values. Network outputs take the real values from a continuous range $[-1;+1]$ to emphasize degrees of similarity of input vector to defined classes.

There are a few algorithms that can also reduce or simplify input data space, e.g. PCA, ICA, roughs sets, SSV trees [7], [9], [11], but they do not transform real inputs into binary inputs, so it is not easy to compare their performance.

The SONN-3 performs many analyses on bipolar binary vectors to optimize a neural network model. It aggregates the most discriminating and most frequent similar values for various training cases, transforms them into an appropriate network architecture and reinforces them in accordance with their discrimination properties, rejecting all useless features for classification and builds a classification model. Thanks this ability, this method is devoided of the curse of the ddimensionality problem.

The SONN-3 analytically computes a network architecture and weight values, so it is also devoided of convergence problems. It always finds a solution that differentiates all defined classes provided that training cases are not contradictory, e.g. two or more cases from different classes can be differentiated neither by a single value nor by their combinations. If training data are not contradictory, SONN-3 always produces a neural solution that always correctly classifies all training data. It also generalizes well. This method does not memorize training data using a huge number of network parameters but builds a very compact neural model (figs. 8, 9) that represents only major ranges of values of real input features for all classes and considers the discrimination of all training cases of different classes.

Chapter 2 describes construction elements and development of a network architecture of the SONN-3. Global Discrimination Coefficients used for optimization processes of SONN-3 are introduced in chapter 3. Chapter 4 describes the construction of a Lossy Binarizing Subnetwork, which is an input subnetwork of a SONN-3 network. Chapter 5 combines the Lossy Binarizing Subnetwork with an Aggregation Reinforcement Subnetwork that aggregates and performes lossless compression of same values and appropriately reinforces inputs. Chapter 6 describes a Maximum Selection Subnetwork that selects maximum outputs for each class and produces an ultimate classification. Comparisons of various soft-computing methods can be found in chapter 7.

2 The SONN-3 Architecture and Construction Elements

The SONN-3 consists of three subnetworks specializing in different tasks:

The first part of the SONN-3 is a single layer Lossy Binarizing Subnetwork (LBS) (figs. 4, 7), consisting of Lossy Binarizing Neurons (LBNs) (fig. 1a), which transform special ranges of real input values into binary ones. The ranges are computed by the ADLBCA algorithm described in the fourth chapter of this paper. The LBN layer is responsible for binarizing and emphasizing important ranges of real input features that will be used to construct a neural model built using the most differentiating features and their ranges of values.

The second subnetwork (Aggregation Reinforcement Subnetwork - ARS) (figs. 6, 7) takes bipolar binary outputs from the first subnetwork LBS. The ARS consists of a number of layers and a number of neurons. The numbers of layers and

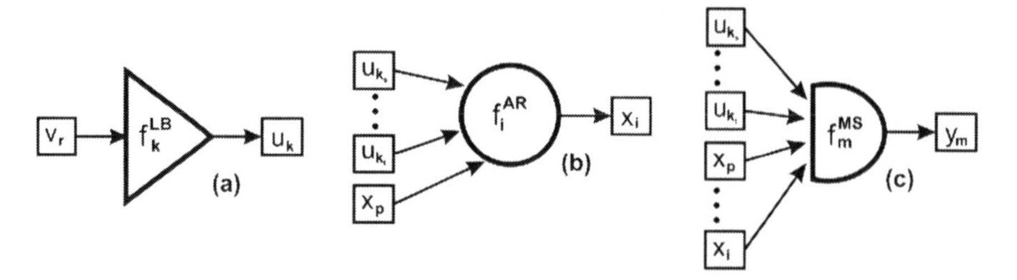

Fig. 1 Three types of SONN-3 neurons: (a) Lossy Binarizing Neuron (LBN) (b) Aggregation Reinforcement Neuron (ARN), (c) Maximum Selection Neuron (MSN)

neurons are data dependent. If data are not very correlated - there is a small number of layers and *vice versa*. This subnetwork is responsible for extraction, counting and aggregation of equal binary values of various cases and various classes and for their appropriate reinforcement depending on their discrimination properties computed after the Global Discrimination Coefficients (GDCs) (1) described in the third chapter of this paper. The aggregation of equal binary values enables the SONN-3 to lossless compress equal binary values of training data and transform these values into single connections and a special architecture of Aggregation Reinforcement Neurons (ARN) (fig. 1b). Each ARN represents a subset of training cases. Each one is determined during the special data division process described in the fifth chapter of this paper. The subset of training cases can consist of training cases of one or of many classes. ARNs that represent training cases of more than a single class are intermediate (hidden) neurons that do not produce their outputs for the next subnetwork (MSS) but only for some other ARNs.

The third subnetwork (Maximum Selection Subnetwork - MSS) (fig. 7) consists of a single layer of Maximum Selection Neurons (MSNs) (fig. 1c). Each defined class is represented by a single MSN. Each MSN is responsible for the selection of a maximum output value for the class it represents. The maximum output value is taken from all outputs of all parentless ARNs that represent a subset of training cases of a single class which is the same as a class for which MSN is created.

3 Global Discrimination Coefficients

The Global Discrimination Coefficients (GDCs) play the most significant role in all optimization processes during network construction and the computation of weights. The GDCs are computed for all LBS bipolar binary outputs (i.e. all ARS bipolar binary inputs) computed for all training cases separately. They precisely determine the discrimination ability of all LBS bipolar binary outputs (i.e. all ARS bipolar binary inputs) and are insensitive to the differences in a number of training cases that represent various defined classes thanks to the normalization factor Q^m (1). They can determine the representativeness of a given bipolar binary value for a given class thanks to the quotients P_k^m/Q^m or N_k^m/Q^m (1). In other words, the more representative for a given class a given bipolar binary value $+1$ or -1 is the bigger value has the quotients P_k^m/Q^m or N_k^m/Q^m respectively (1). The GDCs also include the coefficient defining the differentiating ability of a given bipolar binary value for a given class from other classes represented as the sum normalized by $M-1$ in equation 1. In other words, the more frequent a given bipolar binary value $+1$ or -1 is in other classes the less discriminating is this value for a class of a considered training case.

The GDCs (1) are computed for each k-th bipolar binary feature value v_k for each n-th raw training case v_n of the m-th class C^m:

$$\forall_{m\in\{1,\ldots,M\}}\forall_{u^n\in C^m}\forall_{n\in\{1,\ldots Q\}}\forall_{k\in\{1,\ldots K\}}:$$
$$d_{k+}^n = \begin{cases} \frac{P_k^m}{(M-1)\cdot Q^m}\sum_{h=1\wedge h\neq m}^{M}\left(1-\frac{P_k^h}{Q^h}\right) & if\ f_k^{LB}(v_r^n,R)>0 \wedge v^n\in C^m \\ 0 & if\ f_k^{LB}(v_r^n,R)<0 \wedge v^n\in C^m \end{cases}$$
$$d_{k-}^n = \begin{cases} \frac{N_k^m}{(M-1)\cdot Q^m}\sum_{h=1\wedge h\neq m}^{M}\left(1-\frac{N_k^h}{Q^h}\right) & if\ f_k^{LB}(v_r^n,R)<0 \wedge v^n\in C^m \\ 0 & if\ f_k^{LB}(v_r^n,R)>0 \wedge v^n\in C^m \end{cases} \quad (1)$$

where M denotes a number of defined classes in training data, Q is a number of all training cases, R is a range for which binarized data take value $+1$ (6), K is a number of bipolar binary inputs of the ARS (or a number of bipolar binary outputs of the LBS), V is a set of real training data input vectors $\{v^1, v^2, \ldots, v^n\}$, U is the set of binary transformed training data input vectors $\{u^{1,R}, u^{2,R}, \ldots, u^{n,R}\}$, v_k^n is the k-th real input feature value for the n-th training case, $u_r^{n,R}$ is the r-th bipolar binary input feature value for the range R and for the n-th training case, the f_k^{LB} is the function computing an output value for an LBN and P_k^m, N_k^m, Q^m are defined by the following equations:

$$u_k^{n,R} = f_k^{LB}(v_r^n, R) \quad (2)$$

$$\forall_{m\in\{1,\ldots,M\}}\ Q^m = \left\|\left\{u^{n,R}\in U\bigcap C^m : n\in\{1,\ldots Q\}\right\}\right\| \quad (3)$$

$$\forall_{m\in\{1,\ldots,M\}}\forall_{k\in\{1,\ldots,K\}} : P_k^m = \sum_{u_k^{n,R}\in\left\{u\in U\vee C^m : f_k^{LB}(v_r^n)>0,\, n\in\{1,\ldots,Q\}\right\}} u_k^{n,R} \quad (4)$$

$$\forall_{m\in\{1,\ldots,M\}}\forall_{k\in\{1,\ldots,K\}} : N_k^m = \sum_{u_k^{n,R}\in\left\{u\in U\vee C^m : f_k^{LB}(v_r^n)<0,\, n\in\{1,\ldots,Q\}\right\}} -u_k^{n,R} \quad (5)$$

In order to compute GDCs, all training cases have to be available at the beginning of a construction and training process. It is impossible to compute GDCs if some parts of training cases are not available or when some parts of training cases change during a construction or training process. If training data change or are supplemented then GDCs have to be computed once again and a construction process of

SONN-3 has to be repeated from the very beginning. This drawback is not very significant because training data rarely change during a construction or training process and even if this occurs a construction process of SONNs-3 is so quick that it can be quickly repeated in order to build an improved solution based on a new architecture and new values of weights.

GDCs allow SONN-3 to globally estimate the significance of each bipolar binary value computed for each real input feature value. The GDCs are the basis for constructing an ARS architecture and computing the values of ARN weights.

4 Construction of a Lossy Binarizing Subnetwork

The Lossy Binarizing Subnetwork (LBS) is responsible for a suitable transformation of selected real values of input features into a set of special ranges R that that differentiate classes very well. Next, values of input features inside these ranges are transformed into the value $+1$ for each training case and into the value -1 otherwise (6). The question is how to find these ranges optimally?

The presented algorithm starts its lossy binary conversion from an input data analysis that takes into consideration the following goals:

- the lossy binary conversion ranges should be wide in order to cover important parts of an input data space sufficiently and to achieve good generalization,

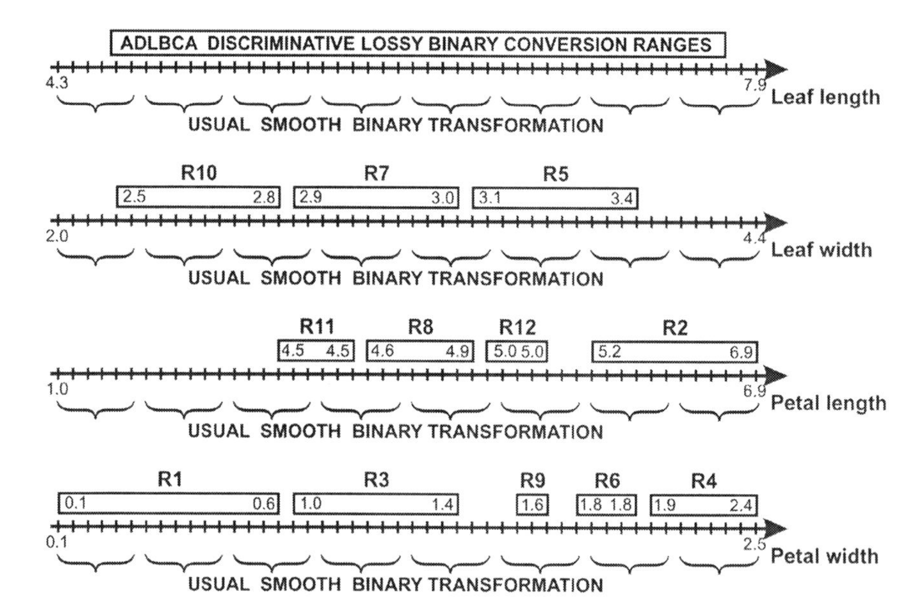

Fig. 2 The comparison of the simple smooth and ADLBCA transformations of real values into binary ones for the Iris data from the ML Repository.

- the number of lossy binary conversion ranges should be as minimal as possible in order to simplify or even reduce the binary input data space and computational cost of the classification method that will use it,
- the computed lossy binary conversion ranges should enable the ARS to discriminate all training cases of all classes provided that TD are not contradictory,
- a discriminative property of lossy binary conversion ranges should be estimated using statistical analysis of training data,
- the computational and memory costs of the method should be as low as possible.

The main goal of this algorithm is to find a possibly minimal set of discriminative lossy binary conversion ranges (DLBCRs) for all real data input features and to convert the values v_t from these ranges $R = [L^t_m; P^t_m]$ into the value $+1$ and other values outside these ranges $R = [L^t_m; P^t_m]$ into the value -1 (6). Moreover, an appropriate selection of these ranges can predifferentiate and prediscriminate some training cases and help the following subnetwork (e.g. ARS - described in the next chapter) ultimately to discriminate them. Real input values from these ranges R are transformed into bipolar binary values $\{+1; -1\}$ using Lossy Binarizing Neurons (LBNs) (fig. 1a) that compute their outputs using equation (6). First, real data inputs have to be sorted and indexed separately for each input feature. Figure 2 illustrates the sorted Iris data after all input features. The heapsort algorithm should be used, which computational cost is always $O(nlogn)$. The stability of the sorting algorithm does not matter for this method. After all input data features are sorted, the algorithm starts to search for a minimal set of DLBCRs taking into account the following criteria:

1. The selected range should contain as many cases of the same class and as few cases of different classes as possible,
2. The ranges containing cases of a smaller number of classes are preferred. The best discriminative ranges contain cases from a single class.
3. The ranges can contain training cases from other classes only if they are discriminated by other ranges.

$$f^{LB}_k(v_r, R) = \begin{cases} +1 & L^k_m \leq v_r \leq P^k_m \; where \; R = [L^k_m; P^k_m] \\ -1 & otherwise \end{cases} \tag{6}$$

These criteria are important to satisfy the requirements mentioned at the beginning of this paper, especially in view of good generalization properties. The algorithm sorts and indexes training data after all input features separately and then looks for optimal ranges R in the following way:

1. First, all training cases are marked as indiscriminated for all classes except the classes they represent.
2. Next, all yet indiscriminated data cases for all input features are looked through in the sorted order and a range containing a maximal number of training cases and the minimum number of classes are sought. Each range is described by an input feature and its range of values.
3. All yet indiscriminated cases for which the range was chosen are marked as discriminated for all classes which this range does not contain.

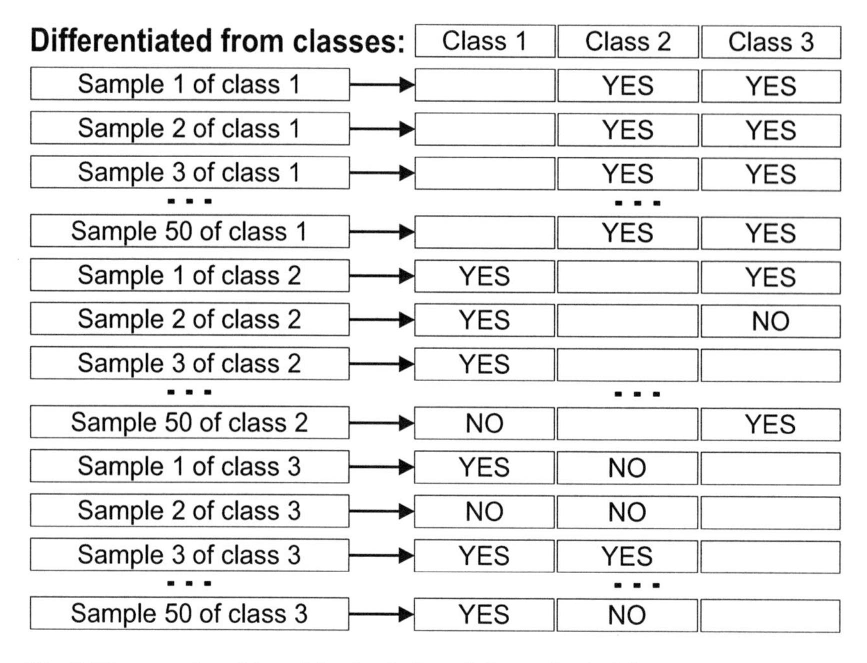

Differentiated from classes:	Class 1	Class 2	Class 3
Sample 1 of class 1		YES	YES
Sample 2 of class 1		YES	YES
Sample 3 of class 1		YES	YES
...		...	
Sample 50 of class 1		YES	YES
Sample 1 of class 2	YES		YES
Sample 2 of class 2	YES		NO
Sample 3 of class 2	YES		
...		...	
Sample 50 of class 2	NO		YES
Sample 1 of class 3	YES	NO	
Sample 2 of class 3	NO	NO	
Sample 3 of class 3	YES	YES	
...		...	
Sample 50 of class 3	YES	NO	

Fig. 3 The exemplar tables of the discriminated classes for the Iris cases.

4. Next, all fully discriminated training cases for all input features are looked through in order to remove their indexes from the sorted index tables.
5. Steps 2, 3 and 4 are repeated until all TD cases are discriminated from all other classes (fig. 3) or there is no more range to consider.
6. If not all training cases are discriminated and no more ranges can be used to carry out their discrimination, all the atomic ranges are chosen for all input features that contain indiscriminated training cases and are added to the previously selected ranges in steps 2, 3 and 4. The atomic ranges always contain a sequence of cases of one class or they may represent a few classes but the range is narrowed to a single value (fig. 2).

If step 6 occurs it means that some training cases are contradictory or they can be differentiated later by a soft-computing algorithm (e.g. the ARS) that can combine these ranges.

5 Construction of an Aggregation Reinforcement Subnetwork

The Aggregation Reinforcement Subnetwork (ARS) is a special kind of constructive ontogenic partially connected multilayer neural network that is able to evaluate input data from the range of $[-1;+1]$. The ARS is not trained, its architecture, connections and weights can be automatically and very quickly constructed, set up and computed using algorithms described in this chapter. This subnetwork neither

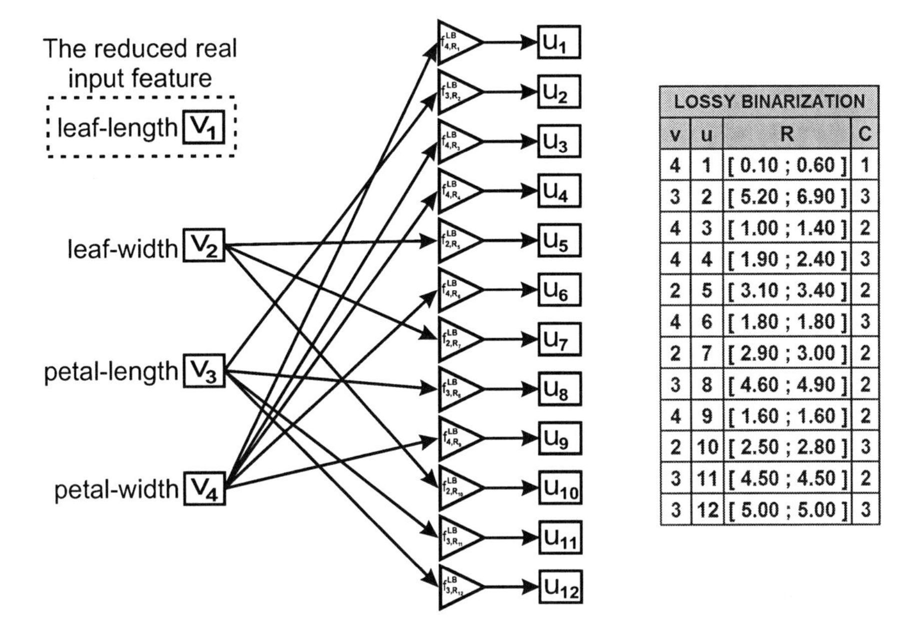

LOSSY BINARIZATION			
v	u	R	C
4	1	[0.10 ; 0.60]	1
3	2	[5.20 ; 6.90]	3
4	3	[1.00 ; 1.40]	2
4	4	[1.90 ; 2.40]	3
2	5	[3.10 ; 3.40]	2
4	6	[1.80 ; 1.80]	3
2	7	[2.90 ; 3.00]	2
3	8	[4.60 ; 4.90]	2
4	9	[1.60 ; 1.60]	2
2	10	[2.50 ; 2.80]	3
3	11	[4.50 ; 4.50]	2
3	12	[5.00 ; 5.00]	3

Fig. 4 The LBS constructed for the Iris data with the specified lossy binarization ranges.

loses nor rounds off any important representative input values but aggregates them and appropriately reinforces them. This algorithm lossless compresses input data and makes it possible to join computations for the same values of input features. It reinforces input features suitable to the values of Global Discrimination Coefficients (GDC) (1) computed for given data and described in the third chapter of this paper. The ARS architecture is always constructed individually for each data set and weights are precisely computed to reflect the discrimination and representative property of any given data. The described ARS automatic configuration can be proceeded only for bipolar binary inputs $\{-1;+1\}$. The ARS construction algorithm can automatically and very quickly find appropriate combinations of binary inputs and automatically simplify or even reduce a bipolar binary input data space. The ARS is placed in the middle part of the SONN-3 architecture (figs. 7,8,9).

Construction of an Aggregation Reinforcement Subnetwork (ARS) begins with computation of GDCs (1) for all bipolar binarized inputs $u_{k_1},...,u_{k_t}$ achieved from binary outputs (6) of the previous LBS described in the previous chapter. The ARS consists of Aggregation Reinforcement Neurons (ARNs) (fig. 1b). These neurons need bipolar binary inputs during their construction and adaptation process. They aggregate inputs of various cases together (fig. 6) if they have the same values (without losing or rounding off any values) and reinforce the values which best discriminate and best represent training cases in-between various classes. All reinforcement factors and weights (9)-(10) are dependent on the appropriate values of Global Discrimination Coefficients (1). The ARNs always produce their outputs in the range of $[-1;+1]$ and are interconnected to the other ARNs or Maximum Selection Neurons

(MSNs) described in the next chapter. ARNs propagate the sum of discrimination coefficient values of all previous connections to ARNs of the next layer (8) during the ARS construction (8). In this way, proper reinforcement is appropriately promoted and propagated through a network without a loss of information about discrimination. The ARNs compute their outputs as an appropriately weighted sum of their inputs (7). The ARNs are connected to a compact subset of bipolar binary ARS inputs (LBS outputs) $\{u_{k_1},...,u_{k_t}\}$ and to a single ARN of a previous layer (if it exists) in order to supplement its discrimination ability (fig. 7). The propagation of information between ARNs never spoils discrimination properties of neurons of previous layers because interneuron weights (9) are computed to keep the influence of GDCs of all previous layers on a computation in the next layers (8).

$$x_i = f_i^{AR}(u_{k_s},...,u_{k_t},x_p) = w_0^{x_p} x_p + \sum_{j \in \{k_s,...,k_t\}} w_j^{x_i} u_j \tag{7}$$

where

$$d_0^{AR_p} = \sum_{j \in J} d_j \tag{8}$$

$$w_0^{AR_p} = \frac{d_0^{AR_p}}{\sum_{j \in \{k_s,...,k_t\}} d_j} \tag{9}$$

$$w_j^{AR_r} = \begin{cases} \frac{u_k^{n,R} d_k^+}{\sum_{j \in \{k_s,...,k_t\}} d_j} & if\ u_k^{n,R} \geq 0 \\ \frac{u_k^{n,R} d_k^-}{\sum_{j \in \{k_s,...,k_t\}} d_j} & if\ u_k^{n,R} < 0 \end{cases} \tag{10}$$

After the GDCs (1) have been computed for all bipolar binary input features of all training cases, it is determined which GDCs will be used as obligatory (fig. 5) to achieve the correct discrimination of all training cases. The goal is to find a minimal subset of GDCs of the largest values that can do this. The larger GDC value is the better discrimination property it represents. Besides obligatory GDCs there are usually many other GDCs with large values that can have values equal to values of the appropriate obligatory GDCs. The GDC with an equal value to the obligatory one is called optional because it can sometimes be represented in the ARS architecture without additional construction elements. All other not null values that are established for the same input features as obligatory ones should be taken into account when discriminating in order to achieve unambiguous discrimination and classification. The determination of the obligatory and optional GDCs proceeds in the following way:

For each training case find the GDCs which have the largest values that discriminate it from other training cases of all other classes in the following way:

1. For each not obligatory GDC of this case compute a number of potentially discriminable training cases of other classes from the indiscriminated cases list and multiply it by the GDC value. This product determines how many training cases can be discriminated using this GDC taking into account the discrimination

property of this bipolar binary input. Only those cases can be discriminated that have an opposite value of that binary input feature to the value of the appropriate binary input of the ARS (e.g. -1 is opposite to $+1$, $+1$ is opposite to -1).

2. Choose a maximal value from the products computed for all not obligatory GDC values.
3. Use this maximal value to discriminate a subset of training cases from the indiscriminated cases list and remove all discriminated training cases from the indiscriminated cases list. If the GDC value for the checked case is null (or its suitable bipolar binary input feature value is the opposite of the bipolar binary input feature value of the given discriminated case) then its training case can be removed from this list.
4. If not all training cases have been discriminated against the given training case then return to step 2.

This algorithm is executed once for each training case and when it finishes all training cases are discriminated against all training cases from all other classes. This guarantees 100% discrimination of all not contradictory training cases and is an important part of the ARS optimization algorithms. Figure 5 presents the obligatory and optional GDCs computed for the Iris data from MLRepository. The black rectangles map the obligatory GDCs that have to be used to totally discriminate all Iris training cases against all cases of all other classes. The dark grey rectangles map the optional GDCs that can be without additional costs included in the ARS. The light grey rectangles map the not null GDC values that have to be taken into

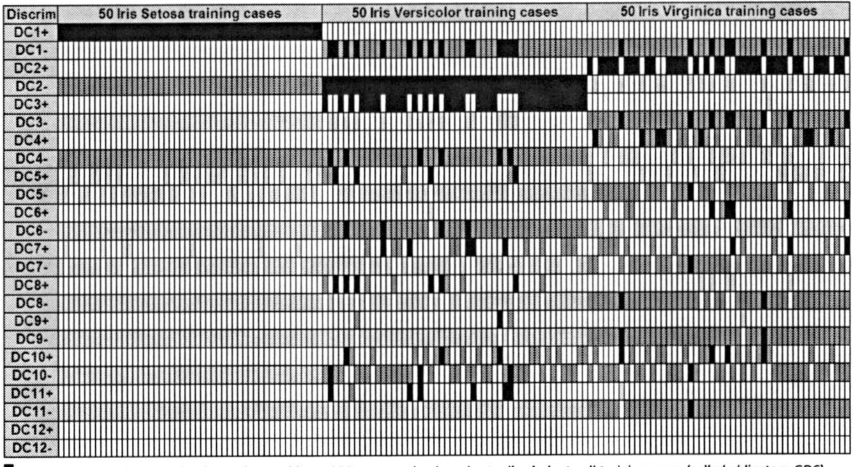

The obligatory major GDC that has to be used in an ARS construction in order to discriminate all training cases (called obligatory GDC).

The optional GDC that can be used during the ARS construction but it is not obligatory to discriminated data (called optional GDC).

The minor GDC that has to be considered when discriminating cases to achieve unambiguous classification (called considered GDC).

The irrelevant, null or minor GDC that can be passed over and is not used in construction of an ARS architecture (called irrelevant GDC).

Fig. 5 The GDC characters computed for the lossy binarized Iris cases.

account in order to achieve unambiguous discrimination and classification. The white rectangles do not play an important role in classification and can be skipped.

$$\varepsilon_j = S_j(C_j - 1) \tag{11}$$

After all GDCs of all training cases are classified as obligatory, optional, considered or irrelevant (fig. 5), the obligatory GDCs of various training cases can be grouped together only if they have equal values for some bipolar binary input. The obligatory and optional GDCs are grouped in such a way as to minimize a number of network construction elements (a number of connections and a number of neurons) that will represent them in the network simultaneously, without limiting exact representation of the obligatory GDCs in this network. The obligatory and optional GDCs that have equal values for the same input features and for different training samples can be grouped together and represented using a single connection. Such grouping and transformation efficiently lossless compress GDC information in the neural network. So called Aggregation Effectiveness Coefficients (AECs) (11) are used to count how many obligatory GDCs can be grouped and aggregated for each obligatory GDC value and the cost of representation of the compressed obligatory GDCs in the network is subtracted. The AECs are used to recursively divide bipolary lossy binarized training cases U into subsets (e.g. $\{U_1, U_2, U_3, U_4, U_5\}$ for the Iris data (fig. 6)) and to create ARNs for these subsets. The maximum value of the AEC in each division step is transformed into an ARN. Each AEC can group many different GDC values (fig. 6) if they are equal for all samples of each input feature and occur for the same samples of various input features simultaneously. The various GDC values are transformed into connections for the ARN created for a given AEC (figs. 6, 7). The AECs also count numbers of saved connections that can be omitted thanks to the aggregations found of GDCs. The AEC (11) is computed as a product of a number of training cases C_j that have equal GDC values for each input k minus one (because the equal GDC values are always represented by a single connection - this is the cost of representation of the aggregated GDCs in the network) and a number S_j of different GDC values that are equal for the same subset found of training cases for various inputs. After AECs are computed for all different GDC values for all inputs, the maximal AEC value is chosen in order to use it to divide the training cases U into two subsets: The first subset consists of the training cases that the obligatory and optional GDC values are equal to and all obligatory GDC values are determined in the maximal AEC. The second subset consists of other training cases that do not belong to the first subset. The division of training cases is demonstrated in fig. 6 for the Iris data. In each division step AECs are computed for one of the divided subsets of training cases until all training cases in subsets belong to single classes and all obligatory GDCs are represented in the ARS connections.

During the process of division of the transformed training cases (fig. 6), the connections are established (fig. 7) and weights are computed (9), (10). The interneuron weights (9) according to the sum (8) of all previously connected GDCs of bipolar binary inputs appropriately reinforce an interneuron input value (fig. 7).

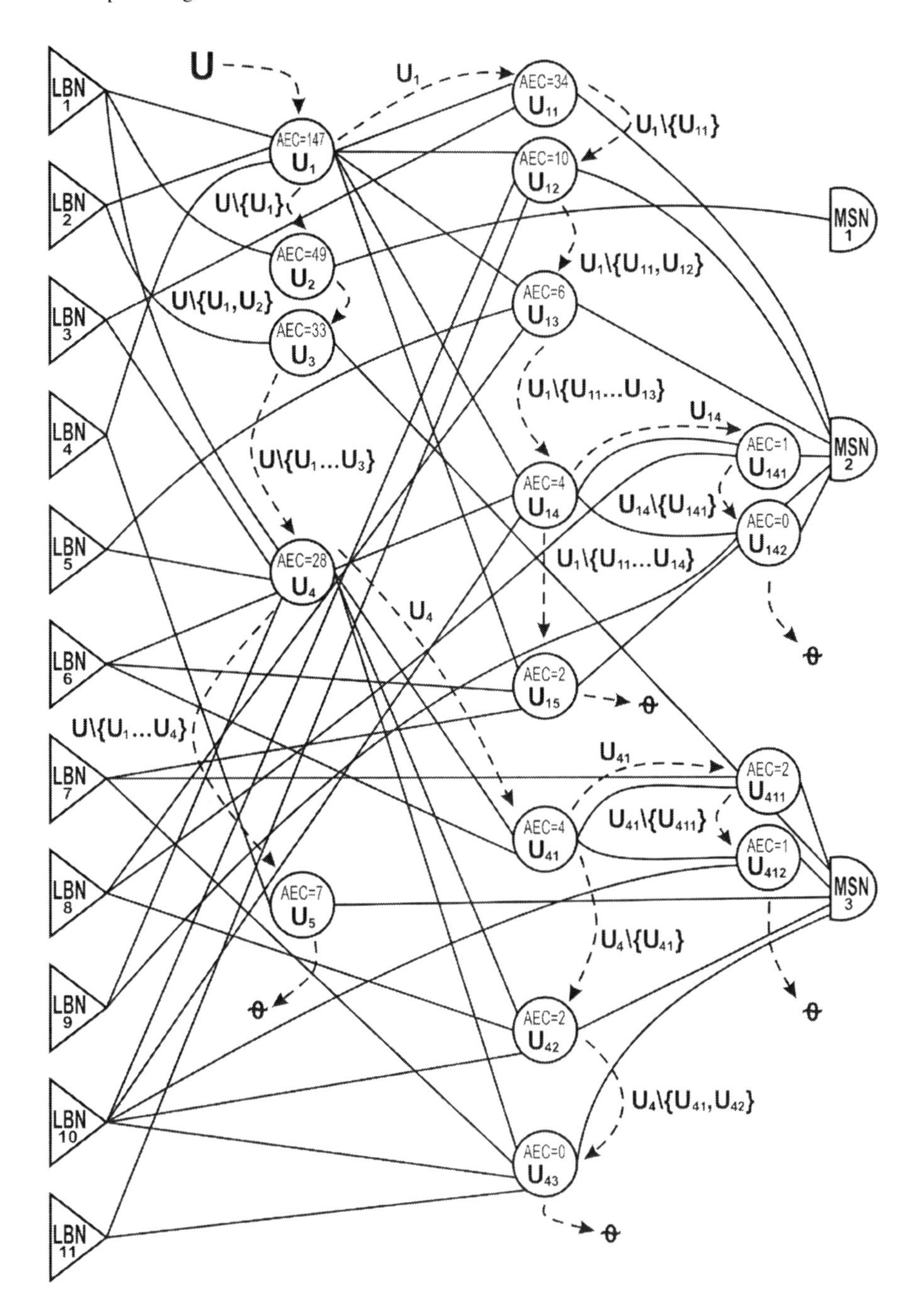

Fig. 6 The division and construction process of the ARS for the Iris data.

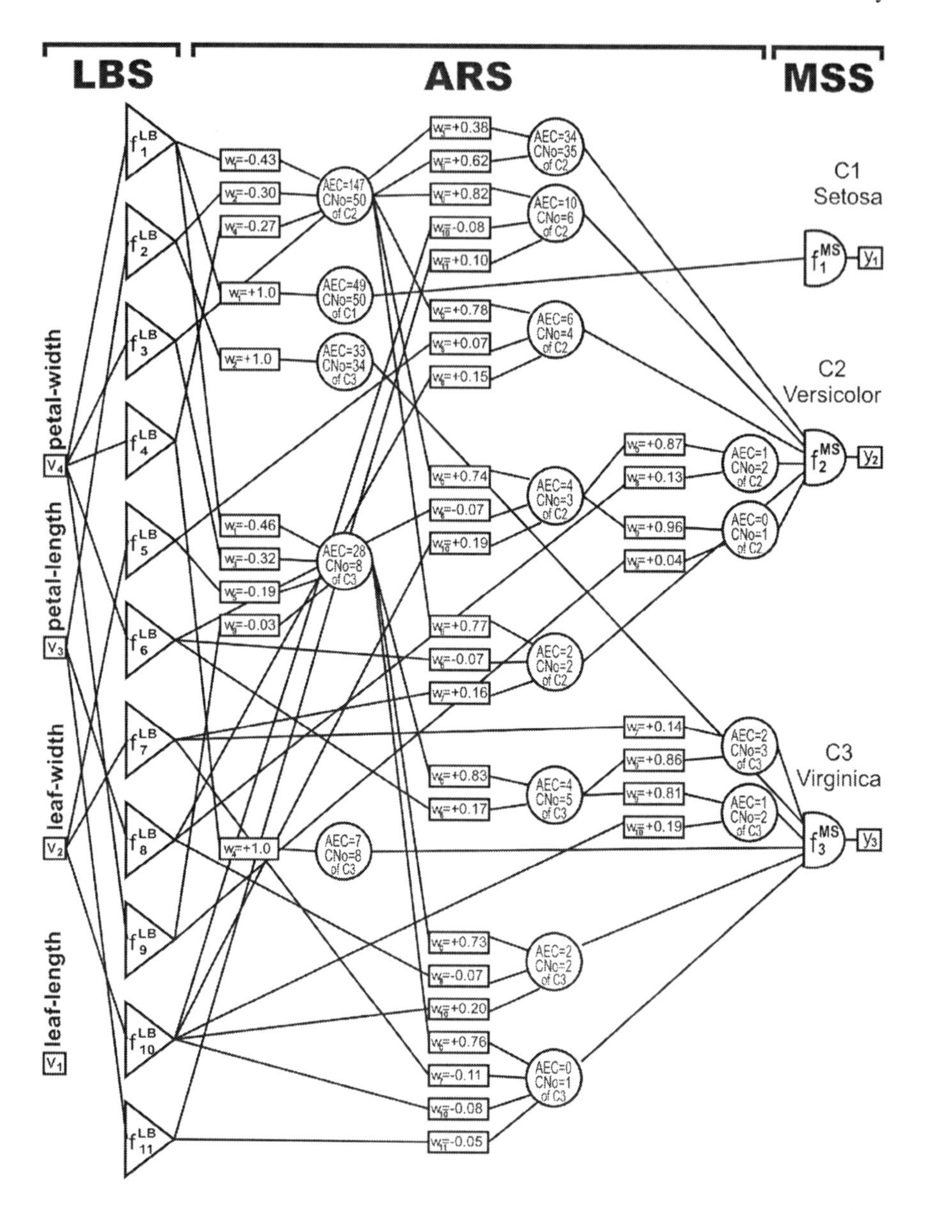

Fig. 7 The architecture of the SONN-3 constructed for the Iris data. The architecture includes all the subnetworks: the LBS, the ARS and the MSS and all computed weights.

6 Construction of a Maximum Selection Subnetwork

The construction of a Maximum Selection Subnetwork (MSS) is carried out during the construction of the ARS. Each subset U_s of training cases (for which all obligatory GDCs have been transformed to the connections) is not farther divided but the

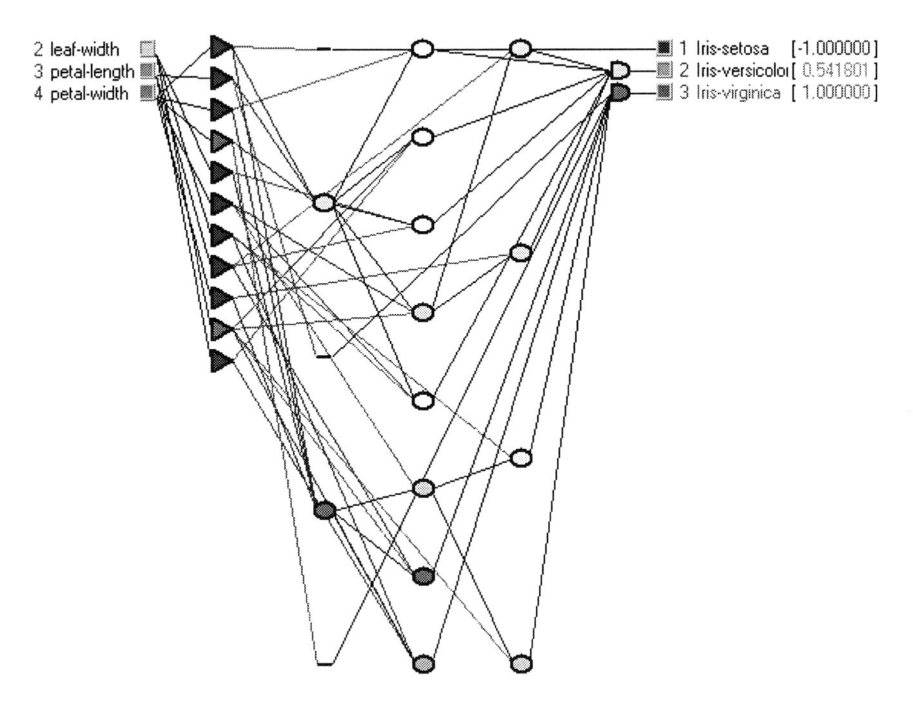

Fig. 8 The SONN-3 architecture (11 LBNs, 14 ARNs, 2 MSNs, 61 connections, 5 layers) automatically constructed for the Iris data.

appropriate ARN (created for the subset U_s) is connected to a Maximum Selection Neuron (MSN) (fig. 1c). This represents the class to which training cases of this subset belong. All training cases of this last undividable subset always belong to a single class.

$$y_m = f_m^{MS}(u_{k_s},...,u_{k_t},x_{i_g},...,x_{i_h}) = max\{ u_{k_s},...,u_{k_t},x_{i_g},...,x_{i_h}\} \tag{12}$$

The MSNs compute the maximum of outputs of all ARNs that represent an undividable subset of training cases of a single class (12). The output value of the MSN can be interpreted as the similarity of an input vector to a considered class. The inputs of MSNs are not weighted (fig. 7). If some MSN has a single input then this MSNs can be reduced. Such situation can occur for some very correlated and easy to discriminate classes of training cases (compare figs. 7 and 8). If some training cases can be discriminated using a single bipolar binary input feature then the MSN can even be connected to this input and a suitable ARN can be reduced (compare figs. 7 and 8).

7 Experiments and Comparisons

The three subnetworks (the LBS, the ARS and the MSS) described in the previous sections constitute the SONN-3 (fig. 7) that can be used as a universal efficient

Table 1 Comparison of the classification results for the Iris data.

Comput. Tool	Name of Method	CONFIGURATION AND TRAINING PARAMETERS			Time	TRAINING RESULTS				
		Function	Other Parameters		sec	Corr.	Incorr.	Unclass.	AVE/RMS	DIM
own	SONN-3	Automatic			1	150	0	0	0	3
Ghost Miner 3.0	Nu SVM	Gaussian	Nu = 0.2	SVM Solver: ON, Stop: 0.001, Shrinking ON, Max iter: 5000000, Alfa: 0.15, Beta 0.15	6	149	1	0	0.006667	4
	C SVM		Auto C & Bias		4	148	2	0	0.013333	4
	C SVM	Polynomial			198	148	1	1	0.013333	4
	IncNet	Gauss/Bi-cent.,Bias:1,Max neur.:20,Growth:0,95,Conf:0.9,Inter:1000			2	147	3	0	0.02	4
	k-NN	Euclidean	Distance measure	Auto k, k <= 20	1	147	3	0	0.02	4
	k-NN	Camberra		Feature weighting, k = 2	2	147	3	0	0.02	4
	FSM	Gaussian, Cent change:0.2, Disp.:0.2, Acc.:0.999, Min. act.:0.2, Tree			1	148	2	0	0.013333	4
	SSV Tree	Gener. strat.: Opt. leaves count, Search strat.: BFS			2	147	3	0	0.02	2
Statistica NN	3L MLP	Automatic designer Unit penalty: 0 Steps: 1 Lookahead: 0		Iterations: 200	862	149	1	0	0.06669	4
	4L MLP				38	146	4	0	0.3652	4
	RBF			Iterations: 2000	41	102	0	48	0.1131	4
	PNN				4	150	0	0	0	4

Table 2 Comparison of the classification results for the Wine data.

Comput. Tool	Name of Method	CONFIGURATION AND TRAINING PARAMETERS			Time	TRAINING RESULTS				
		Function	Other Parameters		sec	Corr.	Incorr.	Unclass.	AVE/RMS	DIM
own	SONN-3	Automatic			1	178	0	0	0	5
Ghost Miner 3.0	Nu SVM	Gaussian	Nu = 0.5	SVM Solver: ON, Stop: 0.001, Shrinking ON, Max iter: 5000000, Alfa: 0.15, Beta 0.15	1	178	0	0	0	13
	C SVM		Auto C & Bias		12	177	0	1	0.0056	13
	C SVM	Polynomial			58	178	0	0	0	13
	IncNet	Gauss/Bi-cent.,Bias:1,Max neur.:20,Growth:0,95,Conf:0.9,Inter:1000			775	105	73	0	0.4101	13
	k-NN	Euclidean	Distance measure	Auto k, k <= 20	1	178	0	0	0	13
	k-NN	Camberra		Feature weighting, k = 1	5	178	0	0	0	13
	FSM	Gaussian, Cent change:0.2, Disp.:0.2, Acc.:0.999, Min. act.:0.2, Tree			1	175	3	0	0.0169	13
	SSV Tree	Gener. strat.: Opt. Pruning degree, Search strat.: Beam search			1	178	0	0	0	8
Statistica NN	3L MLP	Automatic designer Unit penalty: 0 Steps: 1 Lookahead: 0		Iterations: 200	429	178	0	0	0.0001	13
	4L MLP				566	178	0	0	0.0002	13
	RBF				10	178	0	0	0.2603	13
	PNN				3	178	0	0	0	13

classifier for various classification tasks, even to highly non-separable data [1], [8]. The SONN-3 can precisely adjust its architecture and weights to given training data taking into consideration complexity and correlations of them. This adjustment ability of this network is similar to the plasticity processes [5] that take place in natural nervous systems [10].

This chapter compares the classification performance of the SONN-3 with that of other top classification soft-computing methods (tabs. 1-2). The Iris and Wine data from the ML Repository are used to carry out comparisons. Figures 4-8 illustrate the topologies of the SONNs-3 constructed for the above-mentioned training data. The Iris and Wine data from the ML Repository have been used to construct various soft-computing models, i.e. SVM, IncNet, k-NN, FSM, SSV Tree, MLP, RBF, PNN and SONN-3. The 4 dimensional Iris data consists of 150 training cases and 3 classes. The 13 dimensional Wine data consists of 178 training cases and 3 classes. The GhostMiner 3.0 solvers and the Statistica NN automatic designer with 10-fold cross-validation have been used to find the best solutions for these soft-computing methods. Moreover various configurations and parameters have been

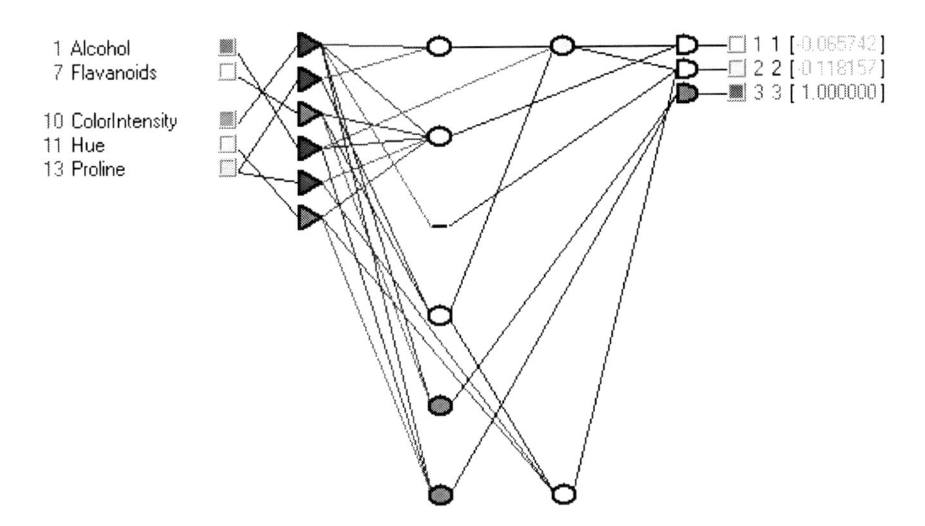

Fig. 9 The SONN-3 architecture (6 LBNs, 7 ARNs, 3 MSNs, 32 connections, 4 layers) automatically constructed for the Wine data.

tested. My own implementation of the SONN-3 has been used to verify assumptions of this method and to construct the solutions (figs. 8-9). Tables 1-2 contain the best results achieved for all tested soft-computing methods mentioned above.

The presented comparisons (tabs. 1-2) confirm that the SONN-3:

- creates very compact architectures,
- is constructed very quickly,
- always classifies all training cases correctly and unambiguously,
- automatically sets up all training parameters,
- achieves results which are competitive with other soft-computing methods,
- generalizes very well [4].

8 Conclusions

The paper deals with the fully automatic construction of the universal ontogenic neural network classifier SONN-3, which is able to automatically adapt itself using binary, integer or real input data without any limitations. It analyses and processes training data very quickly and finds a compact SONN-3 architecture and weights for any given training data set. It can also automatically simplify and reduce an input data space which is very desirable in many practical situations. It is also devoided of the curse of the dimensionality problem.

Moreover, not many soft-computing methods can automatically and effectively select the most discriminative input, so classification results are often influenced by minor or irrelevant parameters that can also spoil them and generalization properties of an achieved soft-computing solution. The SONN-3 reduces original real inputs to a set of the best discriminating inputs and transforms them to bipolar binary ones (figs. 4-8) used in further computations.

The comparison results (tabs. 1-2) show that the SONN-3 is not only very quick and easy to use but it also achieves very good classification and generalization results in comparison with the other popular soft-computing classification methods [4],[6]. Not many other algorithms for training neural networks can effectively compute an architecture and all weights for all training cases using a global analysis of them. Furthermore, the SONN-3 has many interesting features that can be compared with biological neural networks and various neural processes in biological brains [5].

The second main strength of the SONN-3 is that it uses two kinds of information that is very useful for the discrimination: the existence and non-existence of some input values for some classes. This kind of information is rarely used by other soft-computing models. The majority of soft-computing models and methods are limited to using only the information about the existence of some input values for some classes. The SONN methodology expands these abilities and offers better possibilities for a generalization.

The third strength of the SONN-3 is that it is able to construct a compact model without either rounding off or losing any important values of input data. This makes all computations very exact and accurate. Moreover, the SONN-3 can automatically and very accurately estimate discrimination and representative properties of the data. These estimations are used to group and aggregate the most important values of input features in order to produce a compact classification model of any given data using Global Discrimination Coefficients (1). Figures 8 and 9 show how compact and consistent the architectures constructed by the SONN-3 algorithms described in this paper can be.

The fourth strength of the SONN-3 is that it always builds a solution using the most important, well-differentiating and well-discriminating features of all training cases after the global analysis of training data. The SONN-3 also automatically excludes data artifacts because it focuses on the most discriminative features, which are not artifacts.

Finally, the SONN-3 is very quickly, cost-effective and fully automatic. On the other hand, a computer implementation of this method is not easy because of the huge number of optimization algorithms that gradually analyze training data, transform them, develop a final neural network architecture and compute its weights. The interactive website with the implemented SONN-3 algorithms will be published at `http://home.agh.edu.pl/~horzyk` soon.

References

1. Duch, W., Korbicz, J., Rutkowski, L., Tadeusiewicz, R. (eds.): Biocybernetics and Biomedical Engineering. EXIT, Warszawa (2000)
2. Dudek-Dyduch, E., Horzyk, A.: Analytical Synthesis of Neural Networks for Selected Classes of Problems. In: Bubnicki, Z., Grzech, A. (eds.) Knowledge Engineering and Experts Systems, OWPN, Wroclaw, pp. 194–206 (2003)

3. Fiesler, E., Beale, R. (eds.): Handbook of Neural Computation. IOP Publishing Ltd., Oxford University Press, Bristol, New York (1997)
4. Horzyk, A.: Introduction to Constructive and Optimization Aspects of SONN-3. In: Kůrková, V., Neruda, R., Koutník, J. (eds.) ICANN 2008,, Part II. LNCS, vol. 5164, pp. 763–772. Springer, Heidelberg (2008)
5. Horzyk, A.: A New Extension of Self-Optimizing Neural Networks for architecture Optimization. In: Duch, W., Kacprzyk, J., Oja, E., Zadrożny, S. (eds.) ICANN 2005. LNCS, vol. 3696, pp. 415–420. Springer, Heidelberg (2005)
6. Horzyk, A., Tadeusiewicz, R.: Comparison of Plasticity of Self-Optimizing Neural Networks and Natural Neural Networks. In: Mira, J., Alvarez, J.R. (eds.) Proc. of ICANN 2005, pp. 156–165. Springer, Heidelberg (2005)
7. Hyvarinen, A., Karhunen, J., Oja, E.: Independent Component Analysis. John Wiley and Sons, Chichester (2001)
8. Jankowski, N.: Ontogenic neural networks. EXIT, Warszawa (2003)
9. Jolliffe, I.T.: Principal Component Analysis. Springer, Heidelberg (2002)
10. Kalat, J.: Biological Psychology, Thomson Learning Inc. Thomson Learning Inc., Wadsworth (2004)
11. Pawlak, Z.: Rough sets. In: Theoretical Aspects of Reasoning about Data. Kluwer Academic Publishers, Dordrecht (1991)
12. Starzyk, J.A.: Motivation in Embodied Intelligence, Robotics, Automation and Control. I-Tech Education and Publishing (2008)
13. Subirats, J.L., Franco, L., Molina Conde, I., Jerez, J.M.: Active Learning Using a Constructive Neural Network Algorithm. In: Kůrková, V., Neruda, R., Koutník, J. (eds.) ICANN 2008,, Part II. LNCS, vol. 5164, pp. 803–811. Springer, Heidelberg (2008)

M-CLANN: Multiclass Concept Lattice-Based Artificial Neural Network

Engelbert Mephu Nguifo[1,3], Norbert Tsopze[1,2], and Gilbert Tindo[2]

Abstract. Multilayer feedforward neural networks have been successfully applied in different domains. Defining an interpretable architecture of a multilayer perceptron (MLP) for a given problem is still challenging. We propose a novel approach based on concept lattices to automatically design a neural network architecture. The designed architecture can then be trained with the backpropagation algorithm. We report experimental results obtained on different datasets, and then discuss our contribution as a means to provide semantics to each neuron in order to build an interpretable neural network.

1 Introduction

A growing number of real world applications have been tackled with artificial neural networks (ANNs). ANN is an adaptive system that changes its structure based on external or internal information that flows through the network during the learning phase. ANNs offer a powerful and distributed computing architecture, with significant learning abilities and they are able to represent highly nonlinear and multivariable relationships. ANNs have been successfully applied to solve a variety of specific tasks (pattern recognition, function approximation, clustering, feature extraction, optimization, pattern matching

Engelbert Mephu Nguifo and Norbert Tsopze
CRIL CNRS, Artois University, Lens, France
e-mail: tsopze@cril.univ-artois.fr

Norbert Tsopze and Gilbert Tindo
Computer Science Department - University of Yaoundé I,
PO Box 812 Yaoundé - Cameroon
e-mail: tsopze@cril.fr;gtindo@uycdc.uninet.cm

Engelbert Mephu Nguifo
LIMOS CNRS, Université Blaise Pascal, Clermont Ferrand, France
e-mail: mephu@isima.fr

L. Franco et al. (Eds.): Constructive Neural Networks, SCI 258, pp. 103–121.
springerlink.com

and associative memories) of importance to many applications [28, 39]. ANNs are useful especially when data is plentiful and prior knowledge is limited. Different ANN types have been reported in the literature, among which the multilayer feed-forward network, also called multi-layer perceptron (MLP), was the first and arguably simplest type of ANN devised, and is the main concern of this chapter.

MLP networks trained using the backpropagation learning algorithm are limited to search for a suitable set of weights in an apriori fixed network topology. The selection of a network architecture for a specific problem has to be done carefully. In fact there isn't a fixed and efficient method for determining the optimal network topology of a given problem. Too small networks are unable to adequately learn the problem well while overly large networks tend to overfit the training data and consequently result in poor generalization performance. In practice, a variety of architectures are tried out and the one that appears best suited to the given problem is picked. Such a trial-and-error approach is not only computationally expensive but also does not guarantee that the selected network architecture will be close to optimal or will generalize well. An ad-hoc and simple manner deriving from this approach is to use one hidden layer with a number of neurons equal to the average number of neurons in both input and output layers. In the literature, different automatic approaches have been reported to dynamically build the network topology. These works could be divided into two groups:

1- The first group uses prior knowledge (set of implicative rules) of the application to derive the neural network topology [32]. The prior knowledge is provided by an expert of the domain. The main advantage here is that each node in the network represents one variable in the rule set and each connection between two nodes represents one dependency between variables. The obtained neural network is a comprehensible ANN since each node is semantically meaningful, and the ANN's decision is not viewed as deriving from a black-box system, but could easily be explain using a subset of rules from the prior knowledge. The KBANN system (Knowledge-Based ANN) [32] is an example of such an approach. But this solution is limited while the prior knowledge is not available as might be the case in practice.

2- The second group of techniques searches for an optimal network to minimize the number of units in the hidden layers [28, 34]. These techniques bring out a dynamic solution to the ANN topology problem when a priori knowledge is not available. One technique suggests to construct the model by incrementally adding hidden neurons or hidden layers to the network until the obtained network becomes able to better classify the training data set. Another technique is network pruning which begins by training an oversized network and then eliminate weights and neurons that are deemed redundant. An alternative approach consists of using the linear separability [3] approach or the genetic approach [8] even if the latter is computationally expensive. All these (incremental, pruning, genetic) techniques result to neural networks that can be seen as black box systems, since no meaning is associated to each hidden neuron. Their main limitation is the intelligibility of the resulting network (black-box prediction is not satisfactory [1, 11]).

We propose here a novel solution, M-CLANN (Multi-class Concept Lattices-based Artificial Neural Networks), to build a network topology where each node has an associated semantic without using any prior knowledge. M-CLANN is an extended version of the CLANN approach [35]. Both approaches uses formal concept analysis (FCA) theory to build a semi-lattice from which the NN topology is derived and trained by error backpropagation. The main difference between M-CLANN and CLANN are two-folds. First M-CLANN can deal with multi-class classification problems, while CLANN is limited to two-classes. Second, the derived topologies from the semi-lattice are different in both systems.

Our proposed approach presents many advantages: (1) the generated architecture is a multi-layer feed-forward network, such that the use of the backpropagation algorithm is obvious; (2) each neuron has a semantic as it corresponds to a formal concept in the semi-lattice, which is a way to justify the presence of a neuron; (3) each connection (between input neuron and hidden neuron, and between neurons of different hidden layers) in the derived ANN also has a semantic as it is associated to a link in the Hasse diagram of the semi-lattice; (4) the knowledge for other systems (such as expert systems) could be extracted from the training data through the model; (5) Experimental results have shown the efficiency of the approach compared to other well-known techniques.

The rest of this chapter is organized as follows: the next section provides an overview of some related works. Section three recalls some background knowledge on formal concept analysis theory and supervised classification; the fourth section describes our approach M-CLANN (Multiclass CLANN). Experimental studies are reported in section five. Section six discusses the soundness and efficiency of our approach.

2 Related Works

Research works about neural network architecture design could be divided into two groups as mentioned above. The first group uses prior knowledge to propose a MLP topology, while the second group searches for an optimal topology minimizing the number of hidden neurons and layers.

2.1 Defining Neural Topology Using Prior Knowledge

An interesting framework is proposed in [32] to design the ANN topology using the domain theory represented as a set of rules. The derived system

KBANN (Knowledge Based Artificial Neural networks) [32] use a set of rules represented as a set of Horn clauses. From these rules an items hierarchy is defined and the architecture of the neural network is derived from this hierarchy. The hierarchy between items is defined using the following equivalences:

1. Final conclusions $\Leftrightarrow$ output units;
2. Intermediate conclusions $\Leftrightarrow$ internal units;
3. Hypothesis $\Leftrightarrow$ input units;
4. Dependency between items $\Leftrightarrow$ connection links.

The different steps of KBANN are as follows:

1. Rewriting. This step consists of writing the rules such that disjuncts are expressed as a set of rules (each rule has only one antecedent).
2. Mapping. The hierarchy between items is defined and directly mapped to the network.
3. Labeling. Each unit is numbered by its level.
4. Adding new hidden units. In order to make the network able to learn derived features not specified in the initial rule set, it is advised to add new units in the hidden layer.
5. Adding input units. Some relevant features which are referred to by the initial rule set are added.
6. Adding links. Links are added to connect each unit numbered $n-1$ to each unit numbered n. The connection weights of these links are set to 0.
7. Perturbing. A small random number is added to each weight.
8. Initialization of connection weights and ANN training by error backpropagation.

The connection weights and the bias neurons are initialized as follows:

- w for the positive antecedents
- $-w$ for the negated antecedents.
- The bias on the unit corresponding to the rules consequent to $(p-1/2)w$ where p is the number of positive antecedents of the unit.

w is a positive number having 4 (empirically defined) as the default value.

Example 1. *Figure 1 presents a simplified example of defining the neural topology by KBANN approach. In the first column, are the initial rules which are rewritten and presented in the second column. And the final network is presented in the third column.*

2.2 *Defining Neural Topology without Prior Knowledge*

When the prior knowledge is unavailable, it is not possible to use KBANN. To avoid this, there are reported methods that directly define the ANN topology

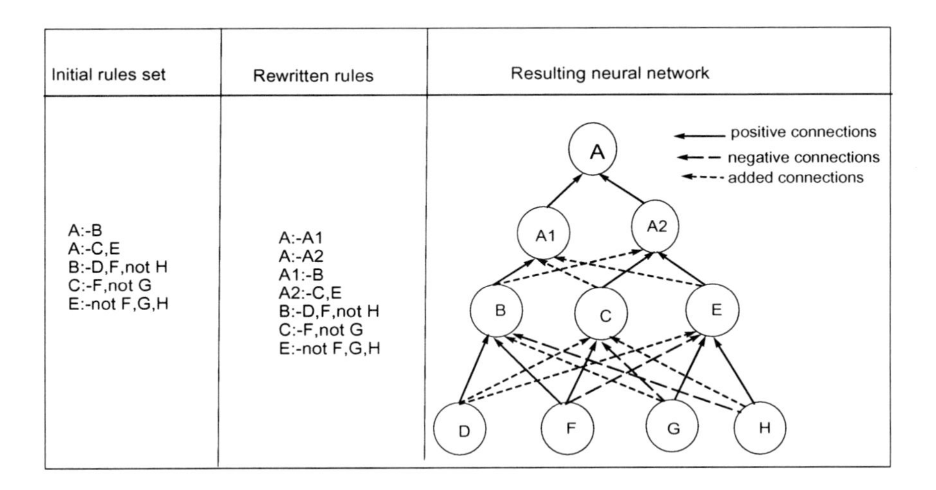

Fig. 1 Example of ANN topology definition with KBANN.

from the data. These methods start with a small network and dynamically grow the network by adding and training neurons as needed until better classification is achieved. These methods can be divided into two subgroups: those with many hidden layers [28] and those with only one hidden layer [38].

2.2.1 Many Hidden Layers

In [28] the authors provide a survey of these methods including MTiling, MUpstart, and MTower. The new added neuron is trained using a variant of perceptron similar to the pocket perceptron with rachet modification [15]. The process adds layers in the existing network until better classification is achieved or the maximum number of layers (user specified value) is attempted.

1. MTiling. It constructs a strictly layered network of threshold neurons. Apart from the most top layer (which is also the output layer) which receives inputs from the layer immediately bellow it and to the inputs neurons, each layer receives input from the layer immediately bellow it. Two kinds of neurons are distinguished: the master unit and the ancillary neurons that are added and trained to ensure a faithful representation of the training data. After training, some ancillary neurons could be pruned to minimize the network structure.
2. MUpstart. The network is constructed as a binary tree. Two kinds of errors are defined: wrongly off ($output = 0$ while the $target = 1$) and wrongly on ($output = 1$ while the $target = 0$). In case of wrongly off (rep. on), one left (resp. right) child neuron is added to the wrong neuron and trained to correct this error.

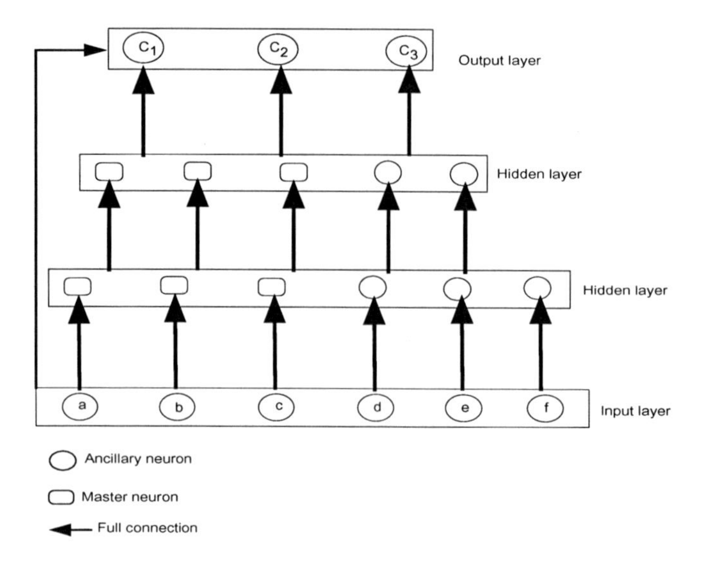

Fig. 2 Example of neural network topology definition by MTiling method.

3. MTower. The resulting network is like a tower. It successively adds new layers in the network until better classification is achieved. The newly added layer is fully connected to the input layer and to the output layer. After connecting this layer, it becomes the new output layer.

Example 2. *Figure 2 is an example of a network constructed by MTiling.*

Recently an approach based on linear separability was introduced in [3] which relies on barycentric correction procedure algorithm for training the individual threshold logic unit.

2.2.2 One Hidden Layer

The Distal method [38] belongs to this category. It builds a 3 layer neural network. Each neuron of the input layer is linked to an attribute. Each neuron of the output layer is associated to a predefined class. The process essentialy consists of defining the hidden layer. Distal clusters training data in disjoint subsets and represents each subset in the hidden layer by one neuron.

Example 3. *Figure 3 presents a neural network defined by Distal. (a) is the initial state of the network and (b) is the generated network.*

There are also in the literature many works which help the user to optimize [37] or prune networks by pruning some connections [23] or by selecting some variables [5] among the entire set of initial variables, or by detecting and filtering noisy examples [34]. These works do not propose an efficient method to build neural network topology, but they can be classified in the second

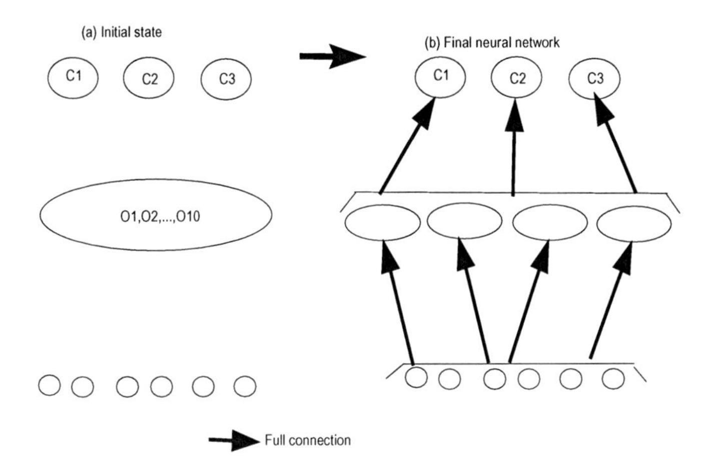

Fig. 3 Example of neural network topology definition by Distal method.

group, since by reducing the number of input neurons, the number of neurons in the hidden layer could also vary.

3 Background - Classification and Formal Concept Analysis

In this section we recall what is supervised classification task, then define basic notions of FCA, and finally presents constraints that can be used to prune the concept lattice in supervised classification.

3.1 Classification

The classification task consists of labelling unknown patterns into a predefined class. The classification process builds a model and trains it for making this model able to affect the unseen patterns to one of the output classes. Here, each known pattern is presented as a pair (x, y) where x is the vector containing different values taken by the pattern on different attributes and y is its class value represented by a particular attribute. The training data is divided into two sets: the training set and the test set. The system operates in two phases: the training phase consists in designing the model while the second step evaluates the trained model.

For instance, in the data table 1, objects or patterns 1 to 6 can be associated to the positive class (+), while patterns 7 to 10 can be associated to the negative class (-).

The model evaluation (or test) consists of calculating its accuracy rate as the ratio between the number of well classified patterns and the total number of patterns. There are many techniques to determine accuracy rate among which:

1. K-fold cross validation. The training data is divided into k disjoint subsets and the model is trained and tested k times. At each iteration i, the ith subset is used to test the model built and trained using the other $k-1$ subsets (all other subsets except i^{th} subset). The accuracy rate is calculated as the average of the different accuracy rates obtained at each iteration. Empirically it is advised to take $k = 10$.
2. Leave-one-out is a variant of k-fold cross validation where k is to the number of patterns on the training set.
3. Holdout. The training set is randomly separated into two disjoint subsets. One of these subsets is used to build and train the model while the other is used for test.

Classification as well as supervised learning are more detailed in [7, 20].

3.2 Formal Concept Analysis

Formal Concept Analysis (FCA) is a mathematical framework that models the world as being composed of objects and attributes, describing an application [40, 16].

Definition 1. *A* ***formal context*** *is a triplet $C = (O, A, I)$ where O is a non empty finite set of objects, A is a non empty finite set of attributes (or items) and I is a binary relation between O and A (formally $I \subseteq O \times A$).*

The formal context (binary) C could be represented as a binary matrix such that $C_{ij} = 1$ if the object represented in row i verifies the attribute represented in column j and 0 if not.

Example 4. *Table 1 is an example of a binary formal context.*
$O = \{1, 2, 3, 4, 5, 6, 7, 8, 9, 10\}$ is the set of objects or patterns while $A = \{a, b, c, d, e, f\}$ is a set of attributes.

The fundamental intuition of FCA relies on the fact that a concept is represented by an intent and an extent.

Definition 2. *Let f and g be two applications defined as follows:*

- $f : 2^O \longrightarrow 2^A$, *s.t.* $f(O_1) = O_1' = \{a \in A \;/\; \forall o \in O_1\;,\; (o, a) \in I\}$, $O_1 \subseteq O$;
- $g : 2^A \longrightarrow 2^O$, *s.t.* $g(A_1) = A_1' = \{o \in O \;/\; \forall a \in A_1\;,\; (o, a) \in I\}$, $A_1 \subseteq A$;

Table 1 Example of a formal context presented as boolean matrix.

O/A	a	b	c	d	e	f
1	1	1		1	1	1
2	1		1		1	1
3		1	1	1	1	
4		1	1	1		
5	1	1		1	1	
6	1		1		1	
7		1		1		1
8		1			1	1
9			1		1	1
10	1			1	1	

A pair (O_1, A_1) *is called* ***formal concept*** *iff* $O_1 = A'_1$ *and* $A_1 = O'_1$. O_1 *(resp.* A_1*) is the extent (resp. intent) of the concept.*

Example 5. *From table 1,* $(\{1,2,5,6,10\},\{a,e\})$ *is a formal concept where* $\{1,2,5,6,10\}$ *is the extent and* $\{a,e\}$ *is the intent. While* $(\{1,2,5\},\{a,e\})$ *is not a formal concept since* $\{1,2,5\}$ *is not the largest set for which each object verifies all attributes of the set* $\{a,e\}$.

Definition 3. *Let* L *be the entire set of concepts extracted from the formal context* C *and* $\leq$ *a relation defined as* $(O_1, A_1) \leq (O_2, A_2) \Rightarrow (O_1 \subset O_2)$ *(or* $A_1 \supset A_2$*). The relation* $\leq$ *defines the order relation on* L *[16].*

If $(O_1, A_1) \leq (O_2, A_2)$ *is verified (without intermediate concept) then the concept* (O_1, A_1) *is called the successor of the concept* (O_2, A_2) *and* (O_2, A_2) *the predecessor of* (O_1, A_1).

The **Hasse diagram** is the graphical representation of the relation *successor/predecessor* on the entire set L of concepts.

The fundamental theorem of FCA [40] states that the set of formal concepts of a formal context forms a complete lattice, called a concept lattice. A complete lattice is a partial order in which the greatest lower bound and least upper bound of any subset of the elements in the lattice must exist.

FCA have shown to be useful in data mining for generating concise representations of implicative rules [19] or association rules [2, 18], but also for supervised classification [14]. More details on FCA could be found in [16, 4].

3.3 Constraints

In order to reduce the size of concept lattice and consequently the time complexity, we introduce some constraints regularly used to select concepts during the learning process.

3.3.1 Frequency of Concept

A concept is frequent if it contains at least α (also refered to as minsupp, is specified by the user) objects. The support s of a concept (X, Y) is the ratio between the cardinality of the set X and the total number of objects ($|O|$) ($s = \frac{100 \times |X|}{|O|}\%$). Frequency is an anti-monotone constraint which helps prune the lattice and reduce its computational complexity. Minimum support is the minimal number of objects that the intent of a concept must verify to be selected.

3.3.2 Validity of Concept

A concept (X, Y) is **complete** if Y recognizes all positive examples. A concept (X, Y) is **consistent** if Y throws back all counter examples or negative examples (formally, the set of consistent concepts is $\{(X, Y)/Y \cap O^- = \{\}\}$ where $O = O^+ \cup O^-$). Both completeness and consistency constraints are restrictive and can lead to overfitting. Other weak constraints are then introduced:

1. **Validity.** A concept (X, Y) is valid if its description recognizes many examples; a valid concept is a frequent concept on the set of examples O^+; formally the set of valid concepts is defined as $\{(X, Y) \,/\, |X^+| \geq \alpha\}$ where $0 < \alpha \leq |O^+|$.
2. **Quasi-consistency.** A concept (X, Y) is quasi-consistent if it is valid and its extent contains few counter examples. Formally, the set of quasi-consistent concepts is defined as $\{(X, Y) \,/\, |X^+| \geq \alpha \text{ and } |X^-| \leq \beta\}$.

3.3.3 Height of a Semi-lattice

The level of a concept c is defined as the minimal number of connections from the supreme concept to c. The height of a semi-lattice is the greatest value of the level of concepts. Using levelwise approach to generate the join semi-lattice, a given constraint can be set to stop concept generation at a fixed level. The height of the lattice could be performed as the depth without considering the cardinality of concept extents (or intents). In fact at each level, concept extents (or intents) do not have the same cardinality. The number of layers of the semi-lattice is a parameter corresponding to the maximum level (height) of the semi-lattice.

4 Concept Lattice-Based Artificial Neural network

We describe in this section the different steps of our new approach, M-CLANN, as shown by figure 4. The process of finding the architecture of neural networks has three steps: (1) build a joint semi-lattice of formal concepts by applying constraints to select relevant concepts [24, 33]; (2) translate

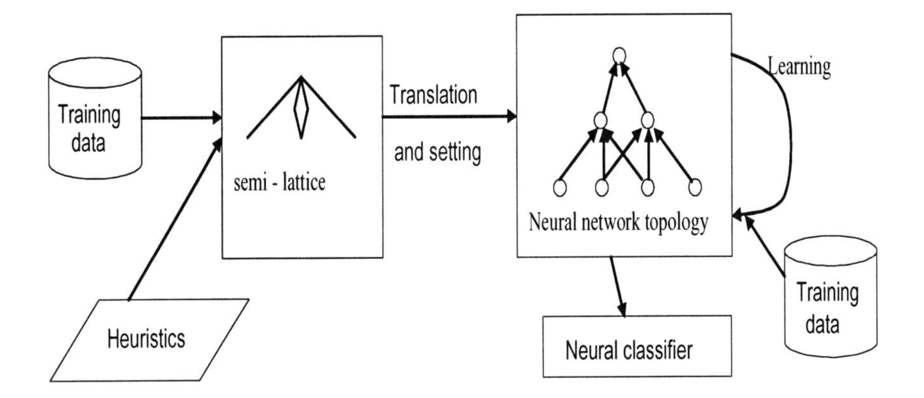

Fig. 4 Neural network topology definition.

the join semi-lattice into a topology of the neural network, and set the initial connections weights; (3) train the neural network.

Variables used in the algorithms defined in this section are : C is a formal context (dataset); L is the semi-lattice built from the training dataset K; c and c' are formal concepts; n is the number of attributes in each training pattern; m is the number of output classes in the training dataset; c a formal concept, element of L; NN is the comprehensive neural network build to classify the data.

4.1 Semi-lattice Construction

There are different algorithms [22] which can be used to generate formal concepts; only a few of them build the Hasse diagram. Lattice could be processed using top-down or bottom-up techniques. In our case, a levelwise approach presents advantage to successively generate concepts of the join semi-lattice and the Hasse diagram. For this reason, we choose to implement the Bordat algorithm [22] which is suitable here. Concepts included in the lattice are only those which satisfy the defined constraints.

In order to prune the concept lattices, we can use one or multiple constraints to select concepts during this step. The constraints used in M-CLANN are frequency of concept and the height of the semi-lattice. For example it is possible to combine frequency and height constraints, or to use only one of them. The semi-lattice construction process starts by finding the supreme element. The process continues by generating the successors of the concepts that belong to the existing set until there are no concepts which satisfies the specified constraints.

Algorithm 1. Modified Bordat algorithm

Require: Binary context C
Ensure: concept lattices (concepts extracted from C) and the Hasse diagram of the order relation between concepts.
1: Init the list L of the concepts $(O, \{\})$ $(L \leftarrow (O, \{\}))$
2: **repeat**
3: **for** concept $c \in L$ such that his successors are not yet been calculated **do**
4: Calculate the successors c' of c.
5: **if** the specified constraint is verified by c' **then**
6: add c' in L as successor of c if c' does not exit in L else connect c' as successor of c.
7: **end if**
8: **end for**
9: **until** no concept is added in L.
10: derive the neural network architecture from the concept semi-lattice.

4.2 Generation of ANN Topology

In the second step, the join semi-lattice is translated into a neural network architecture. Algorithm 2 presents the M-CLANN method to translate the semi-lattice into ANN.

Example 6. *Figure 5 presents an example ANN topology designed with M-CLANN. In this figure, (a) is the semi-lattice while (b) is the corresponding neural network topology.*

Objects used in this algorithm are defined as follows: K is a formal context (dataset); L is the semi-lattice built from the training dataset K; c and c' are formal concepts; n is the number of attributes in each training pattern; m is the number of output classes in the training dataset; c a formal concept, element of L; NN is the comprehensive neural network build to classify the data.

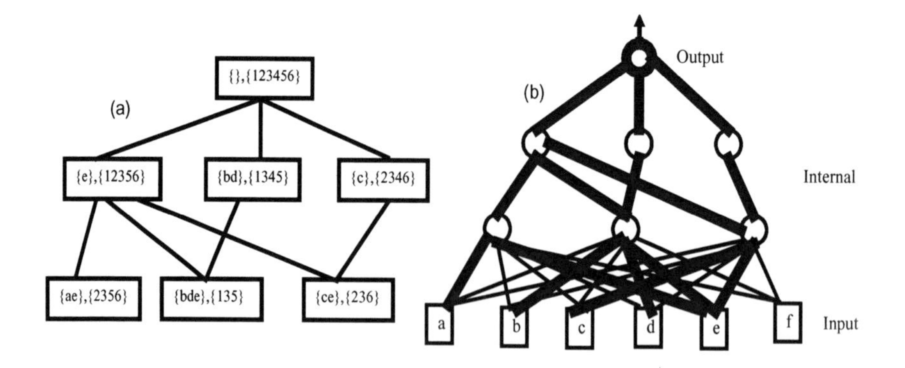

Fig. 5 Example of ANN architecture design using M-CLANN

Algorithm 2. Translation of semi-lattice into ANN topology

Require: L a semi-lattice structure built using specified constraints.
Ensure: NN initial topology obtained from the semi-lattice L

1: **for** each concept $c \in L$ **do**
2: if the set of predecessor of c is empty mark its successor as "last hidden neuron";
3: Else c becomes neurons and add to NN with the successor and predecessor as in L; if the set of successor of c is empty then mark c as "first hidden neuron".
4: Endif
5: **end for**
6: Create a new layer of n neurons and connect each neuron of this layer to the neurons marked as "first hidden neuron" in NN.
7: Create a new layer of m neurons and connect each neuron of this layer to the neurons marked as "last hidden neuron" in NN.
8: Initialize connection weights and train them.

Threshold is zero for all units and the connection weights are initialized as follows:

- Connection weights between neurons derived directly from the lattice is initialized to 1. This implies that when the neuron is active, all its predecessors are active too.
- Connection weights between the input layer and hidden layer are initialized as follows: 1 if the attribute represented by the input appears in the intention Y of the concept associated to the ANN node and -1 otherwise. This implies that the hidden unit connected to the input unit will be active only if the majority of its input (attributes including in its intent) is 1.

4.3 Training the Generated Topology

The last step of M-CLANN is to train the obtained neural network. This is done using the error backpropagation algorithm [30]. This algorithm searches the appropriate connection weights between the different units by propagating the input signals through the network and backpropagating the error from the output units to the input units. This is done by minimizing the quadratic sum of the error.

5 Experimentations and Results

5.1 Data

To examine the practical aspect of the approach presented above, we run the experiments on the data available on the UCI repository [26]. The

Table 2 Experimental data sets

Dataset	#Train	#Test	#Class	#Nom	#Bin
Balance-scale (Bal)	625	0	3	4	20
Chess	3 196	0	2	36	38
Hayes-roth (Hayes)	132	28	3	5	15
Tic-tac-toe (Tic)	958	0	2	9	26
Spect	80	187	2	22	22
Monks1	124	432	2	6	15
Monks2	169	432	2	6	15
Monks3	122	432	2	6	15
Lymphography (lympho)	148	0	3	18	51
Solar-flare1 (Solar1)	323	0	7	12	40
Solar-flare2 (Solar2)	1066	0	7	12	40
Soybean-backup (Soyb)	307	376	19	35	151
Lenses	24	0	3	4	12

characteristics of this data is shown in the table 2 which contains the name of the dataset, the number of training patterns (#Train), the number of test patterns (#Test), the number of output classes (#Class), the initial number of (nominal) attributes in each pattern (#Nom), the number of binary attributes obtained after binarization (#Bin). Attributes were binarized by the Weka [36] binarization procedure "Filters.NominalToBinary". The diversity of this data (from 24 to 3196 training patterns; from 2 to 19 output classes) helps in revealing the behavior of each model in many situations. There are no missing values in these datasets.

Two constraints presented above (frequency and height) have been applied in selecting concepts during experimentation. We first separately use each of them and then we combine them.

5.2 Results

Experimental results are obtained from the model trained by error back-propagation [30] and validated by 10-fold cross-validation or holdout [20]. The learning parameters are the following: as activation function, we use the sigmoid ($f(x) = \frac{1}{1+\exp x}$), 500 iterations in the weight modification process and 1 as learning rate.

Table 3 presents the accuracy rate (percentage) obtained with data in table 2. In table 3, the symbol "-" indicates that no formal concept satisfies the constraints and the process was stopped. The symbol "x" indicates that the classifier CLANN was not applied for those multiclass problem.

In this table, MCL1 is M-CLANN built from a semi-lattice with one level while MCL30 and MCL20 are M-CLANN built using respectively 30 and 20

Table 3 Accuracy rates of MCLANN classifier with some varied input parameters.

Dataset	CLANN	MCL1	MCL2	MCL30	MCL20	MC1-30	MC1-20
Bal	x	99,76	96,23	-	99,89	-	99,89
Chess	93,60	99,87	91,70	93,60	93,78	99,87	99,87
Hayes	x	75,72	76.85	78,58	85,72	78,57	85,71
Tic	94,45	89,64	90.21	99,67	99,86	99,32	100
Spect	93,90	72,74	72,56	92,56	96,73	73,66	77,57
Monks1	82,70	91,67	95,56	91,17	91,17	91,67	91,71
Monks2	78,91	100	98,65	100	100	100	99,67
Monks3	83,61	93,51	100	91,17	93,52	92,59	93,52
Lympho	x	80,78	90,24	84,67	88,91	85,71	92,56
Solar1	x	79,42	78,87	78,67	69,58	71,10	71,10
Solar2	x	75,00	72,62	76,71	70,91	75,34	78,95
Soyb	x	81,33	79,01	89,34	86,95	83,11	84,04
Lenses	x	98,67	90,00	100	99,87	98,67	99,87
Average	86,05	86,62	86,97	89,67	90,53	87,57	90,37

percent as frequency threshold. MC1-30 (respectively MC1-20) is M-CLANN built with a combination of semi-lattice height equals to 1 and 30% (resp. 20%) as frequency threshold. CLANN column represents the precision rate obtained using the original version of CLANN (with lattice height threshold equals to one).

With high minimum support values, sometimes the semi-lattice does not contain sufficient concepts to better classify the data. For instance, with the minimum support value set to 35%, the semi-lattice built from Balance-scale is empty. The best results (accuracy rate) of M-CLANN are obtained with the α value equal to 20% (MCL20). These results are comparable to those of other classifiers as shown in table 4 using some standard machine learning classifiers or some constructive multilayer perceptrons. In table 4, the symbol "x" indicates that the classifier does not converge. The standard classifiers are taken from the WEKA platform [36] and are MLP (a multilayer perceptron classifier), C4.5 (a decision tree based classifier), IB1 (a case based learning classifier model). The constructive multilayer perceptrons are the original versions of author's implementation of Mtiling, Mtower, Mupstart and Distal.

M-CLANN was not compared with KBANN because we have no prior knowledge about this data. The goal of this comparison is to see the behavior (on the supervised classification problems) of M-CLANN regarding those of other standard learning models.

Using different parameters settings, M-CLANN outperfomed standard machine learning classifiers in terms of accuracy on the experimental datasets. MLP is better than C4.5 and IB1.

Table 4 Accuracy rate of other classifiers.

Dataset	MCL20	MLP	C4.5	IB1	MTiling	MUpstart	MTower	Distal
Bal	99,89	98,40	77,92	66,72	94,27	100	95,16	96,77
Chess	93,78	99,30	98,30	89,90	96,24	97,18	96,87	89,74
Hayes	85,72	82,15	89,28	75,00	89,29	90,01	78,57	54,32
Tic	99,86	96,86	93,21	81,63	75,52	73,03	64,21	61,23
Spect	96,73	65,77	66,70	66,31	89,60	83,29	71,40	83,90
Monks1	91,17	100	100	89,35	81,71	77,21	78,01	90,23
Monks2	100	100	70,37	66,89	85,42	82,43	77,87	89,10
Monks3	93,52	93,52	100	81,63	100	89,42	91,21	86,46
Lympho	88,91	81,76	74,32	80,41	85,71	78,57	78,57	86,45
Solar1	69,58	72,79	74,30	68,39	100	100	98,89	x
Solar2	70,91	68,11	69,97	66,56	96,88	93,75	96,88	68,23
Soyb	86,95	92,02	88,83	89,89	83,23	85,45	84,34	x
Lenses	99,87	95,83	91,67	100	99,50	99,00	98,50	99,88
Average	90.59	88,57	84,22	78,67	90,81	87,39	84,43	88.90

Another advantage of M-CLANN over MLP is that each neuron has a semantic as it is associated to each intent of a formal concept. During the experimentations, the running time of MLP and M-CLANN are similar but much higher compared to that of C4.5 and IB1.

MTiling has the best average accuracy rate over the whole dataset, even if this accuracy rate is only slightly greater than that of MCLANN.

6 Discussion

As presented in the previous section, there exists many algorithms which could be used to define the neural network architecture. Each of those algorithms present advantages but they also have issues:

1. **Input data.** Many algorithms could not process other data than numeric. Apart from Distal where the authors have defined the distance between symbolic data, all others only treat numeric data. In addition of the training data, using KBANN method requires a domain theory which is not always available. The choice of the method could hardly be influenced by the input data.
2. **Interpretability of ANN.** It is well known that the ANN is one of the most commonly used methods in classification. As it is seen as 'Black box', it is not used in the domain where result explanations are important. Among the previous methods, only M-CLANN and KBANN present interpretable architectures. So, it could not be advised to use other approaches than M-CLANN and KBANN, while in M-CLANN, each node is associate to one formal concept and each formal concept is formed by

a set of objects (extent) and and a set of attributes (intent) shared by these objects; in KBANN, each node is associated to one variable on the rules set.
3. **Choice of the algorithm's parameters.** One problem with ANN topology design algorithm is the choice of network and training parameters. This problem is avoided in KBANN method where only the maximum number of iterations is needed. In addition to the maximum iterations number (500 as default value), M-CLANN needs to define the constraints value. Other constructive algorithms (except Distal) need to define the maximum layers number, the choice of training algorithm, the maximum iterations number in the training process.

Recently different works showing links between FCA and ANN were reported in the literature. Except from our previous method CLANN, those are different from MCLANN. [12] uses the FCA approach to encode the neural network function, while [31] proposes two ways of directly encoding closure operators on finite sets in a 3 layered feed forward neural network.

7 Conclusion

In this chapter, a new approach of finding the ANN topology is presented. This method is based on concept lattices and is able to define an interpretable ANN topology without any prior domain knowledge. This proposal extends our previous method CLANN in order to treat multi-class supervised classification problems.

Some empirical classification results presented above show its efficiency compared to standard machine learning classification and other constructive multilayer perceptrons.

The extension of this approach will consist of extracting rules from the network after training and the treatment of multivalued context. A more theoretical study of [29] discusses the fact some neural networks compute and others don't. We will explore the link with our proposal.

References

1. Andrews, R., Diederich, J., Tickle, A.: Surevy and critique of techniques for extracting rules from trained artificial neural networks. Knowledge-Based Systems 8(6), 373–389 (1995)
2. Bastide, Y., Pasquier, N., Taouil, R., Stumme, G., Lakhal, L.: Mining minimal non-redundant association rules using frequent closed itemsets. In: Palamidessi, C., Moniz Pereira, L., Lloyd, J.W., Dahl, V., Furbach, U., Kerber, M., Lau, K.-K., Sagiv, Y., Stuckey, P.J. (eds.) CL 2000. LNCS (LNAI), vol. 1861, pp. 972–986. Springer, Heidelberg (2000)

3. Bertini Jr., J.R., do Carmo Nicoletti, M.: MBabCoNN - A Multiclass Version of a Constructive Neural Network Algorithm Based on Linear Separability and Convex Hull. In: Kůrková, V., Neruda, R., Koutník, J. (eds.) ICANN 2008, Part II. LNCS, vol. 5164, pp. 723–733. Springer, Heidelberg (2008)
4. Carpineto, C., Romano, G.: Concept Data Analysis: Theory and Applications. John Wiley and Sons, Chichester (2004)
5. Cibas, T., Fogelman, F., Gallinari, P., Raudys, S.: Variable Selection with Optimal Cell Damage. In: International conference on Artificial Neural Network (ICANN 1994), Part I, pp. 727–730 (1994)
6. Cornuéjols, A., Miclet, L.: Apprentissage Artificiel: Concepts et algorithmes, Eyrolles (2002)
7. Mitchell., T.M.: Machine Learning. McGraw-Hill, New York (1997)
8. Curran, D., O'Riordan, C.: Applying Evolutionary Computation to designing Neural Networks: A study of the State of the art department of Information Technology, Technical report, NUI Galway (2002)
9. Darbari, A.: Rule extraction from trained ANN: A Survey, Technical report, Institute of Artificial Intelligence, Dept. of Computer Science, TU Dresden, Germany (2001)
10. Dreyfus, G., Samuelides, M., Martinez, J.M., Gordon, M., Badran, F., Thiria, S., Hérault, L.: Réseaux de Neurones: Méthodologie et applications. Eyrolles (2002)
11. Duch, W., Setiono, R., Zurada, J.M.: Computational intelligence methods for understanding of data. Proceedings of the IEEE 92(5), 771–805 (2004)
12. Endres, D., Fldiák, P.: An application of formal concept analysis to Neural Decoding. In: Belohlavek, E.R., Kuznetsov, S.O. (eds.) proceedings of sixth Intl. Conf. on Concept Lattices and Applications (CLA), pp. 181–192 (2008)
13. Frean, M.: The Upstart algorithm: A method for constructing and training feed forward neural networks. Neural computation (4), 198–209 (1992)
14. Fu, H., Fu, H., Njiwoua, P., Nguifo, E.M.: A comparative study of FCA-based supervised classification algorithms. In: Eklund, P. (ed.) ICFCA 2004. LNCS (LNAI), vol. 2961, pp. 313–320. Springer, Heidelberg (2004)
15. Gallant, S.: Perceptron based learning algorithms. IEEE Transactions Neural Networks 1, 179–191 (1990)
16. Ganter, B., Wille, R.: Formal Concepts Analysis: Mathematical foundations. Springer, Heidelberg (1999)
17. Garcez d'Avila, A.S., Broda, K., Gabbay, D.M.: Symbolic knowledge extraction from trained neural networks: a sound approach. Artificial Intelligence 125, 155–207 (2001)
18. Gasmi, G., Ben Yahia, S., Mephu Nguifo, E., Slimani, Y.: A new informative generic base of association rules. In: Ho, T.-B., Cheung, D., Liu, H. (eds.) PAKDD 2005. LNCS (LNAI), vol. 3518, pp. 81–90. Springer, Heidelberg (2005)
19. Guigues, J.L., Duquenne, V.: Familles minimales d'implications informatives resultant d'un tableau de donnes binaires. Mathmatiques et sciences sociales 95, 5–18 (1986)
20. Han, J., Kamber, M.: Datamining: Concepts and Techniques. Morgan Kauffman Publishers, San Francisco (2001)
21. Hertz, J., Krogh, A., Palmer, R.G.: Introduction to the theory of neural computation. Lecture Notes, Santa Fe Institute. Addison Wesley Publishing, Reading (1991)
22. Kuznetsov, S., Obiedkov, S.: Comparing Performance of Algorithms for Generating Concept Lattices. JETAI 14(2/3), 189–216 (2002)

23. Le Cun, Y., Denker, J.S., Solla, S.A.: Optimal Brain Damage. In: Advances in Neural Information Processing Systems, vol. 2, pp. 598–605. Morgan Kaufmann Publishers, San Francisco (1990)
24. Mephu Nguifo, E.: Une nouvelle approche base sur le treillis de Galois pour l'apprentissage de concepts. Mathématiques, Informatique, Sciences Humaines 134, 19–38 (1994)
25. Mephu Nguifo, E., Tsopzé, N., Tindo, G.: M-CLANN: Multi-class concept lattice-based artificial neural network for supervised classification. In: Kůrková, V., Neruda, R., Koutník, J. (eds.) ICANN 2008,, Part II. LNCS, vol. 5164, pp. 812–821. Springer, Heidelberg (2008)
26. Newmann, D.J., Hettich, S., Blake, C.L., Merz, C.J.: (UCI)Repository of machine learning databases, Dept. Inform. Comput. Sci. Univ. California, Irvine, CA (1998), `http://www.ics.uci.edu/AI/ML/MLDBRepository.html`
27. Parekh, R., Yang, J., Honavar, V.: Constructive Neural Networks Learning Algorithms for Multi-Category Classification. Department of Computer Science Lowa State University Tech. Report ISU CS TR 95-15 (1995)
28. Parekh, R., Yang, J., Honavar, V.: Constructive Neural-Network Learning Algorithms for Pattern Classification. IEEE Transactions on neural networks 11(2), 436–451 (2000)
29. Piccinini, G.: Some neural networks compute, others dont. Neural Network 21 (special issue), 311–321 (2008)
30. Rumelhart, D.E., Hinton, G.E., Williams, R.J.: Learning representations by backpropagating errors. Nature (323), 318–362 (1986)
31. Rudolph, S.: Using FCA for Encoding Closure Operators into Neural Networks. In: Priss, U., Polovina, S., Hill, R. (eds.) ICCS 2007. LNCS (LNAI), vol. 4604, pp. 321–332. Springer, Heidelberg (2007)
32. Shavlik, W.J., Towell, G.G.: Kbann: Knowledge based articial neural networks. Artificial Intelligence (70), 119–165 (1994)
33. Stumme, G., Taouil, R., Bastide, Y., Pasquier, N., Lakhal, L.: Computing Iceberg concept lattices with TITANIC. Journal on Knowledge and Data Engineering (KDE) 2(42), 189–222 (2002)
34. Subirats, J.L., Franco, L., Molina Conde, I., Jerez, J.M.: Active learning using a constructive neural network algorithm. In: Kůrková, V., Neruda, R., Koutník, J. (eds.) ICANN 2008,, Part II. LNCS, vol. 5164, pp. 803–811. Springer, Heidelberg (2008)
35. Tsopze, N., Mephu Nguifo, E., Tindo, G.: CLANN: Concept-Lattices-based Artificial Neural Networks. In: Diatta, J., Eklund, P., Liquire, M. (eds.) Proceedings of fifth Intl. Conf. on Concept Lattices and Applications (CLA 2007), Montpellier, France, October 24-26, 2007, pp. 157–168 (2007)
36. Witten, I.H., Frank, E.: Data Mining: Practical machine learning tools and techniques. Morgan Kaufmann, San Francisco (2005)
37. Yacoub, M., Bennani, Y.: Architecture Optimisation in Feedforward Connectionist Models. In: Gerstner, W., Hasler, M., Germond, A., Nicoud, J.-D. (eds.) ICANN 1997. LNCS, vol. 1327. Springer, Heidelberg (1997)
38. Yang, J., Parekh, R., Honavar, V.: Distal: An Inter-pattern Distance-based Constructive Learning Algorithm: Intell. Data Anal. 3, 55–73 (1999)
39. Werbos, P.J.: Why neural networks? In: Fiesler, E., Beale, R. (eds.) Handbook of Neural Computation, pp. A2.1:1–A2.3:6. IOP Pub., Oxford University Press, Oxford (1997)
40. Wille, R.: Restructuring Lattice Theory: An Approach Based on Hierarchies of Concepts. In: Rival, I. (ed.) Ordered Sets, pp. 445–470 (1982)

Constructive Morphological Neural Networks: Some Theoretical Aspects and Experimental Results in Classification

Peter Sussner and Estevão Laureano Esmi

Abstract. Morphological neural networks are rooted in mathematical morphology (MM). Several constructive learning algorithms for morphological neural networks have been proposed during the last decade. Since MM can be conducted very generally in the complete lattice setting, MNNs are closely related to other lattice-based neurocomputing models.

This paper reviews and analyzes some important types of constructive morphological neural networks including their learning algorithms from the lattice-theoretical perspective of mathematical morphology. In particular, we present an improved version of the learning algorithm for the morphological perceptron (MP). Moreover, we incorporate competitive nodes into the two variants of the MP and introduce an approach for training these models. Finally, we compare the performance of several constructive morphological models and of conventional multi-layer perceptrons in some classification problems.

1 Introduction

Mathematical Morphology (MM) is a theory that uses concepts from set theory, geometry and topology to analyze geometrical structures in an image [21, 28, 45, 44]. MM has found wide-spread applications over the entire imaging spectrum [7, 19, 20, 26, 32, 47, 48]. Morphological operators were originally developed for binary and grayscale image processing. The subsequent generalization to complete lattices

Peter Sussner
Department of Applied Mathematics, IMECC, University of Campinas,
Campinas, SP 13084 – 970
e-mail: sussner@ime.unicamp.br

Estevão Laureano Esmi
Department of Applied Mathematics, IMECC, University of Campinas,
Campinas, SP 13084 – 970
e-mail: ra050652@ime.unicamp.br

L. Franco et al. (Eds.): Constructive Neural Networks, SCI 258, pp. 123–144.
springerlink.com

is widely accepted today as the appropriate theoretical framework for mathematical morphology [21, 42, 44]. In the complete lattice setting, there are four elementary morphological operators - namely erosion, dilation, anti-erosion, and anti-dilation - which allow for the decomposition of every mapping between complete lattices [4].

Morphological neural networks incorporate morphological operators into the artificial neural network setting. More precisely, a morphological neural network performs a morphological operation at every node. Since the concept of morphological operator is not clearly defined [21], we have suggested to formally define a MNN as an artificial neural network that performs one of the four elementary operators of MM, possibly followed by the application of an activation function, at every node [55].

Several particular morphological models and their respective training algorithms have been proposed in recent years, including morphological perceptrons (MPs) [54], morphological perceptrons with dendrites (MPDs) [38], (fuzzy) morphological associative memories [55, 50, 51, 52, 55], modular morphological neural networks [3], and morphological shared-weight and regularization neural networks [22, 25]. This paper clarifies that fuzzy lattice neural networks (FLNNs) [24] can also be viewed as MNNs. Morphological and hybrid morphological/rank/linear neural networks [30] have been successfully applied to a variety of problems such as pattern recognition [24, 46], prediction [1, 50], automatic target recognition [25], handwritten character recognition [30], control of vehicle suspension [15], self-localization, and hyperspectral image analysis [34, 17].

Although the theory of morphological neural networks (MNNs) and its applications has experienced a steady and consistent growth in the last few years [53], only a brief review and comparison of MNNs has appeared in the literature in the form of a conference paper [29]. The present article focusses on constructive MNNs which automatically update their architecture during the learning phase. Additionally, we provide more background information on MNNs, on the connections between individual models of MNNs, and on the learning algorithms of MPs and FLNNs, the main constructive morphological models [40, 54, 31, 23]. For instance, we present an improved version of the training algorithm for MPs, introduce MPs and MPDs with competitive neurons and show how to train them. Finally, we elaborate on the foundations of morphological perceptrons (MPs) and fuzzy lattice neural networks (FLNNs) in lattice theory [4, 5]. Moreover we explain why fuzzy lattice neural networks can be viewed as morphological models, include some further comparisons with MPs, and provide additional details with respect to the experimental results and the computational effort involved in using morphological models.

The paper is organized as follows. After presenting the lattice background of MNNs, we investigate the most important types of constructive MNNs, namely morphological perceptrons (MPs), morphological perceptrons with dendrites (MPDs), and fuzzy lattice neural networks (FLNNs). Section 4 compares the performances of morphological models and MLPs in some classification problems. We finish the paper with some concluding remarks.

2 Lattice Background for Morphological Neural Networks

Morphological neural networks are geared at merging techniques of artificial neural networks and mathematical morphology. Although mathematical morphology was conceived as a set-theoretic approach to image processing, its theoretical foundations can be found in lattice algebra. This paper concentrates on morphological models of neural networks whose operations can not only be described in terms of set-theoretic ideas but also in terms of the complete lattice framework of mathematical morphology [21, 42, 44].

A partially ordered set $\mathbb{L}$ is called a *lattice* if and only if every finite, non-empty subset of $\mathbb{L}$ has an infimum and a supremum in $\mathbb{L}$. For simplicity, we assume that a partially ordered set is non-empty [18]. A lattice $\mathbb{L}$ is *complete* if every non-empty (finite or infinite) subset has an infimum and a supremum in $\mathbb{L}$ [5]. Every (non-empty) complete lattice has a least element denoted by $0_{\mathbb{L}}$ and a greatest element denoted by $1_{\mathbb{L}}$. The extended real numbers $\bar{\mathbb{R}}$ and the unit interval $[0,1]$ represent specific examples of complete lattices. For any $Y \subseteq \mathbb{L}$, we denote the infimum of Y by the symbol $\bigwedge Y$ and we write $\bigwedge_{j \in J} y_j$ instead of $\bigwedge Y$ if $Y = \{y_j, j \in J\}$ for a index set J. We use similar notations to denote the *supremum* of Y.

If $\mathbb{L}_1, \ldots \mathbb{L}_n$ are lattices, a partial order on $\mathbb{L} = \mathbb{L}_1 \times \ldots \times \mathbb{L}_n$ can be defined by setting

$$(x_1, \ldots, x_n) \leq (y_1, \ldots, y_n) \;\Leftrightarrow\; x_i \leq y_i \quad \forall i \in \{1, \ldots, n\}. \tag{1}$$

The resulting partially ordered set $\mathbb{L}$ is also a lattice and is called the *product lattice* with constituents $\mathbb{L}_1, \ldots \mathbb{L}_n$. If the lattices $\mathbb{L}_1, \ldots \mathbb{L}_n$ are complete then the product lattice $\mathbb{L} = \mathbb{L}_1 \times \ldots \times \mathbb{L}_n$ is complete as well. For notational convenience, the product lattice corresponding to the product of n copies of $\mathbb{L}$ is denoted using the symbol $\mathbb{L}^n$. Suppose that $\mathbb{L}$ and $\mathbb{M}$ are lattices. A function $\varphi : \mathbb{L} \to \mathbb{M}$ that satisfies the following equations for all $x \in \mathbb{L}$ and for all $y \in \mathbb{M}$ is called *lattice homomorphism.*

$$\varphi(x \vee y) = \varphi(x) \vee \varphi(y) \quad \text{and} \quad \varphi(x \wedge y) = \varphi(x) \wedge \varphi(y). \tag{2}$$

A bijective lattice homomorphism is called *lattice isomorphism.* Equivalently, we have that $\varphi : \mathbb{L} \to \mathbb{M}$ is a lattice isomorphism if φ is bijective and order preserving, that is $\varphi(x) \leq \varphi(y)$ for all $x \leq y$.

A central issue in mathematical morphology is the decomposition of mappings between complete lattices in terms of elementary operations.

Definition 1. *Let $\varepsilon, \delta, \bar{\varepsilon}, \bar{\delta}$ be operators from the complete lattice $\mathbb{L}$ to the complete lattice $\mathbb{M}$, and let $Y \subseteq \mathbb{L}$.*

$$\varepsilon \quad \textit{is called erosion} \Leftrightarrow \varepsilon\left(\bigwedge Y\right) = \bigwedge_{y \in Y} \varepsilon(y); \tag{3}$$

$$\delta \quad \textit{is called dilation} \Leftrightarrow \delta\left(\bigvee Y\right) = \bigvee_{y \in Y} \delta(y); \tag{4}$$

$$\bar{\varepsilon} \quad \textit{is called anti-erosion} \Leftrightarrow \bar{\varepsilon}(\bigwedge Y) = \bigvee_{y \in Y} \bar{\varepsilon}(y); \tag{5}$$

$$\bar{\delta} \quad \textit{is called anti-dilation} \Leftrightarrow \bar{\delta}(\bigvee Y) = \bigwedge_{y \in Y} \bar{\delta}(y). \tag{6}$$

The following theorem establishes representations of anti-dilations and anti-erosions in terms of erosions, dilations and negations.

Theorem 1. *Let $\mathbb{L}$ and $\mathbb{M}$ be complete lattices with negations $\nu_{\mathbb{L}}$ and $\nu_{\mathbb{M}}$, respectively.*

- *An operator $\bar{\delta} : \mathbb{L} \to \mathbb{M}$ is an anti-dilation $\Leftrightarrow \bar{\delta} = \varepsilon \circ \nu_{\mathbb{L}}$ or $\bar{\delta} = \nu_{\mathbb{M}} \circ \delta$, where δ is a dilation and ε is a erosion.*
- *An operator $\bar{\varepsilon} : \mathbb{M} \to \mathbb{L}$ is an anti-erosion $\Leftrightarrow \bar{\varepsilon} = \delta \circ \nu_{\mathbb{M}}$ or $\bar{\varepsilon} = \nu_{\mathbb{L}} \circ \varepsilon$, where ε is an erosion and δ is a dilation.*

Banon and Barrera [4] showed that for every mapping $\psi : \mathbb{L} \longrightarrow \mathbb{M}$ there exist erosions ε^i and anti-dilations $\bar{\delta}^i$ for some index set I such that

$$\psi = \bigvee_{i \in I} (\varepsilon^i \wedge \bar{\delta}^i) . \tag{7}$$

Similarly, the mapping ψ can be written as an infimum of supremums of pairs of *dilations* and *anti-erosions*. In the special case that ψ is increasing, ψ can be represented as a supremum of erosions or as an infimum of dilations.

Many models of MNNs can alternatively be defined in terms of certain matrix products in minimax algebra [13, 11]. Minimax algebra is a lattice algebra whose origins lie in the field of operations research and machine scheduling [9, 12, 16, 57].

In minimax algebra, we consider certain algebraic structures called *belts* and *bounded lattice ordered groups*. For our purposes, it is enough to consider the bounded lattice ordered group $(\mathbb{G}, \vee, \wedge, +, +')$, where the symbol $\mathbb{G}$ denotes $\bar{\mathbb{R}} = \mathbb{R} \cup \{-\infty, \infty\}$ or $\bar{\mathbb{Z}} = \mathbb{Z} \cup \{-\infty, \infty\}$. The symbols $\vee$ and $\wedge$ denote the binary operations of maximum and minimum, respectively. The operations $+$ and $+'$ act like the usual sum operation on $\mathbb{G}$ and only differ from each other in the following respect:

$$\infty + (-\infty) = (-\infty) + \infty = \infty \tag{8}$$

$$\infty +' (-\infty) = (-\infty) +' \infty = -\infty \tag{9}$$

There are two types of matrix products with entries in $\mathbb{G}$. Given a matrix $A \in \mathbb{G}^{m \times p}$ and a matrix $B \in \mathbb{G}^{p \times n}$, the matrix $C = A \boxtimes B$, called the *max-product* of A and B, and the matrix $D = A \boxtimes B$, called the *min-product* of A and B, are defined by the following equations:

$$c_{ij} = \bigvee_{k=1}^{p} (a_{ik} + b_{kj}), \quad d_{ij} = \bigwedge_{k=1}^{p} (a_{ik} +' b_{kj}) . \tag{10}$$

Let $A \in \mathbb{G}^{n \times m}$. Consider the following operators ε_A and δ_A:

$$\varepsilon_A(x) = A^t \boxslash x, \tag{11}$$
$$\delta_A(x) = A^t \boxbslash x. \tag{12}$$

Note that the operators ε_A and δ_A represent erosions and dilations from the complete lattice $\mathbb{G}^n$ to the complete lattice $\mathbb{G}^m$, respectively. The theory of minimax algebra includes a theory of conjugation. For more information, we refer the reader to the treatises of Cuninghame-Green [13, 11]. The bounded lattice ordered group $(\mathbb{G}, \vee, \wedge, +, +')$ is self-conjugate. The conjugate of an element $x \in \mathbb{G}$ is denoted using the symbol x^* and is defined as follows:

$$x^* = \begin{cases} -x, & \text{if } x \in \mathbb{G} \setminus \{-\infty, +\infty\} \\ +\infty, & \text{if } x = -\infty \\ -\infty, & \text{if } x = \infty \end{cases} \tag{13}$$

The operator of conjugation gives rise to a negation ν_* on $\mathbb{G}^n$ which maps the i-th component of $\mathbf{x}$ to its conjugate. Formally, we have

$$(\nu_*(\mathbf{x}))_i = (x_i)^* \;\; \forall\, i = 1, \ldots, n. \tag{14}$$

3 Some Constructive Morphological Neural Network Models

Morphological neural networks are equipped with morphological neurons. We speak of a morphological neuron if its aggregation function corresponds to an elementary morphological operation. As mentioned before, the emphasis in this paper is on constructive MNNs. To our knowledge, the class of constructive MNNs consists of morphological perceptrons, morphological perceprons with dendrites, and - as shown in this section - fuzzy lattice neural networks (FLNNs).

This sections provides a new perspective on constructive MNNs by exhibiting the relations between these models. In addition, we introduce a modified version of the training algorithm for morphological perceptron which has led to better experimental results in Section 4 when compared to the original algorithm [54].

3.1 Morphological Perceptron (MP)

Morphological perceptrons [40, 54] grew out of the minimax subalgebra of image algebra [11, 41, 39]. Although MPs have been formulated in terms of matrix products in minimax algebra, it was not until a recent conference paper that MPs were viewed in terms of the complete lattice framework of mathematical morphology [29]. Here, we provide some more details on this issue.

Recall that $\bar{\mathbb{R}}$ and $\bar{\mathbb{R}}^n$ represent complete lattices. Given a vector of inputs $\mathbf{x} \in \bar{\mathbb{R}}^n$ (in practice, we restrict ourselves to input vectors $\mathbf{x} \in \mathbb{R}^n$), a vector of synaptic weights $\mathbf{w} \in \bar{\mathbb{R}}^n$ and an activation function f, a neuron of the morphological perceptron calculates the output y according to one of the following rules:

$$y = f(\varepsilon_{\mathbf{w}}(\mathbf{x})), \text{ where } \varepsilon_{\mathbf{w}}(\mathbf{x}) = \mathbf{w}^t \boxtimes \mathbf{x} = \bigwedge_{i=1}^{n} (x_i + w_i); \tag{15}$$

$$y = f(\delta_{\mathbf{w}}(\mathbf{x})), \text{ where } \delta_{\mathbf{w}}(\mathbf{x}) = \mathbf{w}^t \boxtimes \mathbf{x} = \bigvee_{i=1}^{n} (x_i + w_i); \tag{16}$$

$$y = f(\bar{\varepsilon}_{\mathbf{w}}(\mathbf{x})), \text{ where } \bar{\varepsilon}_{\mathbf{w}}(\mathbf{x}) = \delta_{\mathbf{w}}(\mathbf{x}) \circ \nu_*(\mathbf{x}) = \bigvee_{i=1}^{n} (x_i^* + w_i); \tag{17}$$

$$y = f(\bar{\delta}_{\mathbf{w}}(\mathbf{x})), \text{ where } \bar{\delta}_{\mathbf{w}}(\mathbf{x}) = \varepsilon_{\mathbf{w}}(\mathbf{x}) \circ \nu_*(\mathbf{x}) = \bigwedge_{i=1}^{n} (x_i^* + w_i). \tag{18}$$

Note that $\varepsilon_{\mathbf{w}}$ represents an erosion from the complete lattice $\bar{\mathbb{R}}^n$ to the complete lattice $\bar{\mathbb{R}}$ in the special form of Equation 11. Similarly, $\delta_{\mathbf{w}}$ represents a dilation $\bar{\mathbb{R}}^n \to \bar{\mathbb{R}}$ in the special form of Equation 12. By Theorem 1, the composition of the negation ν_* followed by the erosion $\varepsilon_{\mathbf{w}}$ yields an anti-erosion that we denoted by $\bar{\varepsilon}_{\mathbf{w}}$ and the composition of the negation ν_* followed by the dilation $\delta_{\mathbf{w}}$ yields an anti-dilation that we denoted by $\bar{\delta}_{\mathbf{w}}$.

The values of the morphological perceptron's weights must be determined before it can act as a classifier. More precisely, the weights are determined using a supervised learning algorithm [54] that constructs n-dimensional boxes around sets of points which share the same class value. Convergence occurs in a finite number of steps.

In this paper, we present an improved version of the original training algorithm for MPs that was proposed to solve two-class classification problems [54]. To this end, let us introduce some relevant notations.

The vectors $\mathbf{x}^1, \mathbf{x}^2, \ldots, \mathbf{x}^k \in \mathbb{R}^n$ denote the given training patterns. The set of training patterns belonging to class 0 is denoted using the symbol C_0 and the set of training patterns belonging to class 1 is denoted using the symbol C_1. We define the following index sets:

$$K(0) = \{j \in \{1,\ldots,k\} : \mathbf{x}^j \in C_0\}. \tag{19}$$

$$K(1) = \{j \in \{1,\ldots,k\} : \mathbf{x}^j \in C_1\}. \tag{20}$$

Let P be the hyperbox $\text{box}(\mathbf{p}^\perp, \mathbf{p}^\top) = \{\mathbf{x} \in \mathbb{R}^n : \mathbf{p}^\perp \leq \mathbf{x} \leq \mathbf{p}^\top\}$, where the symbols $\mathbf{p}^\perp$ and $\mathbf{p}^\top \in \bar{\mathbb{R}}^n$ represent the lower vertex and upper vertex, respectively. The symbol P° stands for the interior of P and the symbol ∂P stands for the boundary of P, that is $P^\circ = \{\mathbf{x} \in \mathbb{R}^n : \mathbf{p}^\perp < \mathbf{x} < \mathbf{p}^\top\}$ and $\partial P = P \setminus P^\circ$. Furthermore, let us define the following half-spaces $H_i^+(\mathbf{x})$ and $H_i^-(\mathbf{x})$ for every input pattern $\mathbf{x} \in \mathbb{R}^n$:

$$H_i^+(\mathbf{x}) = \{\mathbf{y} \in \bar{\mathbb{R}}^n : y_i \geq x_i\} \text{ and } H_i^-(\mathbf{x}) = \{\mathbf{y} \in \bar{\mathbb{R}}^n : y_i \leq x_i\}. \tag{21}$$

The training algorithm will automatically produce the arquitecture depicted in Figure 1(**a**). The staircase symbol for the activation function at the output node represents the Heaviside step function given by

$$f(x) = \begin{cases} 1, & \text{if } x \geq 0, \\ 0, & \text{if } x < 0. \end{cases} \tag{22}$$

Given an arbitrary input pattern $\mathbf{x} \in \mathbb{R}^n$, the MP computes the output in terms of $f(g(\mathbf{x}))$, where g is some function $\mathbb{R}^n \to \bar{\mathbb{R}}$. Let $L(0)$ denote the set of indices $j \in \{1,\ldots,k\}$ such that $f(g(\mathbf{x}^j)) \in C_0$ and let $L(1)$ denote the set of indices $j \in \{1,\ldots,k\}$ such that $f(g(\mathbf{x}^j)) \in C_1$. The symbol D refers to the set of indices corrsponding to class 1 patterns that are currently misclassified. Formally, we have $D = \{j \in L(0) \cap K(1)\}$.

We initialize the function $g : \mathbb{R}^n \to \bar{\mathbb{R}}$ by setting $g(\mathbf{x}) = -\infty$ for all $\mathbf{x} \in \mathbb{R}^n$. Thus, initially we have $L(0) = \{1,\ldots,k\}$ and $D = K(1)$.

Step 1 (Find a Hyperbox containing only Class 1 Patterns)

1 *While* $D \neq \emptyset$

1.1 Let $P = \text{box}(\mathbf{p}^{\perp}, \mathbf{p}^{\top})$ where the vertices $\mathbf{p}^{\perp}$ and $\mathbf{p}^{\top}$ satisfy:

$$p_i^{\top} = \bigvee_{j \in D} x_i^j \; \forall i = 1,\ldots,n, \tag{23}$$

$$p_i^{\perp} = \bigwedge_{j \in D} x_i^j \; \forall i = 1,\ldots,n. \tag{24}$$

1.2 If there exists an index i_0 such that $p_{i_0}^{\top} = p_{i_0}^{\perp}$ then perform the following steps.
- (a) If, in addition, the set $P \cap C_0$ is empty then select the pattern $\mathbf{x}^j \in P \cap C^1$ such that j is minimal and set $P = \{\mathbf{x}^j\}$.
- (b) In any event, modify the upper and lower corner of P as follows:

$$p_i^{\perp} = \sup\{x_i < p_i^{\perp} : \mathbf{x} \in C_0\}. \tag{25}$$

$$p_i^{\top} = \inf\{x_i > p_i^{\top} : \mathbf{x} \in C_0\}. \tag{26}$$

Here, the supremum and the infimum are taken in $\bar{\mathbb{R}}$. In particular, we have $\sup\emptyset = -\infty$ and $\inf\emptyset = \infty$.

1.3 Otherwise, proceed as follows. Consider the set $S = C_0 \cap P$.
- (a) If $S = \emptyset$ then use Equations 25 and 26 to expand the hyperbox P.
- (b) If $S \neq \emptyset$ then continue as follows
  - (i) For all $j = 1,\ldots,k$ such that $\mathbf{x}^j \in C_0 \cap P$ execute the following steps. Set $\mathbf{x} = \mathbf{x}^j$. Consider the hyperboxes $H_i^+(\mathbf{x}) \cap P$ and $H_i^-(\mathbf{x}) \cap P$ for $i = 1,\ldots n$. Among these $2n$ hyperboxes, choose the hyperbox $P' \subseteq P$ that contains the largest number of currently misclassified patterns in C_1 such that $\mathbf{x}$ does not belong to P' (if more than one of the hyperboxes $H_i^{\pm}(\mathbf{x}) \cap P$ meets these criteria then randomly select one of the these). Update P by setting $P = P'$.

(ii) Expand the hyperbox P that was obtained in item (a) by applying Equations 25 and 26.

1.4 Determine the smallest hyperbox $B = \text{box}(\mathbf{b}^{\perp}, \mathbf{b}^{\top})$ which contains all the misclassified patterns of class 1 in the interior of P. Formally, we have

$$b_i^{\top} = \bigvee_{\mathbf{x}^j \in P^{\circ}} x_i^j \; \forall i = 1, \ldots, n \tag{27}$$

$$b_i^{\perp} = \bigwedge_{\mathbf{x}^j \in P^{\circ}} x_i^j \; \forall i = 1, \ldots, n \tag{28}$$

If $B = \emptyset$ then choose the pattern $\mathbf{x}^j \in \partial P \cap C_1$ such that j is minimal, redefine P as $P = \{\mathbf{x}^j\}$, and return to Step 1.3(b)(ii) (the next time around, $B \neq \emptyset$).

1.5 Determine an intermediary hyperbox C whose upper and lower corner, denote respectively by $\mathbf{c}^{\top}$ and $\mathbf{c}^{\perp}$, are given by the averages of the corresponding vertices of B and P.

$$c_i^{\top} = \frac{b_i^{\top} + p_i^{\top}}{2}, \tag{29}$$

$$c_i^{\perp} = \frac{b_i^{\perp} + p_i^{\perp}}{2}. \tag{30}$$

Step 2 (Update the Architecture of the Morphological Perceptron) At the end of this step, the patterns in the hyperbox C are assigned to class 1.

2.1 Update the function g as follows:

$$g(\mathbf{x}) = g(\mathbf{x}) \vee \left[\bigwedge_{i=1}^{n} (x_i - c_i^{\top}) \wedge \bigwedge_{i=1}^{n} (c_i^{\perp} - x_i) \right]. \tag{31}$$

2.2 Compute $f(g(\mathbf{x}))$ for all $\mathbf{x} \in D$, update the set D, and return to Step 1.

Note that the function g determines a union of hyperboxes. An arbitrary input pattern $\mathbf{x}$ is assigned to class 1 if and only if it lies in this union of hyperboxes. The term $\bigwedge_{i=1}^{n}(x_i - c_i^{\top}) \wedge \bigwedge_{i=1}^{n}(c_i^{\perp} - x_i)$ of Equation 31 is non-negative if and only if $\mathbf{x}$ is between the upper and lower vertices of C. The term $\bigwedge_{i=1}^{n}(x_i - c_i^{\top})$ corresponds to the erosion $\varepsilon_{\mathbf{v}}$ where $\mathbf{v} = -\mathbf{c}^{\top}$ and the term $\bigwedge_{i=1}^{n}(c_i^{\perp} - x_i)$ corresponds to an anti-dilation $\bar{\delta}_{\mathbf{w}} = \varepsilon_{\mathbf{w}} \circ \nu_*$ where $\mathbf{w} = \mathbf{c}^{\perp}$. During the training phase, pairs of erosive and anti-dilative neurons are added to the hidden layer of the MP that is pictured in Figure 1(**a**). This process ends once all training patterns are classified correctly. After convergence, the output of the MP is determined by the following equation for some $m \in \mathbb{N}$:

$$y = f\Big(\bigvee_{j=1}^{m} (\varepsilon_{\mathbf{v}_j}(\mathbf{x}) \wedge \bar{\delta}_{\mathbf{w}_j}(\mathbf{x}))\Big) \tag{32}$$

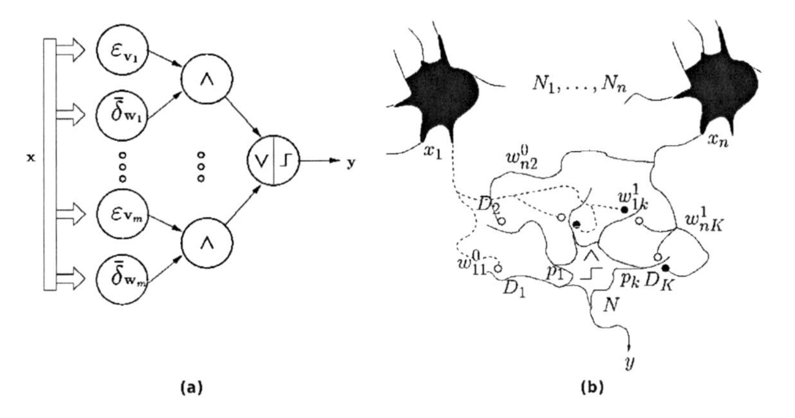

Fig. 1 Architectures of a morphological perceptron (**a**) and of an MPD (**b**), respectively.

An arbitrary input pattern **x** is classified as belonging to class 1 if and only if $y = 1$. According to Equation 34 and Figure 1(**a**), the MP calculates a maximum of pair-wise minimums of erosions and anti-dilations which approximates the decomposition suggested by Banon and Barrera [4] (cf. Equation 7) followed by the application of a hard-limiting function f.

The modifications of the original algorithm proposed in [54] can be found in Steps 1.2, 1.3, and 1.4. We have in particular taken additional measures in order to circumvent situations in which the original algorithm failed to converge. The modified version is guaranteed to converge in a finite number of steps yielding a decision surface that perfectly separates the class 0 and the class 1 training data and led to better results in the classification problems described in Section 4.

In a previous conference paper, we have allocated three morphological neurons to perform a minimum of an erosion $\varepsilon_{\mathbf{v}}$ and an anti-dilation $\bar{\delta}_{\mathbf{w}}$ although one could argue that forming

$$\varepsilon_{\mathbf{v}}(\mathbf{x}) \wedge \bar{\delta}_{\mathbf{w}}(\mathbf{x}) = \bigwedge_{i=1}^{n} (x_i + v_i) \wedge \bigwedge_{i=1}^{n} (x_i^* + w_i) \tag{33}$$

only requires one single morphological neuron or processing element with inputs of the form $(x_1, \ldots, x_n, x_1^*, \ldots, x_n^*)$ and weights in $\bar{\mathbb{R}}^{2n}$. Therefore, we associate only one morphological neuron to the computation of Equation 33. In other words, the number of hidden morphological neurons corresonds to the number of hyperboxes that are generated during the learning phase.

Suppose we have an S-class classification problem. If $S > 2$ then the MP approach has to be adapted so as to be able to deal with *multiple classes*. Let $\bar{C}_s \subseteq \{\mathbf{x}^1, \ldots, \mathbf{x}^k\}$ denote the set of training patterns belonging to the sth class where $s = 1, \ldots, S$. For each $s = 1, \ldots S$, we simply set $C_1 = \bar{C}_s$ and $C_0 = \bigcap_{t \neq s} \bar{C}_t$ and apply the MP training algorithm. This procedure generates weights $\mathbf{v}_j^s$ and $\mathbf{w}_j^s$ for every $s = 1, \ldots S$. Thus, we obtain S MPs with threshold activation functions at their respective output nodes of the form pictured in Figure 1(**a**). Removing the thresholds at the outputs yields S MPs that are given by the following equations for $s = 1, \ldots S$:

$$y_s = \bigvee_{j=1}^{m_s} (\varepsilon_{\mathbf{v}_j^s}(\mathbf{x}) \wedge \bar{\delta}_{\mathbf{w}_j^s}(\mathbf{x})) \tag{34}$$

An MP for multi-class classification problems arises by joining the S MPs and by introducing *competitive output neurons.* In other words, an input pattern $\mathbf{x}$ will be classified as belonging to class $y = \arg\max_s y_s$. For simplicity, we use the acronym MP/C to denote the resulting *morphological perceptron with competitive neurons.*

3.2 Morphological Perceptrons with Dendrites (MPD)

Recent research in neuroscience has given considerable importance to dendritic structures in a single neuron cell [43]. Ritter and Urcid developed a new paradigm for computing with morphological neurons where the process occurs in the dendrites [37, 38]. Figure 1(**b**) provides a graphical representation of an MPD that has a single output neuron N.

The architecture of an MPD is not determined beforehand. During the training phase, the MPD grows new dendrites while the input neurons expand their axonal branches to synapse on the new dendrites. The weight of an axonal branch of input neuron N_i terminating on the k-th dendrite of the output neuron N is denoted by w_{ki}^l where the superscript $l \in \{0, 1\}$ distinguishes between *excitatory* ($l = 1$) and *inhibitory* ($l = 0$) *input* to the dendrite. The k-th dendrite of N will produce either an *excitatory* ($p_k = 1$) or an *inhibitory* ($p_k = -1$) *response* to the total input received from the input neurons N_i. To summarize, the computation performed by the k-th dendrite is given by

$$\tau_k(\mathbf{x}) = p_k \bigwedge_{i=1}^{n} \bigwedge_{l \in L} (-1)^{l+1} (x_i + w_{ki}^l), \tag{35}$$

where $L \subseteq \{0, 1\}$ corresponds to the set of terminal fibers on N_i that synapse on the k-th dendrite of N. After passing the value $\tau_k(\mathbf{x})$ to the cell body, the state of N is given by $\bigvee_{k=1}^{K} \tau_k(\mathbf{x})$, where K denotes the total number of dendrites of N. Finally, an application of the hard limiting function f defined in Equation 22 yields the next state of N, in other words the output y of the MPD depicted in Figure 1(**b**).

$$y = f(\bigwedge_{k=1}^{K} \tau_k(\mathbf{x})) = f(\bigwedge_{k=1}^{K} p_k \bigwedge_{i=1}^{n} \bigwedge_{l \in L} (-1)^{l+1} (x_i + w_{ki}^l)). \tag{36}$$

Leaving the biological motivation aside, the MPD training algorithm that was proposed for binary classification problems [38] resembles the one for MPs [54]. As is the case for MPs, the MPD training algorithm is guaranteed to converge in a finite number of steps and, after convergence, all training patterns will be classified correctly. Learning is based on the construction of n-dimensional hyperboxes. Given two classes C_0 and C_1, an input pattern $\mathbf{x}$ is classified as belonging to class C_1 if and only if $\mathbf{x}$ is contained in one of the constructed hyperboxes. In fact, each dendrite corresponds to a hyperbox. Therefore, we can convert an MPD into an MP and express the computation performed by a dendrite in terms of Equation 33.

When faced with multi-class classification problems, we propose to construct a MPD with competitive output units and to proceed in the same way as we did with MPs at the end of Section 3.1. We will refer to the resulting MNN as *morphological perceptron with dendrites and competitive neurons* (MPD/C).

3.3 *Fuzzy Lattice Neural Network-FLNN*

The theoretical framework of FLNN constitutes a successful combination of fuzzy sets [56], lattice theory [21] and adaptive resonance theory [8]. Figure 2 illustrates the architecture of the FLNN that consists of an *input layer* and a *category layer*. The input layer has N artificial neurons used for storing and comparing input data. The category layer has L artificial neurons that define M classes.

Given a vector of inputs $\mathbf{x}$ and a vector of synaptic weights $\mathbf{w}$, a neuron of an FLNN [31, 24] computes the degree of inclusion of $\mathbf{x}$ in $\mathbf{w}$ in terms of $p(\mathbf{x},\mathbf{w})$ where p is a fuzzy partial order relation. In general, we refer to p as a *fuzzy partial order* on a lattice $\mathbb{L}$ if p is a function $\mathbb{L} \times \mathbb{L} \to [0,1]$ that satisfies the equation $p(\mathbf{x},\mathbf{y}) = 1$ if and only if $\mathbf{x} \leq \mathbf{y}$. (We prefer to speak of a fuzzy partial order instead of a fuzzy membership function or fuzzy inclusion measure because p generalizes the conventional partial order.) A pair $(\mathbb{L}, p)$ consisting of a lattice $\mathbb{L}$ and a fuzzy partial order p is called a *fuzzy lattice*.

In the case of FLNNs, both the input vector $\mathbf{x}$ and the vector of synaptic weights $\mathbf{w}$ are hyperboxes in $\mathbb{L}^N$ where $\mathbb{L}$ is a complete lattice. For the special case where

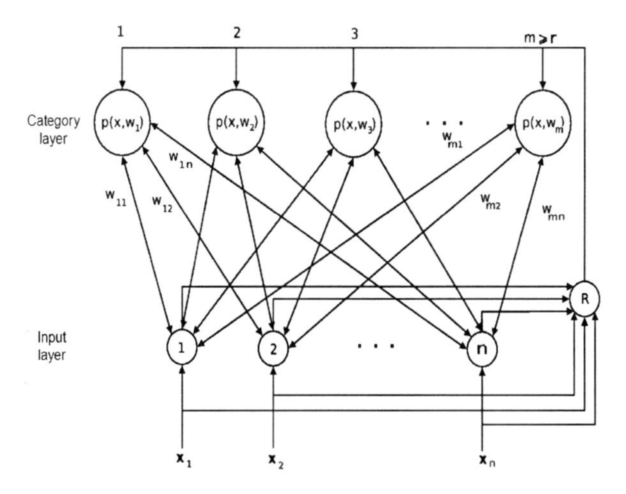

Fig. 2 Architecture of the FLNN.

$N = 1$, a hyperbox in $\mathbb{L}^N$ corresponds to a closed interval in $\mathbb{L}$ and can be written in the form $[a,b]$ where $a,b \in \mathbb{L}$. In particular, if $\mathbb{L} = [0,1]$ we obtain a closed subinterval of the unit interval. The partial order on a given lattice $\mathbb{L}$ induces a partial order on set of intervals $\mathscr{I}_{\mathbb{L}} = \{[\mathbf{a},\mathbf{b}] : \mathbf{a},\mathbf{b} \in \mathbb{L} \text{ and } \mathbf{a} \leq \mathbf{b}\}$ which turns $\mathscr{I}_{\mathbb{L}}$ into a lattice as well:

$$[\mathbf{a},\mathbf{b}] \leq [\mathbf{c},\mathbf{d}] \Leftrightarrow \mathbf{a} \geq \mathbf{c} \text{ and } \mathbf{b} \leq \mathbf{d}\,. \tag{37}$$

Unfortunately, the lattice of the closed intervals is not complete even if $\mathbb{L}$ is complete because $\bigwedge \mathscr{I}_{\mathbb{L}}$ does not exist in $\mathscr{I}_{\mathbb{L}}$. There are however two closely related complete lattices. The first one, called the complete lattice of the *generalized intervals*, is denoted using the symbol $\mathbb{PL}$ and arises by leaving away the restriction $\mathbf{a} \leq \mathbf{b}$. Formally, we have $\mathbb{PL} = \{[\mathbf{a},\mathbf{b}] : \mathbf{a},\mathbf{b} \in \mathbb{L}\}$. If $0_{\mathbb{L}}$ and $1_{\mathbb{L}}$ denote the least element of $\mathbb{L}$ and the greatest element of $\mathbb{L}$, respectively, then the least element of $\mathbb{PL}$ is given by $[1_{\mathbb{L}},0_{\mathbb{L}}]$ and the greatest element of $\mathbb{PL}$ is given by $[0_{\mathbb{L}},1_{\mathbb{L}}]$. The second complete lattice of interest is denoted by $\mathbb{V}_{\mathbb{L}}$ and is given by adjoining $[1_{\mathbb{L}},0_{\mathbb{L}}]$ to $\mathscr{I}_{\mathbb{L}}$. We obtain $\mathbb{V}_{\mathbb{L}} = \mathscr{I}_{\mathbb{L}} \cup \{[1_{\mathbb{L}},0_{\mathbb{L}}]\}$.

The FLNN model employs a fuzzy partial order relation $p : (\mathbb{V}_{\mathbb{L}})^N \times (\mathbb{V}_{\mathbb{L}})^N \rightarrow [0,1]$. In this context, the fuzzy partial order is of a special form. Specifically, the fuzzy partial order relation employed in the FLNN is based on a "function-h" or - as we prefer to call it - a *generating function* [31]. Similar fuzzy lattice models use fuzzy partial order relations based on *positive valuation functions* [5, 24, 23].

In applications of the FLNN to classification tasks such as the ones discussed in Section 4, it suffices to consider - after an appropriate normalization - the complete lattice $\mathbb{U} = [0,1]$, i.e., the unit interval. Thus, the input and weight vectors are N-dimensional hyperboxes in $(\mathbb{V}_{\mathbb{U}})^N$. In this case, we can show that $p(.,\mathbf{w}) : (\mathbb{V}_{\mathbb{U}})^N \rightarrow [0,1]$ represents an elementary operation on mathematical morphology, namely both an anti-erosion and an anti-dilation, if the underlying generating function $h : \mathbb{U} \rightarrow \mathbb{R}$ is continuous. The proof of this result is beyond the scope of this paper since it involves further details on fuzzy lattice neuro-computing models, in particular the construction of a fuzzy partial order from a generating function. Therefore, we postpone the proof to a future paper where we will show a more general result that uses additional concepts of lattice theory. Anyway, the fact that every node in the category layer of an FLNN with a continuous generating function performs an elementary operation of MM has led us to classify FLNNs as belonging to the class of morphological neural networks.

FLNNs can be trained in supervised or unsupervised fashion [31, 24]. Both versions generate hyperboxes that determine the output of the FLNN. In this paper, we focus on the supervised learning algorithm that is used in classification tasks. Here we have a set of n training patterns $\mathbf{x}^1,\ldots,\mathbf{x}^n \in (\mathbb{V}_{\mathbb{L}})^N$ together with their class labels $c_1,\ldots,c_n \in \{1,\ldots,M\}$. Therefore, the basic arquitecture of the FLNN has to be adapted so as to accomodate the class information. This can be achieved by allowing for the storage of a class index in each node of the category layer, by augmenting the input layer by one node that carries the class information c_i corresponding to the pattern $\mathbf{x}^i$ during the training phase, and by fully interconnecting the two layers.

During the training phase, the FLNN successively constructs nodes in the category layer each of which is associated with an N-dimensional hyperbox $\mathbf{w}_i$ - i.e., an element of $(\mathbb{V}_{\mathbb{L}})^N$ - together with its respective class label c_i. Training in the original FLNN model is performed using an exhaustive search and takes $\mathcal{O}(N^3)$ operations [31]. (A modification of the FLNN called *fuzzy lattice reasoning* (FLR) classifier requires $\mathcal{O}(N^2)$ for training if one renounces on optimizing the outcome of training [23].) After training, we obtain L N-dimensional hyperboxes $\mathbf{w}_i \in (\mathbb{V}_{\mathbb{L}})^N$ that correspond to the L nodes that appear in the category layer of the FLNN (cf. Figure 2(**b**)). A class label $c_i \in \{1,\ldots,M\}$ is associated with each hyperbox $\mathbf{w}_i \in (\mathbb{V}_{\mathbb{L}})^N$.

In the testing phase, an input pattern $\mathbf{x} \in (\mathbb{V}_{\mathbb{L}})^N$ is presented to the FLNN and the values $p(\mathbf{x}, \mathbf{w}_i)$ are computed for $i = 1,\ldots,L$. A competition takes place among the L nodes in the category layer and the input pattern $\mathbf{x}$ is assigned to the class c_i that is associated with the hyperbox $\mathbf{w}_i$ exhibiting the highest value $p(\mathbf{x}, \mathbf{w})$. Informally speaking, the degree of inclusion of $\mathbf{x}$ in $\mathbf{w}_i$ is higher than the degree of inclusion of $\mathbf{x}$ in $\mathbf{w}_j$ for all $j \neq i$. In particular, if $\mathbf{x}$ is contained in $\mathbf{w}_i$ but not contained in $\mathbf{w}_j$ for all $j \neq i$, i.e., $p(\mathbf{x}, \mathbf{w}_i) = 1$ and $p(\mathbf{x}, \mathbf{w}_j) < 1$ for all $j \neq i$, the $\mathbf{x}$ is classified as belonging to class c_i.

Occasionally, the training algorithms for FLNNs and its modifications produce overlapping hyperboxes with disparate class memberships although - according to Kaburlasos et al. - this event occurs rarely [23]. In the experiments we conducted in Section 4, this situation did in fact occur as evidenced by the decision surface that is visualized in Figure 5. For more information, we refer the reader to Section 4.

4 Experimental Results

In this section we compare the classification performance of the constructive morphological models and the conventional multi-layer perceptron in a series of experiments on two well known datasets: Ripley's synthetic dataset [35, 36] and the image segmentation dataset that can be found in the UCI Machine Learning Repository [6]. In addition, we have used Ripley's synthetic dataset to visualize the respective decision surfaces. In contrast to our previous conference paper, we have decided to omit morphological models with a fixed architecture such as the modular morphological neural network (MMNN) and the hybrid morphological/rank/linear neural network (MRL-NN) since these models produced poor classification results in our simulations [29].

Tables 1 and 2 display the percentages of the misclassified training and testing patterns. Since the type of operations performed by the individual models in a training epoch varies greatly from one model to another we have also included the average CPU time (on a AMD Athlon 64 X2 Dual Core Processor 4200+ with a processing speed of 2.221 GHz) of each individual model until convergence of the training algorithm. We trained the constructive morphological models until their decision surfaces succeeded in perfectly separating the two classes of training data.

Table 1 Percentage of misclassified patterns for training (E_{tr}) and testing (E_{te}) in the experiments, CPU time in seconds for learning (T_{cpu}) and number of hidden artificial neurons (hyperboxes for MNNs) (H_a).

	Ripley's Synthetic Dataset			
Model	$E_{tr}(\%)$	$E_{te}(\%)$	T_{cpu}	H_a
MP	0.0	11.70	1.1	18
MPD	0.0	17.80	0.58	19
MP/C	0.0	10.80	2.63	38
MPD/C	0.0	13.90	1.26	38
FLNN	0.0	11.4/12.0	75.41	46
MLP	8.41	12.55	144.48	10

Table 2 Percentage of misclassified patterns for training (E_{tr}) and testing (E_{te}) in the experiments, CPU time in seconds for learning (T_{cpu}) and number of hidden artificial neurons (hyperboxes for MNNs) (H_a).

	Image Segmentation Dataset			
Model	$E_{tr}(\%)$	$E_{te}(\%)$	T_{cpu}	H_a
MP/C	0.0	17.38	3.29	88
MPD/C	0.0	13.05	0.21	20
FLNN	0.0	10.00	26.87	17
MLP	15.71	26.81	58.41	20

All the models and algorithms were implemented using MATLAB which favors linear operations over morphological operations. We believe that the CPU times for learning in the constructive MP, MPD, MPD/C, and MPD/C models would be even lower if more efficient implementations of the max-product and min-product were used [33].

For a fair comparison of the number of artificial neurons or processing elements in the constructive MNNs, we have implicitly expressed each individual model as a feedforward model with one hidden layer and competitive output nodes and we have counted the number of hidden nodes or hyperboxes that were constructed during the learning phase. The same sequence of training patterns appearing on the respective internet sites were employed for training the constructive MNNs [36, 6].

4.1 Ripley's Synthetic Problem

Ripley's synthetic dataset [36] consists of data samples from two classes [35, 36]. Each sample has two features. The data are divided into a training set and a test set consisting of 250 and 1000 samples, respectively, with the same number of samples belonging to each of the two classes. Thus, we obtain a binary classification problem in $\mathbb{R}^2$. Figures 3, 4, and 5 provide for more insight into the constructive

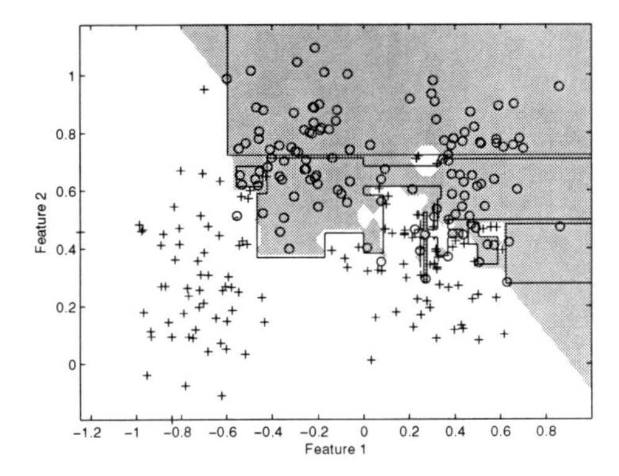

Fig. 3 Decision surfaces of an MP represented by by the continuous line and of an MP with competitive output nodes represented by the difference in shading. Training patterns belonging to class 0 are plotted using "+" symbols. Training patterns belonging to class 1 are plotted using "∘" symbols.

MNNs by visualizing the decision surfaces that are generated by these models after training (Figure 5 also includes the decision surface corresponding to an MLP with ten hidden nodes). Here, we have used the same order in which the training patterns appear on Ripley's internet site. We would like to clarify that the decision surfaces vary slightly depending on the order in which the training patterns are presented to the constructive morphological models.

Recall that the decision surfaces of the constructive morphological models are determined by N-dimensional hyperboxes, i.e., rectangles for $N = 2$. This fact is clearly visible in the decision surfaces of the MP and the MPD with hardlimiting output units, that are pictured by means of the continuous lines in Figures 3 and 4.

In addition, Figures 3, 4, and 5 reveal that the decision surfaces generated by the MP/C, MPD/C, and FLNN models deviate from rectangular appearance of the ones generated by the basic MP and MPD models. In this context, recall that the MP/C and MPD/C models construct separate families of hyperboxes for the training patterns of each class. Each family of hyperboxes is associated to a different class and corresponds to a certain output node. Upon presentation of an input pattern $\mathbf{x}$ to the MP/C or MPD/C model a competition among the output nodes occurs that determines the result of classification.

In the FLNN model a similar competition occurs in the category layer. More precisely, the FLNN uses information on the degrees of inclusion $p(\mathbf{x}, \mathbf{w}_i)$ of an input pattern $\mathbf{x}$ in the hyperboxes $\mathbf{w}_i$ for classification by associating $\mathbf{x}$ to the class of the hyperbox $\mathbf{w}_i$ in which $\mathbf{x}$ exhibits the highest degree of inclusion. This property of the FLNN is evidenced by the diagonal lines in its decision surface (cf. Figure 5).

The training algorithms of all types of constructive MNNs are guaranteed to produce decision surfaces that perfectly separate the training patterns with different class labels. However, the training algorithms of the MP/C, the MPD/C, and the FLNN may result in overlapping hyperboxes with distinct class memberships

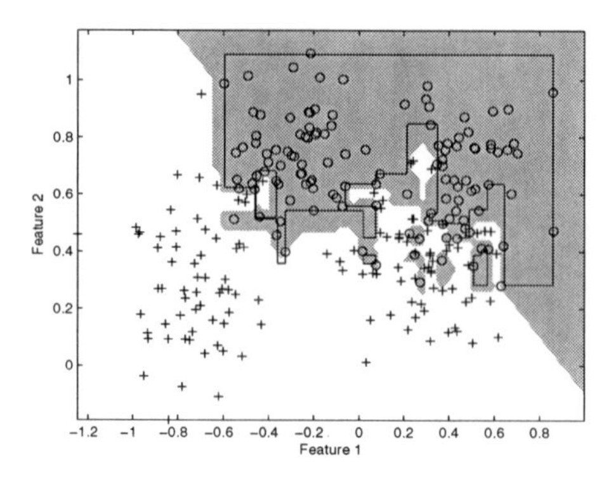

Fig. 4 Decision surfaces of an MPD represented by by the continuous line and of an MPD with competitive output nodes represented by the difference in shading.

although this event did not happen in our simulations with Ripley's synthetic dataset when using the MP/C and MPD/C models. Concerning the FLNN, Figure 5 depicts intersections of rectangles with distinc class labels using a darker shade of gray. These sets of intersection correspond to regions of indecision since a pattern $\mathbf{x}$ that is contained in both $\mathbf{w}_i$ and $\mathbf{w}_j$ with $c_i \neq c_j$ satisfies $p(\mathbf{x},\mathbf{w}_i) = 1 = p(\mathbf{x},\mathbf{w}_j)$.

Table 1 exhibits the results that we obtained concerning the classification performance and the computational effort required by the individual models. The MP training algorithm described in Section 3.1.1 automatically generated 19 hyperboxes corresponding to 19 (augmented) hidden neurons capable of evaluating Equation 33. In a similar manner, training an MPD using the constructive algorithm of Ritter and Urcid [38] yielded 19 dendrites that correspond to 19 hidden computational units. In contrast to the basic MP and MPD models, the MP/C and MPD/C generate one family of hyperboxes for each class of training patterns. Since Ripley's synthetic problem represents a binary classification problem, the number of hidden computational units in the MP/C and MPD/C models is approximately twice as high as in the MP and MPD. The FLNN grew 46 neurons in the category layer during the training phase. Moreover, we compared the morphological models with an MLP with ten hidden nodes that was trained using gradient descent with momentum and adaptive step backpropagation rule (learning rate $\eta = 10^{-4}$, increase and decrease parameters 1.05 and 0.5 respectively, momentum factor $\alpha = 0.9$). In addition, we used 25-fold cross-validation in conjunction with the MLP and chose the weights that led to the least validation error.

Table 1 reveals that the MP and MPD models including their variants with competitive nodes converge rapidly to a set of weights that yield perfect separation of the training data. The FLNN model also produces no training error but the convergence of the training algorithm is slower yet not quite as slow as MLP training. All the models we tested exhibited satisfactory results in classification. The MPD yields the highest classification error for testing which is due to the fact that, in contrast to the MP training algorithm, no expansion of the hyperboxes corresponding to C_1 patterns takes place in MPD training (this is why the MPD learns faster than the MP). This lack of expansion does not cause any problems if competing hyperboxes are constructed for patterns of both classes as is the case for the MPD/C.

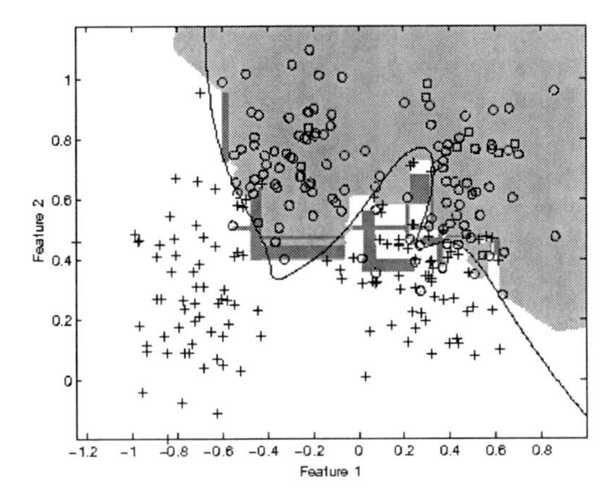

Fig. 5 Decision surfaces of an FLNN represented by the difference in shading (dark regions refer to areas of uncertainty where hyperboxes with different class labels overlap) and of an MLP represented by the continuous line.

As the reader may recall by taking a brief glance at Figure 5, the FLNN produces areas of indecision, i.e., overlapping hyperboxes with distinct class memberships. If these areas of indecision are assigned to either one of the two classes, the percentages of misclassification for testing are 11.40% and 12.00%, respectively. Otherwise, if no decision is taken then we obtain a classification error of 14.2%. In any case, the MP/C model exhibits the best classification performance.

4.2 Image Segmentation Problem

The Image Segmentation Dataset was donated by the Vision Group, University of Massachussets, and is included in the Machine Learning Repository of the University of California, Irvine [6]. This dataset consists of 210 samples for training and 2100 samples for testing. The data have 19 continous attributes. Each sample is decribed by 19 continous attributes and corresponds to a 3×3 region that was randomly drawn from an outdoor image. The images were handsegmented to create a classification for every pixel. The goal is to distinguish between 7 different classes: grass, cement, foliage, brickface, path, sky, and window. The MP and MPD can not be applied directly to such a multi-class problem.

Therefore, we considered the MP/C, the MPD/C, the FLNN, and a MLP. We chose to train an MLP with twenty hidden neurons using the Levenberg-Marquardt algorithm because this algorithm produced the lowest classification error for testing with the Image Segmentation Dataset in a recent paper [10]. Actually, we found a testing error of 26.81% that is slightly lower than the error of 28.13% found by Coskun and Yildirim.

The FLNN yields an excellent recognition rate of 90% of the test patterns without the use of any fine tuning as required by other networks [27]. In this case no action with respect to the region of indecision was taken. The classification error can be lowered to 9.6% by associating a pattern $\mathbf{x}$ contained in the region of indecision with the first class label having value 1. The MLD/C and MP/C also exhibit a better classification performance than the MLP. The high number of hidden neurons grown by the MP/C is probably due to the provisions that were taken to circumvent problems of convergence of the training algorithm. We suspect that these problems are caused by integer-valued attributes of training patterns with different class labels.

5 Conclusions

This paper provides an overview and a comparison of morphological neural networks (MNNs) for pattern recognition with an emphasis on constructive MNNs, which automatically grow hidden neurons during the training phase. We have defined MNNs as models of artificial neural networks that perform an elementary operation of mathematical morphology at every node followed by the application of an activation function. The elementary morphological operations of erosion, dilation, anti-erosion, and anti-dilation can be defined in an arbitrary complete lattice.

In many cases, the underlying complete lattice of choice is $\bar{\mathbb{R}}^n$ which has allowed researchers to formulate morphological neurons (implicitly) in terms of the additive maximum and additive minimum operations in the bounded lattice ordered group $(\bar{\mathbb{R}}, \vee, \wedge, +, +')$ - often without being aware of this connection to minimax algebra [13, 51]. In this setting, the elementary morphological operations can be expressed in terms of maximums (or minimums) of sums, which lead to fast neural computational and easy hardware implementation [33, 38].

As to the resulting models of morphological neurons, recent research results have revealed that the maximum operation lying at the core of morphological neurons is neurobiologically plausible [58]. We have to admit though that there is no neurophysiological justification for summing the inputs and the synaptic weights. This lack of neurobiological plausibility can be overcome by means of the isomorphism between the algebraic structure $(\bar{\mathbb{R}}, \vee, \wedge, +, +')$ and $(\mathbb{R}_{\infty}^{\geq 0}, \vee, \wedge, \cdot, \cdot')$ that transforms additive maximum/minimum operations into multiplicative maximum/minimum operations [52]. In this context, we intend to investigate the connections between MNNs and min-max or adaptive logic networks that combine linear operations with minimums and maximums [2].

In this paper, we have related another lattice based neuro-computing model, namely the fuzzy lattice neural network (FLNN), to MNNs. Specifically, we explained that the FLNN can be considered to be one of the constructive MNN models. Morphological perceptrons (MPs) and morphological perceptrons with dendrites (MPDs) also belong to the class of constructive MNN models. In this paper, we introduced a modified MP training algorithm that only requires a finite number of epochs to converge, resulting in a decision surface that perfectly separates the training data. Furthermore, incorporating competitive neurons into the MP and MPD models led to the MP/C and MPD/C models that can be trained using extensions of the MP and MPD training algorithms. Further research has to be conducted to devise more efficient training algorithms for these new morphological models.

Finally, this article has empirically demonstrated the effectiveness of constructive morphological models in simulations with two well-know datasets for classification [36, 6] by analyzing and comparing the error rates and the computational effort for learning. In general, the constructive morphological models exhibited very satisfactory classification results and - except for the FLNN - extremely fast convergence of the training algorithms. On one hand, the constructive morphological models often require more artificial neurons or computational units than conventional models. On the other hand, morphological neural computations based on max-products or min-products are much less complicated than the usual semi-linear neural computations.

Acknowledgements. This work was supported by CNPq under grant no. $306040/2006-9$ and by FAPESP under grant no. $2006/05868-5$.

References

1. Araújo, R.A., Madeiro, R., Sousa, R.P., Pessoa, L.F.C., Ferreira, T.A.E.: An evolutionary morphological approach for financial time series forecasting. In: Proceedings of the IEEE Congress on Evolutionary Computation, Vancouver, Canada, pp. 2467–2474 (2006)
2. Armstrong, W.W., Thomas, M.M.: Adaptive Logic Networks. In: Fiesler, E., Beale, R. (eds.) Handbook of Neural Computation, vol. C1.8, pp. 1–14. IOP Publishing, Oxford University Press, Oxford United Kingdom (1997)
3. Araújo, R.A., Madeiro, F., Pessoa, L.F.C.: Modular morphological neural network training via adaptive genetic algorithm for design translation invariant operators. In: Proceedings of the IEEE International Conference on Acoustics, Speech and Signal Processing, Toulouse, France (May 2006)
4. Banon, G., Barrera, J.: Decomposition of Mappings between Complete Lattices by Mathematical Morphology, Part 1, General Lattices. Signal Processing 30(3), 299–327 (1993)
5. Birkhoff, G.: Lattice Theory, 3rd edn. American Mathematical Society, Providence (1993)
6. Asuncion, A., Newman, D.J.: UCI Repository of machine learning databases, University of California, Irvine, CA, School of Information and Computer Sciences (2007), http://www.ics.uci.edu/~mlearn/MLRepository.html
7. Braga-Neto, U., Goutsias, J.: Supremal multiscale signal analysis. SIAM Journal of Mathematical Analysis 36(1), 94–120 (2004)
8. Carpenter, G.A., Grossberg, S.: ART 3: Hierarchical search using chemical transmitters in self-organizing pattern recognition architectures. In: Neural Networks, vol. 3, pp. 129–152 (1990)
9. Carré, B.: An algebra for network routing problems J. Inst. Math. Appl. 7, 273–294 (1971)
10. Coskun, N., Yildirim, T.: The effects of training algorithms in MLP network on image classification. Neural Networks 2, 1223–1226 (2003)

11. Cuninghame-Green, R.: Minimax Algebra and Applications. In: Hawkes, P. (ed.) Advances in Imaging and Electron Physics, vol. 90, pp. 1–121. Academic Press, New York (1995)
12. Cuninghame-Green, R., Meijer, P.F.J.: An Algebra for Piecewise-Linear Minimax Problems. Discrete Applied Mathematics 2(4), 267–286 (1980)
13. Cuninghame-Green, R.: Minimax Algebra: Lecture Notes in Economics and Mathematical Systems, vol. 166. Springer, New York (1979)
14. Davey, B.A., Priestley, H.A.: Introduction to lattices and order. Cambridge University Press, Cambridge (2002)
15. Gader, P.D., Khabou, M., Koldobsky, A.: Morphological Regularization Neural Networks. Pattern Recognition, Special Issue on Mathematical Morphology and Its Applications 33(6), 935–945 (2000)
16. Giffler, B.: Mathematical solution of production planning and scheduling problems. IBM ASDD, Tech. Rep (1960)
17. Graña, M., Gallego, J., Torrealdea, F.J., D'Anjou, A.: On the application of associative morphological memories to hyperspectral image analysis. In: Mira, J., Álvarez, J.R. (eds.) IWANN 2003. LNCS, vol. 2687, pp. 567–574. Springer, Heidelberg (2003)
18. Grätzer, G.A.: Lattice theory: first concepts and distributive lattices. W. H. Freeman, San Francisco (1971)
19. Haralick, R.M., Shapiro, L.G.: Computer and Robot Vision, vol. 1. Addison-Wesley, New York (1992)
20. Haralick, R.M., Sternberg, S.R., Zhuang, X.: Image Analysis Using Mathematical Morphology: Part I. IEEE Transactions on Pattern Analysis and Machine Intelligence 9(4), 532–550 (1987)
21. Heijmans, H.: Morphological Image Operators. Academic Press, New York (1994)
22. Hocaoglu, A.K., Gader, P.D.: Domain learning using Choquet integral-based morphological shared weight neural networks. Image and Vision Computing 21(7), 663–673 (2003)
23. Kaburlasos, V.G., Athanasiadis, I.N., Mitkas, P.A.: Fuzzy lattice reasoning (FLR) classifier and its application for ambient ozone estimation. International Journal of aproximate reasoning 45(1), 152–188 (2003)
24. Kaburlasos, V.G., Petridis, V.: Fuzzy lattice neurocomputing (FLN) models. Neural Networks 13, 1145–1170 (2000)
25. Khabou, M.A., Gader, P.D.: Automatic target detection using entropy optimized shared-weight neural networks. IEEE Transactions on Neural Networks 11(1), 186–193 (2000)
26. Kim, C.: Segmenting a low-depth-of-field image using morphological filters and region merging. IEEE Transactions on Image Processing 14(10), 1503–1511 (2005)
27. Kwok, J.T.: Moderating the Outputs of Support Vector Machine Classifiers. IEEE Transactions on Neural Networks 10(5), 1018–1031 (1999)
28. Matheron, G., Random Sets and Integral Geometry. Wiley, New York, (1975).
29. Monteiro, A.S., Sussner, P.: A Brief Review and Comparison of Feedforward Morphological Neural Networks with Applications to Classification. In: Kůrková, V., Neruda, R., Koutník, J. (eds.) ICANN 2008,, Part II. LNCS, vol. 5164, pp. 783–792. Springer, Heidelberg (2008)
30. Pessoa, L.F.C., Maragos, P.: Neural networks with hybrid morphological/rank/linear nodes: a unifying framework with applications to handwritten character recognition. Pattern Recognition 33, 945–960 (2000)
31. Petridis, V., Kaburlasos, V.G.: Fuzzy lattice neural network (FLNN): a hybrid model for learning. IEEE Transactions on Neural Networks 9(5), 877–890 (1998)

32. Pitas, I., Venetsanopoulos, A.N.: Morphological Shape Decomposition. IEEE Transactions on Pattern Analysis and Machine Intelligence 12(1), 38–45 (1990)
33. Porter, R., Harvey, N., Perkins, S., Theiler, J., Brumby, S., Bloch, J., Gokhale, M., Szymanski, J.: Optimizing digital hardware perceptrons for multispectral image classification. Journal of Mathematical Imaging and Vision 19(2), 133–150 (2003)
34. Raducanu, B., Graña, M., Albizuri, X.F.: Morphological Scale Spaces and Associative Morphological Memories: Results on Robustness and Practical Applications. Journal of Mathematical Imaging and Vision 19(2), 113–131 (2003)
35. Ripley, B.D.: Pattern Recognition and Neural Networks. Cambridge University Press, Cambridge (1996)
36. Ripley, B.D.: Datasets for Pattern Recognition and Neural Networks, Cambridge, United Kingdom (1996), `http://www.stats.ox.ac.uk/pub/PRNN/`
37. Ritter, G.X., Iancu, L., Urcid, G.: Morphological perceptrons with dendritic structure. In: The 12th IEEE International Conference, Fuzzy Systems, May 2003, vol. 2, pp. 1296–1301 (2003)
38. Ritter, G.X., Urcid, G.: Lattice Algebra Approach to Single-Neuron Computation. IEEE Transactions on Neural Networks 14(2), 282–295 (2003)
39. Ritter, G.X., Wilson, J.N.: Handbook of Computer Vision Algorithms in Image Algebra, 2nd edn. CRC Press, Boca Raton (2001)
40. Ritter, G.X., Sussner, P.: Morphological Perceptrons Intelligent Systems and Semiotics, Gaithersburg, Maryland (1997)
41. Ritter, G.X., Wilson, J.N., Davidson, J.L.: Image Algebra: An Overview. Computer Vision, Graphics, and Image Processing 49(3), 297–331 (1990)
42. Ronse, C.: Why Mathematical Morphology Needs Complete Lattices. Signal Processing 21(2), 129–154 (1990)
43. Segev, I.: Dendritic processing. In: Arbib, M.A. (ed.) The handbook of brain theory and neural networks, MIT Press, Cambridge (1998)
44. Serra, J.: Image Analysis and Mathematical Morphology. In: Theoretical Advances, vol. 2, Academic Press, New York (1998)
45. Serra, J.: Image Analysis and Mathematical Morphology. Academic Press, London (1982)
46. Khabou, M.A., Gader, P.D., Keller, J.M.: LADAR target detection using morphological shared-weight neural networks. Machine Vision and Applications 11(6), 300–305 (2000)
47. Sobania, A., Evans, J.P.O.: Morphological corner detector using paired triangular structuring elements. Pattern Recognition 38(7), 1087–1098 (2005)
48. Soille, P.: Morphological Image Analysis. Springer, Berlin (1999)
49. Sousa, R.P., Pessoa, L.F.C., Carvalho, J.M.: Designing translation invariant operations via neural network training. In: Proceedings of the IEE International Conference on Image Processing, Vancouver, Canada, pp. 908–911 (2000)
50. Sussner, P., Valle, M.E.: Morphological and Certain Fuzzy Morphological Associative Memories with Applications in Classification and Prediction. In: Kaburlasos, V.G., Ritter, G.X. (eds.) Computational Intelligence Based on Lattice Theory. Studies in Computational Intelligence, vol. 67, pp. 149–173. Springer, Heidelberg (2007)
51. Sussner, P., Valle, M.E.: Gray Scale Morphological Associative Memories. IEEE Transactions on Neural Networks 17(3), 559–570 (2006)
52. Sussner, P., Valle, M.E.: Implicative Fuzzy Associative Memories. IEEE Transactions on Fuzzy Systems 14(6), 793–807 (2006)

53. Sussner, P., Grana, M.: Guest Editorial: Special Issue on Morphological Neural Networks. Journal of Mathematical Imaging and Vision 19(2), 79–80 (2003)
54. Sussner, P.: Morphological Perceptron Learning. In: Proceedings of IEEE ISIC/CIRA/ISAS Joint Conference, Gaithersburg, MD, September 1998, pp. 477–482 (1998)
55. Valle, M.E., Sussner, P.: A General Framework for Fuzzy Morphological Associative Memories. Fuzzy Sets and Systems 159(7), 747–768 (2008)
56. Zadeh, L.A.: Fuzzy Sets. Information and Control 8(3), 338–353 (1965)
57. Zimmermann, U.: Linear amd Combinatorial Optimization of Ordered Algebraic Structures. North-Holland, Amsterdam (1981)
58. Yu, A.J., Giese, M.A., Poggio, T.: Biophysiologically Plausible Implementations of the Maximum Operation. Neural Computation 14(12), 2857–2881 (2002)

A Feedforward Constructive Neural Network Algorithm for Multiclass Tasks Based on Linear Separability

João Roberto Bertini Jr. and Maria do Carmo Nicoletti

Abstract. Constructive neural network (CoNN) algorithms enable the architecture of a neural network to be constructed along with the learning process. This chapter describes a new feedforward CoNN algorithm suitable for multiclass domains named MBabCoNN, which can be considered an extension of its counterpart BabCoNN, suitable for two-class classification tasks. Besides describing the main concepts involved in the MBabCoNN proposal, the chapter also presents a comparative analysis of its performance *versus* the multiclass versions of five well-known constructive algorithms, in eight knowledge domains, as empirical evidence of the MBabCoNN suitability and efficiency for multiclass classification tasks.

Keywords: Constructive neural network algorithm, LS-discriminant learning, Barycentric Correction Procedure, Multiclass classification.

1 Introduction

There are many different methods that allow the automatic learning of concepts, as can be seen in [1] and [2]. One particular class of relevant machine learning methods is based on the concept of linear separability (LS).

The concept of linear separability permeates many areas of knowledge and based on the definition given in [3] it can be stated as: Let E be a finite set of N distinct patterns $\{E_1, E_2, \ldots, E_N\}$, each pattern E_i ($1 \leq i \leq N$) described as $E_i = \langle x_1,\ldots,x_k \rangle$, where k is the number of attributes that defines a pattern. Let the patterns of E be classified in such a way that each pattern in E belongs to only one of the M classes C_j ($1 \leq j \leq M$). This classification divides the set of patterns E into

João Roberto Bertini Jr.
Institute of Mathematics and Computer Science, University of São Paulo,
Av. Trabalhador São Carlense 400, São Carlos, Brazil
bertini@icmc.usp.br

Maria do Carmo Nicoletti
Department of Computer Science, Federal University of São Carlos,
Via Washington Luiz, km. 238, São Carlos, Brazil
carmo@dc.ufscar.br

L. Franco et al. (Eds.): Constructive Neural Networks, SCI 258, pp. 145–169.
springerlink.com

the subsets EC_1, EC_2, …, EC_M, such that each pattern in EC_i belongs to class C_i, for i = 1, …, *M*. If a linear machine can classify the patterns in *E* into the proper class, the classification of *E* is a linear classification and the subsets EC_1, EC_2, …, EC_M are linearly separable. Stated another way, a classification of *E* is linear and the subsets EC_1, EC_2, …, EC_M, are linearly separable if and only if linear discriminant functions g_1, g_2, …, g_M exist such that

$$g_i(E) > g_j(E) \qquad \text{for all } E \in EC_i$$
$$j = 1, \ldots, M, j \neq i \qquad \text{for all } i = 1, \ldots, M$$

Since the decision regions of a linear machine are convex, if the subsets EC_1, EC_2, …, EC_M are linearly separable, then each pair of subsets EC_i, EC_j, i, j = 1, …, *M*, i ≠ j, is also linearly separable. That is, if EC_1, EC_2, …, EC_M, are linearly separable, then EC_1, EC_2, …, EC_M, are also pairwise linearly separable.

According to Elizondo [4], linearly separable based learning methods can be divided into four groups. Depending on their main focus they may be based on linear programming, computational geometry, neural networks or quadratic programming.

This chapter describes a new neural network algorithm named MBabCoNN (Multiclass Barycentric-based Constructive Neural Network) suitable for multiclass classification problems. The algorithm incrementally constructs a neural network by adding hidden nodes that linearly separate sub-regions of the feature space. It can be considered a multiclass version of the two-class CoNN named BabCoNN (Barycentric-based Constructive Neural Network) proposed in [5].

The chapter is an extended version of an earlier paper [6] and is organized as follows. Section 2 stresses the importance of CoNN algorithms and discusses the role played by the algorithm used for training individual Threshold Logic Units (TLU), particularly focusing on the BCP algorithm [7]. Section 3 highlights the main characteristics of the five well-known CoNN multiclass algorithms used in the empirical experiments described in Section 5. Section 4 initially outlines the basic features of the two-class BabCoNN algorithm briefly presenting the main concepts and strategies used by BabCoNN when learning and classifying and,presents a detailed description of the multiclass MBabCoNN algorithm divided into two parts: learning the neural network and using the network learnt for classifying previously unseen patterns. Section 5 presents and discusses the results of 16 algorithms; four of them are versions of PRM and BCP for multiclass tasks and the other 12 are variants of the basic multiclass algorithms used, namely: MTower, MPyramid, MUpstart, MTiling, MPerceptron-Cascade and MBabCoNN in eight knowledge domains from the UCI Repository [8]. The Conclusion section ends the chapter by presenting a summary of the main results highlighting a few possible research lines to investigate, aiming at improving the MBabCoNN algorithm.

2 Constructive NN and the Relevance of TLU Training

Whereas conventional neural network (NN) training algorithms such as the Backpropagation algorithm require the NN architecture to be defined before learning can begin, constructive neural network (CoNN) algorithms allow the network architecture to be constructed simultaneously with the learning process; both subprocesses, learning and constructing the network, are interdependent.

Constructive neural network (CoNN) algorithms do not assume fixed network architecture before training begins. The main characteristic of a CoNN algorithm is the dynamic construction of the network's hidden layer(s), which occurs simultaneously with training. A description of a few well-known CoNN algorithms can be found in [9] and [10]; among the most well-known are: Tower and Pyramid [11], Tiling [12], Upstart [13], Perceptron-Cascade [14], Pti and Shift [15].

2.1 Training Individual TLUs

Usually the basic function performed by a CoNN algorithm is the addition to the network architecture of a new TLU and its subsequent training. For this reason CoNN algorithms are very dependent on the TLU training algorithm used. For training a TLU, a constructive algorithm generally employs the Perceptron or any of its variants, such as Pocket or Pocket with Ratchet Modification (PRM) [11]. Considering that CoNN algorithms depend heavily on an efficient TLU training algorithm, there is still a need for finding new and better methods, although some of the Perceptron variants (especially the PRM) have been widely used with good results.

The Barycentric-based Correction Procedure (BCP) algorithm [7] [16], although not widely adopted, has performed well when used for training individual TLUs (see [17] for instance) and has established itself as a good competitor compared to the PRM when used by CoNN algorithms (see [18] for a performance comparison). Good results have also been obtained by allowing both algorithms (BCP and PRM) to compete for training the next neuron to be added to the network; the proposal of this hybrid constructive algorithm and its results can be found in [19]. In spite of BCP being poorly explored in the literature, its good performance motivated the choice of this algorithm as the TLU's training algorithm embedded in both the BabCoNN and its multiclass version MBabCoNN, described in this chapter.

The BCP is based on the geometric concept of the barycenter of a convex hull and the algorithm (for a two-class problem) iteratively calculates the barycenters of the regions defined by the positive and the negative training patterns. Unlike Perceptron based algorithms, this algorithm calculates the weight vector and the bias separately. The BCP defines the weight vector as the vector that connects two points: the barycenter of the convex hull of positive patterns (class +1) and the barycenter of the convex hull defined by negative patterns (class −1). The convex hull of an n-dimensional set of points E is the intersection of all convex sets containing E, and the barycenter stands for its center of mass [20]. It follows a brief overview of the BCP algorithm.

Let $E = E1 \cup E2$ be a training set such that E1 is the subset of training patterns with class 1 and E2 the set of training patterns with class −1, and let $|E1| = k_1$ and $|E2| = k_2$. The barycenters b_1 and b_2 represent the center of mass of the convex hull formed by patterns belonging to each class, respectively. In the algorithm they are defined as the weighted averages of patterns in E1 and E2 respectively as described by eq. (1),

$$b_1 = \frac{\sum_{i=1}^{k_1} \alpha_i E1^i}{\sum_{i=1}^{k_1} \alpha_i} \quad \text{and} \quad b_2 = \frac{\sum_{i=1}^{k_2} \mu_i E2^i}{\sum_{i=1}^{k_2} \mu_i} \tag{1}$$

where α and μ are weight vectors, $\alpha = \langle \alpha_1, \alpha_2, \ldots, \alpha_{k_1} \rangle$ and $\mu = \langle \mu_1, \mu_2, \ldots, \mu_{k_2} \rangle$, responsible for modifying the position of the barycenters. For the experiments described in Section 5, both weight vectors were randomly initialized in the range [1,2], as recommended in [17]. They are used to vary the barycenters, at each execution, increasing the probability of finding a better weight vector.

In the BCP procedure the weight vector is defined as $W = b_1 - b_2$ and the hyperplane it defines is given by $W.x + \theta = 0$, where θ is the bias term. Once W is determined, the bias term θ is separately defined according to the following procedure. Let p be a pattern and consider the function $V: R^n \rightarrow R$ given by eq. (2).

$$V(p) = -W.p \tag{2}$$

Consider subsets $V1 = \{V(p) \mid p \in E1\}$ and $V2 = \{V(p) \mid p \in E2\}$ and let $V = V1 \cup V2$. The greatest and the smallest values of V1 and V2 are then determined. If $\max(V1) < \min(V2)$, the training set is linearly separable and θ is chosen such that $\max(V1) < \theta < \min(V2)$ and the algorithm ends. Otherwise either the set is not linearly separable or the current weight vector is not correctly positioned.

If $\max(V1) \geq \min(V2)$ the chosen value for θ should minimize the misclassifications. To do so, consider the set Ex = {ext1, ext2, ext3, ext4} whose values correspond to the smallest and biggest values of V1 and V2 respectively. Consider $P_- = [ext1, ext2) \cap V$; $P_+ = [ext3, ext4) \cap V$ and $P_{ov} = [ext2, ext3) \cap V$. As sets P_- and P_+ have patterns belonging to only one class, they are called exclusion zones. Since P_{ov} has patterns belonging to both classes it is called the overlapping zone. To choose an appropriate bias, the algorithm iteratively establishes its value as the arithmetic mean of two consecutive values in the overlapping zone. The value that correctly classifies the greatest number of patterns is chosen as bias.

At a certain, iteration let R and S be the sets of patterns belonging to classes 1 and –1 respectively and let b_1 and b_2 be the barycenters of region R and S respectively (calculated as in eq. (1)). Let $RE \subset R$ and $SE \subset S$ be the subsets of misclassified patterns. The algorithm determines the barycenters be_1' and be_2' of RE and SE respectively and then, creates two vectors $e_1 = b_1 - be_1'$ and $e_2 = b_2 - be_2'$. The two vectors are then multiplied by random values from [0,1], say r_1 and r_2, giving rise to the new barycenters $b_1' = r_1.e_1$ and $b_2' = r_2.e_2$. A new weight vector W' (and consequently the new hyperplane H'), is then obtained by connecting the new barycenters. The process continues while wrongly classified patterns remain or the number of iterations has not reached its predefined value. Due to its geometric approach, the BCP ends after a few iterations and the final hyperplane tends to be a good separator between the two classes, even in situations where the training set is not linearly separable.

3 Reviewing Five Well-Known Multiclass CoNN Algorithms

Multiclass classification tasks are common in pattern recognition. Frequently a classification task with M (> 2) classes is treated as M two-class tasks. Although this approach may be suitable for some applications, there is still a need for more effective ways of dealing with multiclass problems. CoNNs have proved to be a good alternative for two-class tasks and have the potential to become good alternatives for multiclass domains as well.

Multiclass constructive algorithms start by training as many output neurons as there are classes in the training set; generally two different strategies can be employed for the task, the *independent* (I) and the *winner-takes-all* (WTA). As stated in [21] in the former strategy each output neuron is trained independently of the others. The WTA strategy, however, explores the fact that the membership of a pattern in one class prevents its belonging to any other class. Using the WTA strategy, for any pattern, the output neuron with the highest net input is assigned an output of 1 and all other neurons are assigned outputs of -1. In the case of a tie for the highest net input all neurons are assigned an output of -1, thereby rendering the pattern incorrectly classified.

The main goal of this section is to provide a brief overview of the five well-known multiclass CoNN algorithms used in the experiments (for a more detailed description see [9]) described in Section 5 and to describe the new multiclass algorithm MBabCoNN. So far, multiclass problems have not been the main focus of CoNN research and consequently most of the multiclass algorithms available are extensions of their two-class counterparts.

The multiclass MTower algorithm was proposed in [22] and can be considered a direct extension of the two-class Tower algorithm. The Tower creates a NN with only one TLU per hidden layer. In a Tower network [11] each new hidden neuron introduced is connected to all the input neurons and to the hidden neuron previously created – this causes the network to resemble a tower. Similarly to the two-class Tower, the MTower adds TLUs to the network; instead of one at a time, like the Tower, it adds as many hidden neurons as there are classes.

For an M-class problem, the MTower adds M hidden neurons per hidden layer. Each one of the M neurons in a certain hidden layer has connections with all the neurons in the input layer as well as connections with all the M neurons of the previously added hidden layer. The addition of new layers to the network ends when any of the following stopping criteria is satisfied: (1) the current network correctly classifies all the training patterns; (2) the threshold on the number of layers has been reached; (3) the current network accuracy is worse than the accuracy of the previous network (i.e., the current network without the addition of the last hidden layer). If (3) happens the algorithm removes the last layer added and ends the process, returning the network constructed so far.

The multiclass MPyramid, also proposed in [22], is a direct extension of its two-class counterpart Pyramid algorithm, described in [11]. MPyramid extends the Pyramid simply by adding M hidden neurons per layer (corresponding to the existing M classes in the training set) instead of only one at each step. The difference between the Tower and Pyramid algorithms (and consequently between their

M-class versions) lies on the connections. In a Pyramid network each newly added hidden neuron has connections with all the previously added hidden ones as well as with the input neurons.

The two-class Upstart algorithm [13] constructs the neural network as a binary tree of TLUs and it is governed by the addition of new hidden neurons, specialized in correcting *wrongly-on* or *wrongly-off* errors made by the previously added neurons. A natural extension of this algorithm for multiclass tasks would be an algorithm that constructs *M* binary trees, each one responsible for the learning of one of the *M* classes found in the training set. This approach, however, would not take into account a possible relationship that might exist between the *M* different classes. The MUpstart proposal, described in [23], tries to circumvent the problem by grouping the created hidden neurons in a single hidden layer. Each hidden neuron is created aiming at correcting the most frequent error (*wrongly-on* or *wrongly-off*) committed by a single neuron among the *M* output neurons. The hidden neurons are trained with patterns labeled with two classes only and they can fire 0 or 1. Each hidden neuron is directly connected to every neuron in the output layer. The input layer is connected to the hidden neurons as well as to the output neurons.

The Tiling algorithm [12] constructs a neural network where hidden nodes are added to a layer in a similar fashion to laying tiles. Each hidden layer in a Tiling network has a master neuron and a few ancillary neurons. The output layer has only one master neuron. Tiling constructs a neural network in successive layers such that each new layer has a smaller number of neurons than the previous layer. Similarly to this approach, the MTiling method, as proposed in [24], constructs a multi-layer neural network where the first hidden layer has connections to the input layer and each subsequent hidden layer has connections only to the previous hidden layer. Each layer has master and ancillary neurons with the same functions they perform in a Tiling network i.e., the master neurons are responsible for classifying the training patterns and the ancillary ones are responsible for making the layer faithful. The role played by the ancillary neurons in a hidden layer is to guarantee that the layer does not produce the same output for any two training patterns belonging to different classes. In the MTiling version the process of adding a new layer is very similar to the one implemented by Tiling. However, while the Tiling algorithm adds only one master neuron per layer, the MTiling adds *M* master neurons (where *M* is the number of different classes in the training set).

The Perceptron Cascade algorithm [14] is a neural constructive algorithm that constructs a neural network with an architecture resembling the one constructed by the Cascade Correlation algorithm [25] and it uses the same approach for correcting the errors adopted by the Upstart algorithm [13]. Unlike the Cascade Correlation however, the Perceptron Cascade uses the Perceptron (or any of its variants) for training individual TLUs). Like the Upstart algorithm, the Perceptron Cascade starts the construction of the network by training the output neuron and hidden neurons are added to the network similarly to the process adopted by the Cascade Correlation: each new neuron is connected to both the output and input neurons and has connections with all hidden neurons previously added to the network. The MPerceptron-Cascade version, proposed in [22], is very similar to the

MUpstart described earlier in this section, the main difference between them being that the neural network architecture induced by both. The MPerceptron-Cascade adds the new hidden neurons in new layers while the MUpstart adds them in one single layer.

4 The Multiclass MBabCoNN Proposal

The MBabCoNN proposal can be considered an extension of the two-class algorithm called BabCoNN [5], suitable for classification tasks involving $M > 2$ classes. In order to present and discuss the MBabCoNN proposal, this section initially presents a brief description of the most important features of the BabCoNN algorithm, paying particular attention to the mechanism employed by the hidden neurons for firing their outputs, since the MBabCoNN shares the same strategy. To facilitate the understanding of MBabCoNN, the learning and the classification processes implemented by the algorithm are approached separately; the trace of both processes is shown via an example.

4.1 The Two-Class BabCoNN Algorithm

BabCoNN is a new proposal that borrows some ideas from the BCP to build a neural network. Like Upstart, Perceptron Cascade (PC) and Shift, BabCoNN also constructs the network beginning with the output neuron. However, it creates only one hidden layer; each hidden neuron is connected to the input layer as well as to output neuron, like the Shift algorithm [15]. Unlike Shift however, the connections created by BabCoNN do not have an associated weight. The Upstart, PC and Shift algorithms construct the network by adding new hidden neurons specialized in correcting *wrongly-on* or *wrongly-off* errors. The BabCoNN, however, employs a different strategy to add new hidden neurons to the network.

Network construction starts by training the output neuron, using the BCP. Next, the algorithm identifies all the misclassified training patterns; if there are none, the algorithm stops, otherwise it starts adding neurons (one at a time) to the single hidden layer of the network, in order not to have misclassified patterns. A hidden neuron added to the hidden layer will be trained with the training patterns that were misclassified by the last added neuron; the first hidden neuron will be trained with the patterns that were misclassified by the output neuron; the second hidden neuron will be trained with the set containing the patterns that the first hidden neuron was unable to classify correctly, and so on. The process continues up to the point where no training patterns remain or all the remaining patterns belong to the same class.

The process of building the network architecture is described by the pseudocode given in Fig. 1, where $E = \{E_1, E_2, \ldots E_N\}$ represents the training set and each training pattern is described as $E_i = \langle x_1, x_2, \ldots x_k, C\rangle$, i.e., k attributes and an associated class $C \in \{-1, 1\}$.

In Fig. 1 the variables *output* and *hiddenLayer[ ]* define the neural network. The variable *output* represents a single neuron, and *hiddenLayer[ ]* is a vector representing the hidden neurons. The function *bcp( )* stands for the BCP algorithm, used for training individual neurons. The function *removeClassifiedPatterns( )* removes from the training set the patterns that were correctly classified by the last added neuron and *bothClasses( )* is a Boolean function that returns 'true' if the current training set still has patterns belonging to both classes and 'false' otherwise.

Due to the way the learning phase is conducted by BabCoNN, each hidden neuron of the network is trained using patterns belonging to a region of the training space (i.e., the one defined by the patterns that were misclassified by the previous hidden neuron added to the network). This particular aspect of the algorithm has the effect of introducing an undesirable 'redundancy', in the sense that a pattern may be correctly classified by more than one hidden neuron. This has been sorted out by implementing a classification process where the neurons of the hidden layer have a particular way of firing their output.

```
procedure BabCoNN_learner(E)
begin
   output ← bcp(E)
   nE ← removeClassifiedPatterns(E)
   h ← 0
   while bothClasses(nE) do
    begin
      h ← h + 1
      hiddenLayer[h] ← bcp(nE)
      nE ← removeClassifiedPatterns(nE)
    end
end procedure.
```

Fig. 1 BabCoNN algorithm for constructing a neural network.

Given an input pattern to be classified by a BabCoNN network, each hidden neuron has three possible outputs: 1, when the input pattern is classified as positive; −1, when the pattern is classified as negative and 0, when the pattern is classified as undetermined. Aiming at stressing the classification power of the hidden neurons, as well as providing a way for them to deal with unknown patterns, a limited 'region of action' is assigned to each hidden neuron. The region is limited by two thresholds associated to each hidden neuron, one for the positive class and the other for the negative class. The threshold values are determined as the largest Euclidean distance between the barycenter of a given class and the patterns of the same class are used to train the current neuron.

Figure 2 illustrates the process. The two-dimensional patterns used for training the hidden neuron are represented by '+' (positive) and '–' (negative); b_1 and b_2 are the barycenters of the regions defined by the '+' and the '–' patterns respectively; W is the weight vector after the training and H is the hyperplane defined by both, W and the bias. For each class, the region is defined as the hypersphere whose radius is given by the largest distance between all correctly classified patterns and the corresponding barycenter of the region.

To exemplify how a hidden neuron behaves during the classification phase, let each $Y_i = \langle y_{i_1}, y_{i_2} \rangle$, i = {1, 2 ,3 ,4} be a given pattern to be classified. As can be seen in Figure 2, four situations may occur:

(1) The new pattern (Y_1) is in the positive classification region of the hidden neuron. The pattern Y_1 is classified as positive by the neuron, which fires +1;
(2) The new pattern (Y_2) is in the positive region, but now lying on the other side of the hyperplane; this would make the neuron classify Y_2 as negative. However, the neuron will fire the value 0 since there is no guarantee that the pattern is negative;
(3) The new pattern (Y_3) is not part of any region; in this case the neuron fires the value 0 independently of the classification given by the hyperplane it represents;
(4) The new pattern (Y_4) is in the negative classification region of the hidden neuron. The pattern Y_4 is classified as negative and the neuron fires the value −1. Note that the regions may overlap with each other and, eventually, a pattern may lie in both regions. When that happens, the hidden neuron (as implemented by the version used in the experiments described in Section 5) assigns the pattern the class is given by the hyperplane.

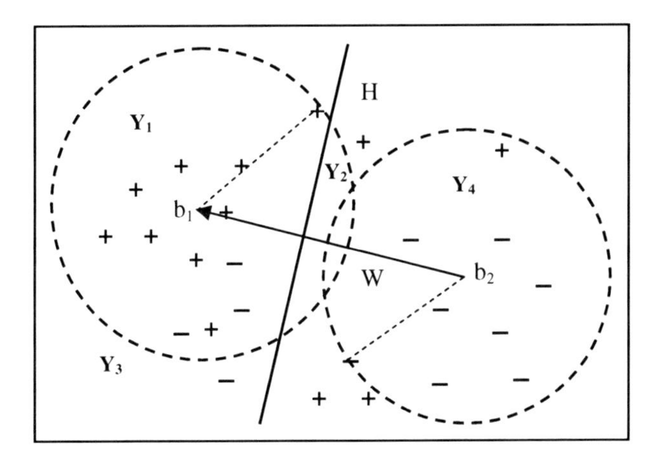

Fig. 2 BabCoNN hidden neuron firing process.

The pseudocode of the classification procedure is described in Fig. 3. After each hidden neuron fires its output, the output neuron decides which class the given pattern belongs to. The decision process is based on the sum of all the responses; if the resulting value is positive, the pattern is classified as positive, otherwise, as negative. If the sum result is 0, the output node is in charge of classifying the pattern.

The function *classification( )* returns the neuron classification (1 or –1), this is the usual classification that uses the weight vector and bias. Both functions *belongsToPositive( )* and *belongsToNegative( )* are Boolean functions. The first returns 'true' if the pattern lies in the positive region and 'false' otherwise. The second returns 'true' if the pattern lies in the negative region and 'false' otherwise. The *Hlc[ ]* vector stores the classifications given by all hidden neurons, for a given pattern and the last conditional command in the classification algorithm defines the class associated with the input pattern X.

```
procedure BabCoNN_classifier(X)
{X is the pattern to be classified}
begin
  for i ← 1 to h do
   begin
     C ← classification(hiddenLayer[i], X)
     Bp ← belongsToPositive(hiddenLayer[i], X)
     Bn ← belongsToNegative(hiddenLayer[i], X)
     if (C = 1 and Bp) then Hlc[i] ← 1
                      else if (C = –1 and Bn)
                           then Hlc[i] ← –1
                           else Hlc[i] ← 0
   end
  sum ← 0
  for j ← 1 to h do
     sum ← Hlc[j] + sum
  if sum ≠ 0 then sum / |sum|
           else classification(output,X)
end procedure.
```

Fig. 3 BabCoNN classification process.

4.2 The MBabCoNN Learning Algorithm

Figure 4 presents the pseudocode of the algorithm that implements the MBabCoNN learning process; the input to the algorithm is the training set E with patterns belonging to $M > 2$ classes. MBabCoNN can deal with Boolean, integer and real-valued tasks.

MBabCoNN constructs the network beginning with the output layer containing as many neurons as there are classes in the training set (each output neuron is associated to a class). The algorithm is flexible enough to allow the output neurons to be trained using any TLU algorithm combined with either strategy, independent or WTA.

After adding and training the M output neurons using procedure *MTluTraining( )*, the algorithm identifies the misclassifications the current network makes on the training set, via procedure *evaluateNetwork( )*, and starts to add neurons to its single hidden layer in order to correct the classification mistakes made by the output neurons.

In a MBabCoNN neural network each hidden neuron can be considered a two-class BabCoNN-like hidden neuron, i.e. it only fires 1, −1 or 0 values. In order to add a hidden neuron, MBabCoNN first finds which output neuron (class) is responsible for the greatest number of misclassifications in relation to patterns belonging to all the other output classes, via *highest_wrongly-on_error( )*, detailed in Figure 5.

A hidden neuron is then added to the hidden layer and is trained with a set containing patterns of two classes only: those belonging to the class the output neuron represents (which are relabeled as class −1) and those belonging to the misclassified class (which are relabeled as class 1).

Each newly added hidden neuron is then connected only to the two output neurons whose classes it separates. The connection to the neuron responsible for the misclassifications has weight 1 and the other −1. In fact the classes' labels can be arbitrarily chosen, the only proviso is that the weight must correspond to the relabeled class of the output neuron in question, e.g. the connection associated with a neuron recently represented by label 1 must be 1.

```
procedure MBabCoNNLearner(E)
begin
  currentAccuracy ← 0,
  previousAccuracy ← 0
  output ← MTluTraining(E)  {output layer with M neurons for a M-class problem}
  currentAccuracy ← evaluateNetwork(E)
  h ← 0 {hidden neuron index}
  while (currentAccuracy > previousAccuracy) and (currentAccuracy < 1) do
    begin
      highest_wrongly-on_error(E,WrongNeuron,Wrongly-onClass)
      twoClassesE ← createTrainingSet(WrongNeuron,Wrongly-onClass,E)
      h ← h + 1
      hiddenLayer[h] ← bcp(twoClassesE)           {hidden BabCoNN neuron}
      previousAccuracy ← currentAccuracy
      currentAccuracy ← evaluateNetwork(E)
    end
  if currentAccuracy ≠ 1 then begin
                                remove(hiddenLayer,h)
                                h ← h − 1
                              end
end procedure.
```

Fig. 4 Pseudocode of the MBabCoNN learning procedure.

As mentioned before in situations of uncertainty, BabCoNN neurons fire 0; this is convenient in a multiclass situation because it causes no side effects concerning the other patterns that do not belong to either two classes responsible for the hidden neuron creation. After a hidden neuron is added, the whole training set is input to the network grown so far and the classification process is repeated again. Depending on the classification accuracy, new hidden neurons may be added to the network in a similar fashion as the one previously described. If with the addition of a new hidden neuron the accuracy of the network decreases, the new hidden neuron is removed and the learning process ends. The other trivial stopping criteria is the convergence of the network i.e., when the network makes no mistakes).

```
procedure highest_wrongly-on_error(E,WrongNeuron,Wrongly-onClass)
begin
{initializing error matrix}
 for i ← 1 to M do
  for j ← 1 to M do
    outputErr[i,j] ← 0
{collecting errors made by output neurons in training set E={E1,E2,...,EN} }
 for i ← 1 to N do
  begin
    predClass ← MBabCoNN(Ei)
    if predClass ≠ class(Ei)
       then outputErr[predClass,class(Ei)] ← outputErr[predClass,class(Ei)] + 1
  end
{identifying which neuron makes the highest number of wrongly-on errors within a class}
highWrong ← 0
highErr ← 0
highWrongly-onClass ← 0
for i ← 1 to M do
   for j ← 1 to M do
   if outputErr[i,j] > highErr
    then begin
            highErr ← outputErr[i,j]
            highWrong ← i
            highWrongly-onClass ← j
          end
WrongNeuron ← highWrong
Wrongly-onClass ← highWrongly-onClass
end procedure.
```

Fig. 5 Pseudocode for determining the neuron responsible for the highest number of *wrongly-on* misclassifications as well as for the corresponding misclassified class.

4.3 An Example of the MBabCoNN Learning Algorithm

This section shows a simple example of the MBabCoNN learning algorithm according to the pseudocode described in Fig. 4. The example considers a training set with six training patterns identified by numbers 1 to 6 describing three classes identified by numbers 1 to 3. The figure on the left shows a MBabCoNN network after training the output neurons and, on the right, the four misclassifications it makes. The following figures show the evolution of the network implemented by MBabCoNN in the process of correcting the misclassifications.

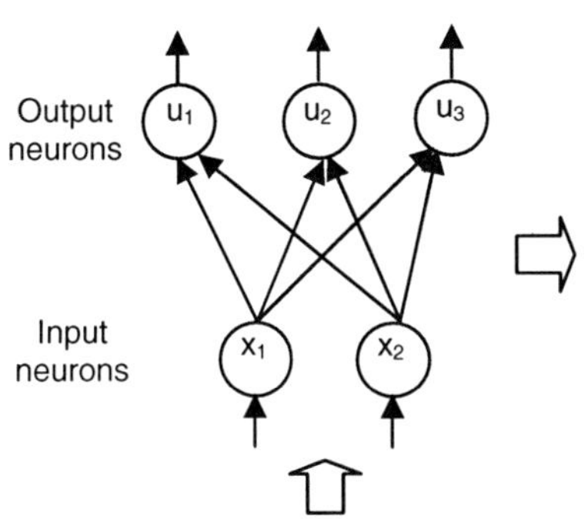

Training set: {#1, #2, #3, #4, #5, #6}

#P	Output neurons u_1	u_2	u_3	Class	
#1	1	–1	–1	1	C
#2	–1	–1	1	2	W
#3	–1	–1	1	2	W
#4	–1	–1	1	1	W
#5	–1	1	–1	3	W
#6	–1	–1	1	3	C

#P: pattern id; C: correctly classified; W: wrongly classified

Number of *wrongly-on* errors by u_1: 0
Number of *wrongly-on* errors by u_2: 1
Number of *wrongly-on* errors by u_3: 3 (patterns #2, #3 (class 2) and #4 (class 1))
u_3: has the highest number of *wrongly-on* errors within a class (misclassifies #2 and #3 from class 2). A new hidden neuron (h_1) is added to the network and trained with all patterns belonging to class 2 and class 3 i.e., hidden neuron h_1 is trained with E = {#2,#3,#5,#6}.

	class	new class label
#2	2	1
#3	2	1
#5	3	–1
#6	3	–1

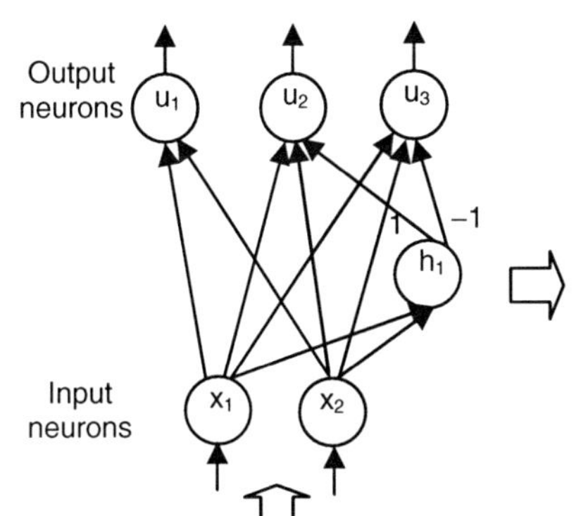

Training set: {#1, #2, #3, #4, #5, #6}

#P	Output neurons u_1	u_2	u_3	Class	
#1	1	–1	–1	1	C
#2	–1	**1**	**–1**	2	**C**
#3	–1	**1**	**–1**	2	**C**
#4	–1	–1	1	1	W
#5	–1	1	–1	3	W
#6	–1	–1	1	3	C

After the addition of h_1, patterns #2 and #3 are correctly classified.

Number of *wrongly-on* errors by u_1: 0
Number of *wrongly-on* errors by u_2: 1
Number of *wrongly-on* errors by u_3: 1 (pattern #4 (class 1))
u_2 and u_3 have the highest number of *wrongly-on* errors within a class. Randomly choose one of them; u_3 for example. A new hidden neuron (h_2) is added to the network and trained with all patterns belonging to class 1 and class 3 i.e., trained with E = {#1,#4,#5,#6}.

	class	new class label
#1	1	1
#4	1	1
#5	3	–1
#6	3	–1

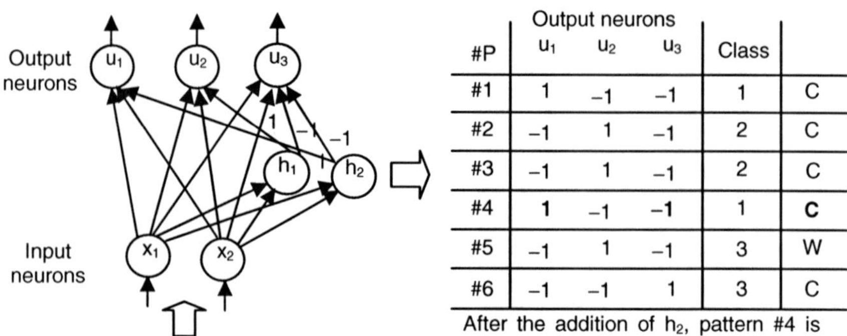

#P	Output neurons u1	u2	u3	Class	
#1	1	−1	−1	1	C
#2	−1	1	−1	2	C
#3	−1	1	−1	2	C
#4	**1**	−1	**−1**	1	**C**
#5	−1	1	−1	3	W
#6	−1	−1	1	3	C

Training set: {#1, #2, #3, #4, #5, #6}

After the addition of h_2, pattern #4 is correctly classified.

Number of *wrongly-on* errors by u_1: 0
Number of *wrongly-on* errors by u_2: 1 (pattern #5 (class 3))
Number of *wrongly-on* errors by u_3: 0
u_2 has the highest number of *wrongly-on* errors within a class. A new hidden neuron (h_3) is added to the network and trained with all patterns belonging to class 3 and class 2 i.e., trained with E = {#5, #6, #2, #3}.

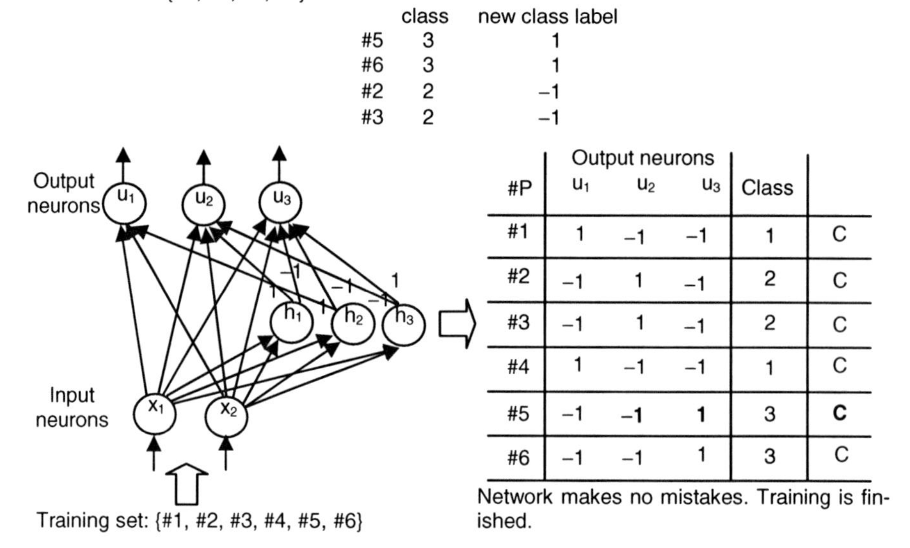

	class	new class label
#5	3	1
#6	3	1
#2	2	−1
#3	2	−1

#P	Output neurons u1	u2	u3	Class	
#1	1	−1	−1	1	C
#2	−1	1	−1	2	C
#3	−1	1	−1	2	C
#4	1	−1	−1	1	C
#5	−1	**−1**	**1**	3	**C**
#6	−1	−1	1	3	C

Training set: {#1, #2, #3, #4, #5, #6}

Network makes no mistakes. Training is finished.

4.4 The MBabCoNN Classifying Algorithm

The MBabCoNN classifying algorithm is described in Fig. 6. For the classification process an output neuron that has any connections to hidden neurons is said to have dependencies.

In the pseudocode of Fig. 6, the procedure *classification( )* approaches the network as constituted by a single output node indexed by *output[ ]* and no hidden neurons; the procedure gives as result the classification of the input pattern by the output node.

The classification process promotes the lack of dependency; if an output neuron fires 1 and has no dependencies then the class given to the pattern being classified is the class the neuron represents. Intuitively, an output neuron that does not create dependencies reflects the fact that it has not been associated with misclassifications during training. This can be an indication that the class represented by this particular neuron is reasonably easy to identify from the others (i.e., is linearly separable from the others). Figure 7 shows an example of this situation.

```
procedure MBabCoNN_classifier(X)
{X: new pattern}
{MBabCoNN network with M output neurons}
begin
 result ← 0, counter ← 0, neuronIndex ← 0
 j ← 1
 while (j ≤ M) and (counter < 2) do
   begin
    OutputClassification[j] ← classification(output[j],X)  {retrieves 1 or -1}
    if OutputClassification[j] = 1 then
         begin
           counter ← counter + 1
           neuronIndex ← j
         end
    j ← j + 1
   end
 if ( (counter = 1) and not hasDependencies(output[neuronIndex]) )
   then result ← class(neuronIndex)
   else
    begin
      for j  ← 1 to M do
        begin
        sum ← 0
        for k ← 1 to h do
            begin   {h is the number of  hidden neurons}
             if isConected(k,j)  then   {verifies connection between hidden neuron k and output j}
                   sum ← sum + classification(hiddenLayer[k], X) {BabCoNN-like neuron}
              hiddenClassification[j] ← sum
           end
       end
     result ← class(greatest(hiddenClassification))
     {returns the class associated with the  index of the greatest value in hiddenClassification}
    end
end  procedure.




end procedure.
```

Fig. 6 Pseudocode of the MBabCoNN procedure for classifying a new pattern.

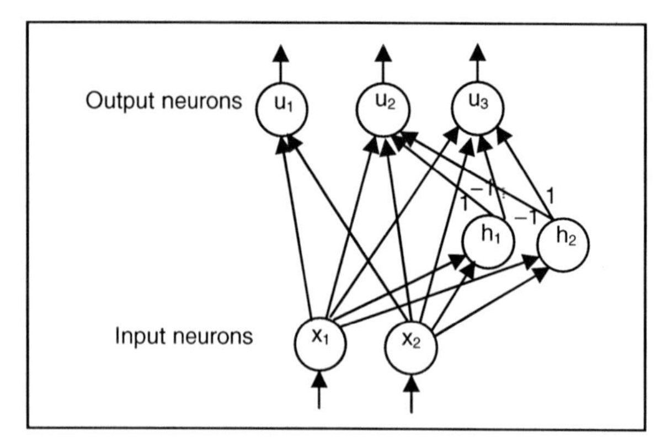

Fig. 7 Two out of three output neurons have dependencies.

In Fig. 7, two of the output neurons, u_2 and u_3 have dependencies (have connections with hidden neurons) and neuron u_1 does not have dependencies. If a new pattern is to be classified, the classification procedure checks if the output given by the node(s) that has (have) no dependencies (in this example, the u_1) is +1; if that is the case the new pattern is assigned the class represented by u_1 otherwise, the classification procedure takes into consideration only the sum of the outputs by the hidden neurons.

In cases where hidden neurons fire value 0, the classification procedure ignores the hidden neurons and takes into account the information given by the output neurons only. If, however, the output neuron that classifies the pattern has dependencies, the output result will be the sum of the outputs of all hidden neurons. If the sum is 0 the output neuron will be in charge of classifying the pattern.

The three output neurons, u_1, u_2 and u_3, in the MBabCoNN network of Fig. 8 have dependencies. Each output node has two connections with the added hidden neurons. A pattern to be classified will result in three outputs, one from each of the three hidden nodes (+1, −1 or 0), which will be multiplied by the connection

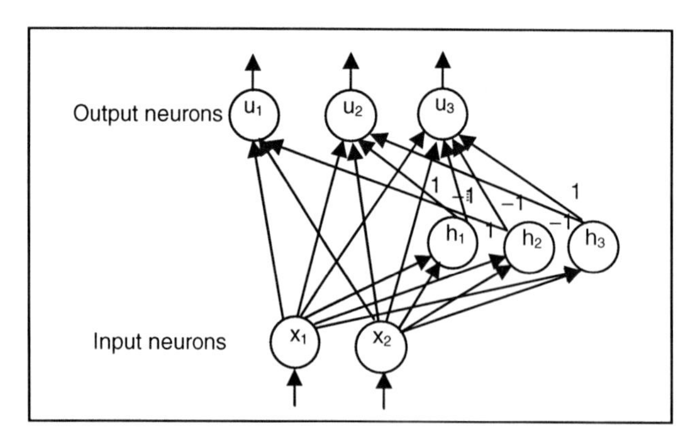

Fig. 8 MBabCoNN network where the three output neurons have dependencies.

weight, producing values +1, −1 or 0. Each output neuron will sum the input received from the hidden neurons and the pattern will be assigned the class represented by the output neuron with the highest score.

5 Experimental Results and Discussion

This section presents and compares the results of using MBabCoNN and the five other multiclass CoNN algorithms previously described, when learning from eight multiclass knowledge domains. Each algorithm was implemented in Java using two different algorithms for training individual TLUs, namely, the PRM and the BCP, identified in tables 2 to 9 by the suffixes P and B added to their names respectively.

Also for comparative purposes, the results of running a multiclass version of PRM and BCP, each implemented in two versions, WTA and independent(I), are presented. The eight knowledge domains used in the experiments have been downloaded from the UCI-Repository [8] and are summarized in Table 1.

Taking into consideration the MBabCoNN proposal (implemented in two versions: MBabCoNNP and MBabCoNNB), the two different versions implemented for each of the five algorithms (MTower, MPyramid, MUpstart, MTiling and MPerceptron-Cascade) and the two different strategies employed for implementing the MPRM and the MBCP, a total of 16 different algorithms have been implemented and evaluated.

Versions MBabCoNNP and MBabCoNNB differ in relation to the algorithm used for training their output neurons, the PRMWTA and the BCPWTA respectively, since both versions use the BCP for training hidden neurons. In the experiments, the accuracy of each neural network is based on the percentage of successful predictions on test sets for each domain. For each of the eight datasets the experiments consisted of performing a ten-fold cross-validation process with each of the 16 algorithms. The results are the average of the ten runs followed by their standard deviation.

Table 1 Domain Specifications

Domain	# PATTERNS	# ATTRIBUTES	# CLASSES
Iris	150	4	3
E. coli	336	7	8
Glass	214	9	6
Balance	625	4	3
Wine	178	13	3
Zoo	101	17	7
Car	1,728	6	4
Image Segmentation	2,310	19	7

Runs with the various learning procedures were carried out on the same training sets and evaluated on the same test sets. The cross-validation folds were the same for all the experiments in each domain. For each domain, each learning procedure was run considering one, ten, a hundred and a thousand iterations; only the best test accuracy among these iterations for each algorithm is presented. All the results obtained with MBabCoNN and the other algorithms (and their variants) are presented in tables 2 to 9, organized by knowledge domain.

The following abbreviations were adopted for presenting the tables: #I: number of iterations, TR training set, TE testing set. The accuracy (Acc) is given as a percentage followed by the standard deviation value. The 'Absolute Best' (AB) column gives the best performance of the learning procedure (in TE) over the ten runs and the 'Absolute Worst' (AW) column gives the worst performance of the learning procedure (in TE) over the ten runs; #HN represents the number of hidden nodes; AB(HN) gives the smallest number of hidden nodes created and AW(HN) gives the highest number of hidden nodes created.

Obviously the PRMWTA, PRMI, BCPWTA and BCPI do not have values for #HN, AB(HN) and AW(HN) because the networks they create only have input and output layers.

In relation to the results obtained in the experiments shown in tables 2 to 9, it can be said that as far as accuracy in test sets is concerned, MBabCoNNP has shown the best performance in four out of eight domains, namely the Iris, E. Coli, Wine and Zoo. In the Balance domain, although its result is very close to the best

Table 2 Iris

Algorithm	#I	Acc(TR)	Acc(TE)	#HN	AB(TE)	AW(TE)	AB(HN)	AW(HN)
MBabCoNNP	10^3	98.7~0.4	98.0~3.2	3.1~0.3	100.0	93.3	3.0	4.0
MBabCoNNB	10^2	94.2~3.1	94.0~4.9	4.9~0.6	100.0	86.7	6.0	4.0
PRMWTA	10^2	98.7~0.4	95.3~8.9	–	100.0	7.3.3	–	–
PRMI	10^2	89.3~2.6	86.7~18.1	–	100.0	46.7	–	–
BCPWTA	10^2	87.9~1.4	84.0~13.4	–	100.0	53.3	–	–
BCPI	1	85.0~7.3	72.7~41.3	–	100.0	6.7	–	–
MTowerP	10^2	98.9~0.4	96.7~6.5	3.6~1.3	100.0	80.0	3.0	6.0
MTowerB	10^2	87.8~1.6	83.3~14.5	3.3~0.9	100.0	53.3	3.0	6.0
MPyramidP	10^2	98.8~0.4	96.0~8.4	3.3~0.9	100.0	73.3	3.0	6.0
MPyramidB	10^2	88.3~1.4	82.7~13.8	3.6~1.3	100.0	53.3	3.0	6.0
MUpstartP	10^2	98.9~0.4	93.3~12.9	3.3~0.0	100.0	60.0	3.0	3.0
MUpstartB	10^2	88.6~1.6	80.7~13.9	3.6~0.5	100.0	53.3	3.0	4.0
MTilingP	10	98.2~0.8	95.3~8.9	3.0~0.0	100.0	73.3	3.0	3.0
MTilingB	10^2	89.6~4.7	84.0~14.5	7.0~8.8	100.0	53.3	3.0	28.0
MPCascadeP	10^2	98.8~0.4	95.3~8.9	3.0~0.0	100.0	73.3	3.0	3.0
MPCasdadeB	10^2	88.5~1.4	80.7~13.5	3.3~0.5	100.0	53.3	3.0	4.0

Table 3 E. Coli

Algorithm	#I	Acc(TR)	Acc(TE)	#HN	AB(TE)	AW(TE)	AB(HN)	AW(HN)
MBabCoNNP	10^2	90.6~0.6	83.9~5.3	8.2~0.6	91.2	75.8	8.0	9.0
MBabCoNNB	10^2	85.1~1.5	81.0~5.2	9.1~0.7	88.2	73.5	10.0	8.0
PRMWTA	10^2	90.8~1.3	77.8~18.9	–	100.0	52.9	–	–
PRMI	10^2	87.2~2.6	73.8~24.0	–	100.0	42.4	–	–
BCPWTA	10^2	76.5~3.0	69.1~15.0	–	97.1	42.4	–	–
BCPI	10	85.1~3.3	72.0~28.2	–	100.0	27.3	–	–
MTowerP	10	90.2~1.5	78.4~22.2	27.8~14.1	100.0	30.3	6.0	56.0
MTowerB	10^2	76.6~2.9	69.2~14.4	9.2~3.2	97.1	48.5	8.0	16.0
MPyramidP	10	90.1~1.3	79.6~18.3	29.1~10.3	100.0	42.4	14.0	48.0
MPyramidB	10^2	76.6~2.8	69.5~16.2	8.4~2.1	97.1	36.4	6.0	14.0
MUpstartP	10^2	90.7~1.6	76.9~19.7	8.1~1.3	100.0	50.0	6.0	10.0
MUpstartB	10^2	80.5~2.8	75.2~10.4	8.9~1.3	97.1	60.7	6.0	11.0
MTilingP	10	87.9~2.2	76.3~20.7	7.7~0.7	100.0	38.2	6.0	8.0
MTilingB	10^2	76.3~3.3	67.3~18.0	25.3~55.4	97.1	33.3	6.0	183.0
MPCascadeP	10	88.8~1.2	82.9~16.2	8.6~1.3	100.0	57.8	8.0	11.0
MPCasdadeB	10^2	79.6~2.7	70.1~15.3	9.0~1.7	97.0	42.4	6.0	12.0

Table 4 Glass

Algorithm	#I	Acc(TR)	Acc(TE)	#HN	AB(TE)	AW(TE)	AB(HN)	AW(HN)
MBabCoNNP	10^3	64.7~2.5	60.3~12.7	7.5~0.7	68.4	60.6	85.7	42.9
MBabCoNNB	10^3	63.2~2.2	56.1~8.1	7.4~0.5	66.8	60.9	66.7	40.9
PRMWTA	10^3	55.7~1.4	48.1~14.3	–	57.5	52.3	71.4	23.8
PRMI	10^3	53.9~2.6	46.8~14.0	–	57.3	50.0	81.0	28.6
BCPWTA	10^2	54.3~2.7	49.6~10.2	–	59.6	48.7	66.7	36.4
BCPI	10^3	58.8~5.2	54.2~5.8	–	66.7	51.6	66.7	47.6
MTowerP	10^3	61.2~2.5	56.1~10.8	21.6~8.6	65.8	58.0	71.4	33.3
MTowerB	10^3	56.1~1.9	50.6~6.8	18.6~10.4	60.4	53.4	61.9	36.4
MPyramidP	10^2	64.9~5.0	56.3~13.9	33.0~17.7	71.0	56.5	77.3	28.6
MPyramidB	10^2	55.1~2.3	51.5~11.0	17.4~4.4	59.1	52.1	66.7	36.4
MUpstartP	10^3	72.2~1.5	63.6~6.8	8.3~0.9	74.6	69.3	76.2	52.4
MUpstartB	10^2	64.3~4.7	54.7~8.4	7.9~1.0	68.2	52.8	66.7	42.9
MTilingP	10^2	79.9~16.5	55.9~15.2	81.5~63.1	92.7	54.9	72.7	23.8
MTilingB	10^3	54.6~1.5	50.0~8.2	6.0~0.0	57.5	52.1	61.9	38.1
MPCascadeP	10^3	71.4~1.8	62.7~10.4	7.6~1.2	73.6	68.2	81.0	42.9
MPCasdadeB	10^2	57.6~5.2	54.4~13.4	7.2~1.2	65.8	48.7	76.2	36.4

Table 5 Balance

Algorithm	#I	Acc(TR)	Acc(TE)	#HN	AB(TE)	AW(TE)	AB(HN)	AW(HN)
MBabCoNNP	10	92.1~0.8	91.4~2.3	3.5~0.7	93.7	87.1	3.0	5.0
MBabCoNNB	10	92.1~0.9	89.3~2.5	5.3~0.5	93.5	85.5	6.0	5.0
PRMWTA	10^2	92.2~0.5	90.1~3.7	–	96.8	84.1	–	–
PRMI	10	89.1~1.6	89.3~4.8	–	98.4	84.1	–	–
BCPWTA	10^2	80.1~4.4	77.9~9.9	–	91.9	65.1	–	–
BCPI	10	89.5~1.7	88.0~3.5	–	93.5	82.3	–	–
MTowerP	10	94.8~1.1	90.6~6.2	20.4~5.8	98.4	79.4	12.0	30.0
MTowerB	10^2	83.6~4.8	80.8~8.0	9.0~3.5	91.9	65.1	6.0	15.0
MPyramidP	10	95.1~0.9	90.1~6.3	24.0~6.2	96.8	76.2	15.0	33.0
MPyramidB	10^2	83.3~4.1	83.1~5.8	6.2~2.2	93.5	76.2	3.0	9.0
MUpstartP	10	91.8~0.4	91.2~4.7	3.4~0.9	98.4	85.5	3.0	6.0
MUpstartB	10^2	82.1~4.9	81.1~6.9	4.0~0.8	91.9	69.8	3.0	5.0
MTilingP	10^2	95.5~2.9	92.3~3.3	28.1~21.8	96.8	88.9	3.0	49.0
MTilingB	10^2	79.8~4.5	76.4~8.4	3.0~0.0	91.9	66.7	3.0	3.0
MPCascadeP	10^2	92.1~0.6	90.0~4.5	3.2~0.4	98.4	84.1	3.0	4.0
MPCasdadeB	10^2	81.6~4.7	78.6~9.9	4.0~0.8	91.9	66.7	3.0	5.0

Table 6 Wine

Algorithm	#I	Acc(TR)	Acc(TE)	#HN	AB(TE	AW(TE)	AB(HN)	AW(HN)
MBabCoNNP	10^3	94.2~0.9	92.7~5.5	3.3~0.5	95.0	93.1	100.0	82.4
MBabCoNNB	10^2	76.2~2.2	75.8~8.8	4.4~0.5	81.9	73.8	88.9	58.8
PRMWTA	10^3	85.8~3.3	81.9~10.4	–	90.1	78.9	100.0	64.7
PRMI	10^3	92.0~1.4	88.3~8.9	–	94.4	90.0	100.0	66.7
BCPWTA	10^3	74.0~1.4	70.8~7.1	–	75.6	70.8	82.4	61.1
BCPI	10	73.7~1.7	71.9~7.3	–	75.8	70.0	77.8	58.8
MTowerP	10^3	86.5~3.5	83.1~13.5	4.2~2.1	90.0	81.2	100.0	61.1
MTowerB	10^2	73.6~0.9	69.1~7.6	4.2~1.5	74.5	71.9	83.3	61.1
MPyramidP	10^3	87.3~2.7	81.5~17.1	8.1~4.9	90.7	83.1	100.0	50.0
MPyramidB	10^3	73.6~0.9	69.1~7.6	4.2~15	74.5	71.9	83.3	61.1
MUpstartP	10^3	94.5~0.5	91.6~6.0	3.1~0.3	95.6	93.8	100.0	83.3
MUpstartB	10^2	74.1~1.6	71.4~11.3	3.2~0.4	76.2	70.8	94.1	55.6
MTilingP	10^2	86.7~5.0	82.5~11.2	13.1~31.9	99.4	81.9	94.4	61.1
MTilingB	10^2	73.8~2.1	74.8~8.7	3.0~0.0	76.2	69.6	88.9	61.1
MPCascadeP	10^3	92.1~0.7	84.9~9.0	3.1~0.3	93.1	91.2	94.1	66.7
MPCasdadeB	10	75.7~1.2	74.1~9.2	3.2~0.4	77.5	73.3	88.9	61.1

Table 7 Zoo

Algorithm	#I	Acc(TR)	Acc(TE)	#HN	AB(TE)	AW(TE)	AB(HN)	AW(HN)
MBabCoNNP	10^2	100.0~0.0	97.0~4.8	7.0~0.0	100.0	100.0	100.0	90.0
MBabCoNNB	10^2	89.5~5.1	85.1~12.7	9.3~0.9	95.6	77.8	100.0	60.0
PRMWTA	10^2	100.0~0.0	95.2~6.7	–	100.0	100.0	100.0	81.8
PRMI	10^2	100.0~0.0	95.0~7.1	–	100.0	100.0	100.0	80.0
BCPWTA	10^3	78.4~7.5	71.3~16.6	–	89.0	65.6	90.0	40.0
BCPI	10	64.9~16.0	65.5~16.1	–	94.5	36.3	90.0	40.0
MTowerP	10^2	100.0~0.0	95.0~7.1	7.0~0.0	100.0	100.0	100.0	80.0
MTowerB	10^2	79.0~5.3	70.4~7.7	9.1~3.4	86.8	72.5	80.0	60.0
MPyramidP	10^2	100.0~0.0	95.0~8.5	7.0~0.0	100.0	100.0	100.0	80.0
MPyramidB	10^2	78.2~7.9	74.3~15.0	9.8~4.9	86.7	64.8	100.0	50.0
MUpstartP	10^3	100.0~0.0	96.0~7.0	7.0~0.0	100.0	100.0	100.0	80.0
MUpstartB	10^2	81.0~6.6	74.4~13.1	7.9~0.7	90.1	70.3	90.0	50.0
MTilingP	10^3	100.0~0.0	97.0~4.8	7.0~0.0	100.0	100.0	100.0	90.0
MTilingB	10^3	78.5~7.9	72.4~12.8	13.3~10.2	90.1	65.9	90.0	50.0
MPCascadeP	10^3	100.0~0.0	95.1~7.0	7.0~0.0	100.0	100.0	100.0	80.0
MPCasdadeB	10^2	77.8~6.2	72.4~15.2	7.4~0.7	85.7	68.1	100.0	50.0

Table 8 Car

Algorithm	#I	Acc(TR)	Acc(TE)	#HN	AB(TE)	AW(TE)	AB(HN)	AW(HN)
MBabCoNNP	10^3	81.2~0.7	80.1~2.8	4.7~0.5	82.8	80.4	83.8	76.3
MBabCoNNB	10^2	78.3~3.6	77.5~4.4	5.8~0.9	80.5	68.2	84.4	70.5
PRMWTA	10^2	80.5~0.5	79.7~1.7	–	81.4	79.9	82.7	78.0
PRMI	10^2	79.1~0.5	78.9~4.0	–	80.2	78.3	83.8	72.1
BCPWTA	10^3	68.7~0.6	68.5~3.1	–	69.4	67.3	73.8	63.0
BCPI	10^2	77.3~1.1	76.3~2.7	–	79.4	75.8	81.5	73.4
MTowerP	10^2	82.7~1.2	81.5~2.4	26.8~18.4	85.1	81.6	86.1	78.5
MTowerB	10^3	73.8~2.1	73.3~3.7	5.2~2.7	75.4	68.2	78.6	65.7
MPyramidP	10	83.6~1.3	80.4~3.8	32.8~14.2	85.7	81.1	86.7	73.8
MPyramidB	10^2	68.6~1.2	68.5~4.1	5.2~2.7	71.2	67.2	78.0	64.7
MUpstartP	10^2	81.9~1.7	80.8~2.3	5.3~1.6	85.3	80.3	85.5	77.5
MUpstartB	10	75.3~0.3	74.8~3.3	4.3~0.5	76.1	74.9	79.2	69.4
MTilingP	10^2	89.2~3.0	83.0~3.9	88.1~30.3	91.1	80.6	86.7	73.4
MTilingB	10	74.9~0.5	74.7~2.3	4.0~0.0	75.3	73.6	79.8	72.1
MPCascadeP	10^3	82.8~1.4	81.2~4.4	6.3~1.7	84.3	80.2	87.9	74.4
MPCasdadeB	10^3	74.8~1.7	73.8~4.9	4.7~0.8	75.8	70.1	79.2	62.2

Table 9 Image Segmentation

Algorithm	#I	Acc(TR)	Acc(TE)	#HN	AB(TE)	AW(TE)	AB(HN)	AW(HN)
MBabCoNNP	10^3	92.2~1.6	83.3~7.2	7.9~0.7	95.2	89.4	95.2	71.4
MBabCoNNB	10^3	74.0~4.7	70.0~9.5	8.8~0.9	81.5	68.8	85.7	52.4
PRMWTA	10^3	91.2~0.9	83.8~10.6	–	92.6	89.4	100.0	66.7
PRMI	10^3	88.9~1.5	83.8~6.4	–	91.5	86.8	90.5	71.4
BCPWTA	10^2	62.2~2.3	60.5~6.8	–	65.6	57.1	71.4	52.4
BCPI	10^2	67.6~5.9	63.8~15.9	–	78.8	60.3	81.0	33.3
MTowerP	10^2	91.7~1.2	82.4~9.0	14.0~8.1	93.1	88.9	95.2	71.4
MTowerB	10^2	67.9~3.7	63.8~14.1	11.9~4.7	72.0	61.9	85.7	42.9
MPyramidP	10^3	92.0~1.1	81.4~4.7	11.2~4.9	93.1	89.9	85.7	71.4
MPyramidB	10^3	61.3~3.0	60.5~8.7	9.1~3.4	68.3	57.7	81.0	52.4
MUpstartP	10^3	94.8~0.7	85.2~10.9	7.1~0.3	95.8	93.7	100.0	66.7
MUpstartB	10^3	64.7~2.9	59.5~8.5	7.3~0.7	68.8	58.2	71.4	47.6
MTilingP	10^2	93.4~4.0	82.4~7.1	35.5~65.9	100.0	89.9	90.5	71.4
MTilingB	10^2	64.7~5.0	63.8~8.8	7.0~0.0	70.9	54.0	76.2	52.4
MPCascadeP	10^2	88.8~1.2	86.2~6.1	7.3~0.5	90.5	87.3	90.5	76.2
MPCasdadeB	10^3	66.7~4.0	64.8~10.3	7.4~0.7	70.4	57.1	81.0	47.6

result (obtained with MTilingP), it is worth noting that MTilingP created 28.1 hidden neurons on average while MBabCoNN created only 3.5. A similar situation occurred in the Car domain.

All the algorithms with a performance higher than 80% induced networks bigger than the network induced by MBabCoNNP, especially the MTilingP, the MPyramid and the MTowerP, although the accuracy values of the three were very close to those of MBabCoNN. In the Glass domain, MBabCoNNP is ranked third considering only accuracy; however, when taking into account the standard deviation as well, it can be said that the MBabCoNNP and MPCascadeP (second position in the rank) are even. In the last domain, Image Segmentation, MBabCoNNP accuracy was average while the best performance was obtained with the MPCascadeP.

In relation to the versions that used BCPWTA for training the output neuron, MBabCoNNB outperformed all the other algorithms in the eight domains. The results however were inferior to those obtained using the PRMWTA for training the output nodes. This fact is due to the particular characteristics of the two training approaches; in general the BCPWTA is not a good match for the PRMWTA. Future work concerning BCPWTA needs to be done in order for this algorithm to be considered an option for network construction.

Now, considering the PRMWTA versions of all algorithms, it is easy to see that the test accuracies are more standardized. In order to have a clearer view of the algorithm performances, Table 10 presents the values for the average ranks considering the test accuracy for all PRMWTA based CoNN algorithms. In the table the results are ranked in ascending order of accuracy. Ties in accuracy were sorted out by averaging the corresponding ranks.

According to Demšar [26], average ranks provide a fair comparison of the algorithms. Taking into account the values obtained with the average ranks it can be said that, as far as the eight datasets are concerned, MBabCoNN is the best choice among the six algorithms. As can be seen in Table 10, MBabCoNNP obtained the smallest value, followed by MPCascadeP and MUpstartP. MBabCoNNP was ranked last in only one domain (Car); in the Car domain, however, the test accuracies among the six algorithms were very close, i.e. the maximum difference was about 3.0%.

It is worth noticing that the three algorithms ranked first in the average rankings, construct the network by first adding the output neurons and then starting to correct their misclassifications by adding two-class hidden neurons. The good performance may be used to corroborate the efficiency of this technique. Based on the empirical results obtained, it can be said that the MBabCoNN algorithm is a good choice among the multiclass CoNN algorithms available.

Table 10 Average rank over PRMWTA based algorithms concerning test accuracy

Domain	MBabCoNNP	MTowerP	MPyramidP	MUpstartP	MTilingP	MPCascadeP
Iris	98.0~3.2 **(1)**	96.7~6.5**(2)**	96.0~8.4**(3)**	93.3~12.9**(6)**	95.3~8.9**(4.5)**	95.3~8.9**(4.5)**
EColi	83.9~5.3**(1)**	78.4~22.2**(4)**	79.6~18.3**(3)**	76.9~19.7**(5)**	76.3~20.7**(6)**	82.9~16.2**(2)**
Glass	60.3~12.7**(3)**	56.1~10.8**(5)**	56.3~13.9**(4)**	63.6~6.8**(1)**	55.9~15.2**(6)**	62.7~10.4**(2)**
Balance	91.4~2.3**(2)**	90.6~6.2**(4)**	90.1~6.3**(5)**	91.2~4.7**(3)**	92.3~3.3**(1)**	90.0~4.5**(6)**
Wine	92.7~5.5**(1)**	83.1~13.5**(4)**	81.5~17.1**(6)**	91.6~6.0**(2)**	82.5~11.2**(5)**	91.0~6.1**(3)**
Zoo	97.0~4.8**(1.5)**	95.0~7.1**(5.5)**	95.0~8.5**(5.5)**	96.0~7.0**(3)**	97.0~4.8**(1.5)**	95.1~7.0**(4)**
Car	80.1~2.8**(6)**	81.5~2.4**(2)**	80.4~3.8**(5)**	80.8~2.3**(4)**	83.0~3.9**(1)**	81.2~4.4**(3)**
Image	83.3~7.2**(3)**	82.4~9.0**(4.5)**	81.4~4.7**(6)**	85.2~10.9**(2)**	82.4~7.1**(4.5)**	86.2~6.1**(1)**
Average Rank	2.312	3.875	4.687	3.25	3.687	3.187

5 Conclusions

This chapter proposes the multiclass version, MBabCoNN, of a recently proposed constructive neural network algorithm named BabCoNN, which is based on the geometric concept of convex hull and uses the BCP algorithm for training individual TLUs added to the network during learning. The chapter presents the accuracy results of learning experiments conducted in eight multiclass knowledge domains, using the MBabCoNN implemented in two different versions: MBabCoNNP and MBabCoNNB, *versus* five well-known multiclass algorithms (each implemented in two versions as well). Both versions of the MBabCoNN use the BCP for training the hidden neurons and differ from each other in relation to the algorithm used for training their output neurons (PRMWTA and BCPWTA respectively).

As far as results in eight knowledge domains are concerned, it can (easily) be observed that all algorithms performed better when using PRMWTA for training the output neurons. This may occur because BCPWTA is not a good strategy for training M (>2) classes. Now considering the PRMWTA versions, it can be said the MBabCoNNP version has shown superior average performance in relation to both accuracy in test sets and the size of the induced neural network. This work had established MBabCoNN as a good option among other CoNNs for multiclass domains.

Acknowledgments. To CAPES and FAPESP for funding the work of the first author and for the project financial help granted to the second author, and to Leonie C. Pearson for proofreading the first draft of this chapter.

References

[1] Mitchell, T.M.: Machine learning. McGraw-Hill, USA (1997)
[2] Duda, R.O., Hart, P.E., Stork, D.G.: Pattern classification. John Wiley & Sons, USA (2001)
[3] Nilsson, N.J.: Learning machines. McGrall-Hill Systems Science Series, USA (1965)
[4] Elizondo, D.: The linear separability problem: some testing methods. IEEE Transactions on Neural Networks 17(2), 330–344 (2006)
[5] Bertini Jr., J.R., Nicoletti, M.C.: A constructive neural network algorithm based on the geometric concept of barycenter of convex hull. In: Rutkowski, L., Tadeusiewiza, R., Zadeh, L.A., Zurada, J. (eds.) Computational Intelligence: Methods and Applications, 1st edn., vol. 1, pp. 1–12. Academic Publishing House EXIT, Warsaw (2008)
[6] Bertini Jr., J.R., Nicoletti, M.C.: MBabCoNN - a multiclass version of a constructive neural network algorithm based on linear separability and convex hull. In: Kůrková, V., Neruda, R., Koutník, J. (eds.) ICANN 2008, Part II. LNCS, vol. 5164, pp. 723–733. Springer, Heidelberg (2008)
[7] Poulard, H.: Barycentric correction procedure: A fast method for learning threshold unit. In: WCNN 1995, Washington, DC, vol. 1, pp. 710–713 (1995)
[8] Asuncion, A., Newman, D.J.: UCI machine learning repository. University of California, School of Information and Computer Science, Irvine (2007), `http://www.ics.uci.edu/~mlearn/MLRepository.html`
[9] Parekh, R.G.: Constructive learning: inducing grammars and neural networks. Ph.D. Dissertation, Iowa State University, Ames, Iowa (1998)
[10] Nicoletti, M.C., Bertini Jr., J.R.: An empirical evaluation of constructive neural network algorithms in classification tasks. Int. Journal of Innovative Computing and Applications (IJICA) 1, 2–13 (2007)
[11] Gallant, S.I.: Neural network learning & expert systems. The MIT Press, Cambridge (1994)
[12] Mézard, M., Nadal, J.: Learning feedforward networks: the tiling algorithm. J. Phys. A: Math. Gen. 22, 2191–2203 (1989)
[13] Frean, M.: The upstart algorithm: a method for constructing and training feedforward neural networks. Neural Computation 2, 198–209 (1990)
[14] Burgess, N.: A constructive algorithm that converges for real-valued input patterns. International Journal of Neural Systems 5(1), 59–66 (1994)
[15] Amaldi, E., Guenin, B.: Two constructive methods for designing compact feedforward networks of threshold units. International Journal of Neural System 8(5), 629–645 (1997)
[16] Poulard, H., Labreche, S.: A new threshold unit learning algorithm, Technical Report 95504, LAAS (December 1995)
[17] Poulard, H., Estèves, D.: A convergence theorem for barycentric correction procedure, Technical Report 95180, LAAS-CNRS, Toulouse (1995)

[18] Bertini Jr., J.R., Nicoletti, M.C., Hruschka Jr., E.R.: A comparative evaluation of constructive neural networks methods using PRM and BCP as TLU training algorithms. In: Proceedings of the IEEE International Conference on Systems, Man and Cybernetics, pp. 3497–3502. IEEE Press, Los Alamitos (2006)
[19] Bertini Jr., J.R., Nicoletti, M.C., Hruschka Jr., E.R., Ramer, A.: Two variants of the constructive neural network Tiling algorithm. In: Proceedings of The Sixth International Conference on Hybrid Intelligent Systems (HIS 2006), pp. 49–54. IEEE Computer Society, Washington (2006)
[20] de Berg, M., van Kreveld, M., Overmars, M., Schwarzkopf, O.: Computational geometry: algorithms and applications, 2nd edn. Springer, Berlin (2000)
[21] Parekh, R.G., Yang, J., Honavar, V.: Constructive neural network learning algorithms for multi-category real-valued pattern classification. TR ISU-CS-TR97-06, Iowa State University, IA (1997)
[22] Parekh, R.G., Yang, J., Honavar, V.: Constructive neural network learning algorithm for multi-category classification. TR ISU-CS-TR95-15a, Iowa State University, IA (1995)
[23] Parekh, R.G., Yang, J., Honavar, V.: MUpstart – a constructive neural network learning algorithm for multi-category pattern classification. In: ICNN 1997, vol. 3, pp. 1920–1924 (1997)
[24] Yang, J., Parekh, R.G., Honavar, V.: MTiling – a constructive network learning algorithm for multi-category pattern classification. In: Proc. of the World Congress on Neural Networks, pp. 182–187 (1996)
[25] Fahlman, S., Lebiere, C.: The cascade correlation architecture. In: Advances in Neural Information Processing Systems, vol. 2, pp. 524–532. Morgan Kaufmann, San Mateo (1990)
[26] Demšar, J.: Statistical comparisons of classifiers over multiple data sets. Journal of Machine Learning Research 7, 1–30 (2006)

Analysis and Testing of the m-Class RDP Neural Network

David A. Elizondo, Juan M. Ortiz-de-Lazcano-Lobato, and Ralph Birkenhead

Abstract. The Recursive Deterministic Perceptron (RDP) feed-forward multilayer neural network is a generalisation of the single layer perceptron topology. This model is capable of solving any two-class classification problem unlike the single layer perceptron which can only solve classification problems dealing with linearly separable sets. For all classification problems, the construction of an RDP is done automatically and convergence is always guaranteed. A generalisation of the 2-class Recursive Deterministic Perceptron (RDP) exists. This generalisation always allows the deterministic separation of m-classes. It is based on a new notion of linear separability and it arises naturally from the 2 valued RDP. The methods for building 2-class RDP neural networks have been extensively tested. However, no testing has been done before on the m-class RDP method. For the first time, a study on the performance of the m-class method is presented. This study will allow the highlighting of the main advantages and disadvantages of this method by comparing the results obtained while building m-class RDP neural networks with other more classical methods such as Backpropagation and Cascade Correlation in terms of level of generalisation and topology size. The networks were trained and tested using the following standard benchmark classification datasets: Glass, Wine, Zoo, Iris, Soybean, and Wisconsin Breast Cancer.

1 Introduction

The RDP for 2-class classification problems was introduced in [12]. This topology is a generalisation of the single layer perceptron topology (SLPT) developed

David A. Elizondo and Ralph Birkenhead
School of Computing, De Montfort University, The Gateway, Leicester,
LE1 9BH, United Kingdom
e-mail: {elizondo, rab}@dmu.ac.uk

Juan M. Ortiz-de-Lazcano-Lobato
School of Computing, University of Málaga, Bulevar Louis Pasteur, 35, Málaga, Spain
e-mail: jmortiz@lcc.uma.es

L. Franco et al. (Eds.): Constructive Neural Networks, SCI 258, pp. 171–192.
springerlink.com

by Rosenblatt [11]. This generalisation is capable of transforming any non-linearly separable (NLS) 2-class classification problem into a linearly separable (LS) one, thus making it possible for the SLPT to find a solution to the problem. An extension of the RDP algorithm to m-class problems (with $m \geq 2$) was introduced in [13]. This extension is based on a new notion of linear separability and, it evolves naturally from the 2-valued RDP.

1.1 Preliminaries

We use the following standard notions:
• $\mathcal{S}_m$ stands for the set of permutations of $\{1, ..., m\}$.
• If $\mathbf{u} = (u_1, ..., u_d), \mathbf{v} = (v_1, ..., v_d) \in I\!R^d$, then $\mathbf{u}^T\mathbf{v}$ stands for $u_1v_1 + ... + u_dv_d$; and $\mathbf{u}(j) = u_j$ (i.e. $\mathbf{u}(j)$ is the *j-th* component of $\mathbf{u}$).
• $\Pi_{\{i_1,...,i_k\}}\mathbf{u} = (u_{i_1}, ..., u_{i_k})$ and by extension,
if $S \subset I\!R^d$ then $\Pi_{\{i_1,...,i_k\}}S = \{\Pi_{\{i_1,...,i_k\}}\ \mathbf{x} \mid \mathbf{x} \in S\}$.
• Let $r \in I\!R$, $Adj(\mathbf{u}, r) = (u_1, ..., u_d, r)$ and by extension,
if $S \subset I\!R^d$, $Adj(S, r) = \{Adj(\mathbf{x}, r) \mid \mathbf{x} \in S\}$.
• $Im(E, F) = \{(x_1, ..., x_d, x_{d+1}) \in F \mid (x_1, ..., x_d) \in E\}$ is defined for $E \subset I\!R^d$ and $F \subset I\!R^{d+1}$.
• $\mathcal{P}(\mathbf{w}, t)$ stands for the hyperplane $\{\mathbf{x} \in I\!R^d \mid \mathbf{w}^T\mathbf{x} + t = 0\}$ of $I\!R^d$.

1.2 Some Definitions and Properties

In this section, we introduce the notions of convex hull(CH), [10], and of linear separability. A discussion on the different methods for testing linear separability can be found in [12].

Definition 1. Let S be a sub-set of $I\!R^d$,$CH(S) = \{t_1\mathbf{x}_1 + \ldots + t_k\mathbf{x}_k \mid \mathbf{x}_1, \ldots, \mathbf{x}_k \in S,$ $t_1, \ldots, t_k \in [0, 1]$ and $t_1 + \ldots + t_k = 1\}$.
Thus, if S is finite, then there exists $\mathbf{a}_1, ..., \mathbf{a}_k \in I\!R^d$ and $b_1, ..., b_k \in I\!R$ such that $CH(S) = \{\mathbf{x} \in I\!R^d \mid \mathbf{a}_i^T\mathbf{x} \geq b_i\ for\ 1 \leq i \leq k\}$.

Definition 2. Two subsets X and Y of $I\!R^d$ are said to be linearly separable if there exists a hyperplane $\mathcal{P}(\mathbf{w}, t)$ of $I\!R^d$, such that ($\forall \mathbf{x} \in X$, $\mathbf{w}^T\mathbf{x} + t > 0$ and $\forall \mathbf{y} \in Y$, $\mathbf{w}^T\mathbf{y} + t < 0$). In the following we will denote the fact that X and Y are LS by $X \mid\mid Y$ or $X \mid\mid Y$ $(\mathcal{P}(\mathbf{w}, t))$ if we want to specify the hyperplane which linearly separates X and Y.

This paper is divided into four sections. The m-class generalisation of the RDP neural network, based on a notion of linear separability for m classes, is presented in section two. In this section also, some of the notions used throughout this paper are introduced. In section three, the procedure used to evaluate the generalisation of the m-class RDP model is presented. Six machine learning benchmarks (Iris, Soybean, and Wisconsin Breast Cancer) were used [3] and datasets were generated using cross validation. The method is compared with Backpropagation and Cascade Correlation in terms of their level of generalisation. A summary and some conclusions are presented in section four.

2 The m-Class RDP Algorithm

The m-class RDP algorithm is an adaptation of the 2-class RDP based on the following notion of linear separability for m classes ($m > 2$).

Definition 3. Let $X_1, ..., X_m \subset \mathbb{R}^d$ and $a_0 < a_1 < ... < a_m$, $X_1, ..., X_m$ are said to be linearly separable relatively to the ascending sequence of real numbers $a_0, ..., a_m$ if
$\exists \sigma \in \mathcal{S}_m, \exists \mathbf{w} \in \mathbb{R}^d, \exists t \in \mathbb{R}$ *such that* $\forall i, \forall \mathbf{x} \in X_{\sigma(i)},\ a_{i-1} < \mathbf{w}^T\mathbf{x} + t < a_i$.

Remarks
Let $X_1, ..., X_m \subset \mathbb{R}^d$ and $a_0 < a_1 < a_2 < ... < a_m$,
• $X_1, ..., X_m$ are linearly separable relatively to $a_0, ..., a_m$ iff $CH(X_1), ..., CH(X_m)$ are linearly separable relatively to $a_0, ..., a_m$.
• Let $\sigma \in \mathcal{S}_m$.
Put: $X^\sigma = Adj(X_{\sigma(1)}, -a_0) \cup Adj(X_{\sigma(2)}, -a_1) ... \cup Adj(X_{\sigma(m)}, -a_{m-1})$,
$Y^\sigma = Adj(X_{\sigma(1)}, -a_1) \cup Adj(X_{\sigma(2)}, -a_2) ... \cup Adj(X_{\sigma(m)}, -a_m)$, then, $X_1, ...,$ X_m are linearly separable relatively to $a_0, ..., a_m$ by using σ iff $X^\sigma \parallel Y^\sigma$. In other words, we reduce the problem of linear separability for m classes to the problem of linear separability for 2 classes. We do this by augmenting the dimension of the input vectors with the ascending sequence $a_0, ..., a_m$.
• If $X_1 \parallel X_2$ $(\mathcal{P}(\mathbf{w}, t))$ and $\alpha = Max(\{|\mathbf{w}^T\mathbf{x} + t| \ ; \ \mathbf{x} \in (X_1 \cup X_2)\}$, then X_1, X_2 are linearly separable relatively to $-\alpha, 0, \alpha$.

Proposition 1. *Let* $X_1, ..., X_m \subset \mathbb{R}^d$, $a, b \in \mathbb{R}, h, k > 0$ *and let* $a_i = a + ih, b_i = b + ik$, *for* $0 \le i \le m$, *then* $X_1, ..., X_m$ *are linearly separable relatively to* $a_0, ..., a_m$ *iff they are linearly separable relatively to* $b_0, ..., b_m$. *In other words, the linear separability between* m *classes is independent of the arithmetic sequence.*

Proof. Let $\sigma \in \mathcal{S}_m$ represent a class, and let $\mathbf{w} \in \mathbb{R}^d, t \in \mathbb{R}$ such that $\forall i, \forall \mathbf{x} \in X_{\sigma(i)},\ a_{i-1} < \mathbf{w}^T\mathbf{x} + t < a_i$.
Thus, $\forall i, \forall \mathbf{x} \in X_{\sigma(i)},\ b_{i-1} < \frac{k}{h}\mathbf{w}^T\mathbf{x} + \frac{k}{h}(t-a) + b < b_i$ □

Definition 4. $X_1, ..., X_m \subset \mathbb{R}^d$ are said to be linearly separable if there exists $a \in \mathbb{R}, h > 0$ such that $X_1, ..., X_m$ are linearly separable relatively to $a, a + h, ...,$ $a + mh$.

Definition 5. A m-SLPT with the weight $\mathbf{w} \in \mathbb{R}^d$, the threshold $t \in \mathbb{R}$, the values $v_1, v_2, ..., v_m \in \mathbb{R}$ and the characteristic $(c, h) \in \mathbb{R} \times \mathbb{R}^+$ (*c represents the value corresponding to the starting hyperplane, and h a chosen distance between a hyperplane which we will call the step size*), has the same topology as the 2-class SLPT.

The only difference is that the function corresponding to a m-SLPT is a $m-$valued function f defined by : $\forall \mathbf{y} \in \mathbb{R}^d$

$$f(\mathbf{y}) = \begin{cases} v_1 & \text{if } \ \mathbf{w}^T\mathbf{y} + t < c + h \\ v_i & \text{if } \ c + (i-1)h < \mathbf{w}^T\mathbf{y} + t < c + ih, \text{ for } 1 < i < m \\ v_m & \text{if } \ \mathbf{w}^T\mathbf{y} + t > c + (m-1)h \end{cases} \quad (1)$$

2.1 The Specialised NLS to LS Transformation Algorithm for m Classes

A specialised version of the transformation algorithm , from two to m classes, was proposed in [12]. This extension is based on the notion of linear separability for m classes described above.

Let $c \in \mathbb{R}, h > 0, m$ be the number of classes and $b = -(m - \frac{3}{2})h$,for $1 \leq i < m$ $b_i = c + (m-i)b + (\frac{(m-1)(m-2)}{2} - \frac{(i-1)(i-2)}{2})h$ and $b_m = c$.

Table 1 shows the specialized NLS to LS transformation algorithm for m classes. We proceed as in the 2-class specialized transformation algorithm. That is to say, at each step we select a LS sub-set which belongs to a single class and add an artificial variable to the entire input data set. To this artificial variable we assign a value b_i for all the input vectors belonging to the selected LS sub-set and a value b_j to the rest of the set of input vectors, where $b_i \neq b_j$. Two cases for assigning the values to the artificial inputs are possible depending on the class to which the LS sub-set belongs:

1. If the selected LS sub-set belongs to the jth class, with $j < m$, we add to its input vector a new component with value b_j and we add to the rest of the input vector a new component with value b_{j+1}.
2. If the selected LS sub-set belongs to the last class *(mth class)*, we add to its input vector a new component with value b_m and we add to the rest of the input vector a new component with value b_{m-1}.

In the following theorem we prove the correctness and the termination of the algorithm presented in table 1 which allows the construction of an $m-$RDP for linearly separating any given m classes.

Theorem 1. *If $X_1^i, ..., X_m^i$ are not linearly separable, then there exists Z_i such that ($Z_i \subset X_1'^i$ or ... or $Z_i \subset X_m'^i$), $Z_i \neq \emptyset$ and $Z_i \parallel (S_i \setminus Z_i)$.*

Proof. We will prove that, there exists $\mathbf{x} \in X_1'^i \cup ... \cup X_m'^i$ such that $\{\mathbf{x}\} \parallel (S_i \setminus \{\mathbf{x}\})$.

Assume that $\forall \mathbf{x} \in X_1'^i \cup ... \cup X_m'^i$, $\{\mathbf{x}\}$ and $(S_i \setminus \{\mathbf{x}\})$ are not linearly separable, then $X_1'^i \cup ... \cup X_m'^i \subset CH(S_i \setminus (X_1'^i \cup ... \cup X_m'^i))$.

Table 1 Specialized NLS to LS transformation algorithm for m classes.

SNLS2LS($X_1, .., X_m, X_{0i}, .., X_{mi}$)
– data: m data set vectors, $X_0, .., X_m$ representing m NLS classes
– result: A m-RDP $[(\mathbf{w}_0, t_0, a_0, h_0, b_{0,1}, b_{0,2}), ..., (\mathbf{w}_{i-1}, t_{i-1}, a_{i-1}, h_{i-1}, b_{i-1,1}, b_{i-1,2}), (\mathbf{w}_i, t_i, a_i, h_i, b_1, ..., b_m)]$ which linearly separates $X_1, ..., X_m$.
INITIALIZE : Let $i := 0$; $X_1^0 := X_1; ...; X_m^0 := X_m$; ${X'_1}^0 := X_1; ...; {X'_m}^0 := X_m$; $S_0 = X_1 \cup ... \cup X_m$;

WHILE $(X_1^i, ..., X_m^i)$ are not linearly separable
BEGIN
SELECT : Select a non-empty sub-set Z_i from ${X'_1}^i$ or ... or from ${X'_m}^i$ (if it exists) such that Z_i, $(S_i \setminus Z_i)$ are linearly separable
(i.e. $(Z_i \subset {X'_1}^i$ *or ... or* $Z_i \subset {X'_m}^i)$ *and* $Z_i \parallel (S_i \setminus Z_i))$ $(\mathcal{P}(\mathbf{w}_i, t_i))$;
CASE :

Case $Z_i \subset {X'_1}^i$:

$S_{i+1} := Adj(Z_i, b_1) \cup Adj(S_i \setminus Z_i, b_2)$;
${X'_1}^{i+1} := Im({X'_1}^i, S_{i+1}) \setminus Im(Z_i, S_{i+1})$;
${X'_2}^{i+1} := Im({X'_2}^i, S_{i+1}); ...$;
${X'_2}^{i+1} := Im({X'_m}^i, S_{i+1})$;
$X_1^{i+1} := Im(X_1^i, S_{i+1})$;
$X_2^{i+1} := Im(X_2^i, S_{i+1}); ...$;
$X_m^{i+1} := Im(X_m^i, S_{i+1})$;
$i := i + 1$;

....

Case $Z_i \subset {X'_j}^i$:

$S_{i+1} := Adj(Z_i, b_j) \cup Adj(S_i \setminus Z_i, b_{j+1})$;
${X'_1}^{i+1} := Im({X'_1}^i, S_{i+1}); ...$;
${X'_{j-1}}^{i+1} := Im({X'_{j-1}}^i, S_{i+1})$;
${X'_j}^{i+1} := Im({X'_j}^i, S_{i+1}) \setminus Im(Z_i, S_{i+1})$;
${X'_{j+1}}^{i+1} := Im({X'_{j+1}}^i, S_{i+1}); ...$;
${X'_m}^{i+1} := Im({X'_m}^i, S_{i+1})$;
$X_1^{i+1} := Im(X_1^i, S_{i+1})$;
$X_2^{i+1} := Im(X_2^i, S_{i+1}); ...$;
$X_m^{i+1} := Im(X_m^i, S_{i+1})$;
$i := i + 1$;

....

Case $Z_i \subset {X'_m}^i$:

$S_{i+1} := Adj(Z_i, b_m) \cup Adj(S_i \setminus Z_i, b_{m-1})$;
${X'_1}^{i+1} := Im({X'_1}^i, S_{i+1}); ...$;
${X'_{m-1}}^{i+1} := Im({X'_{m-1}}^i, S_{i+1})$;
${X'_m}^{i+1} := Im({X'_m}^i, S_{i+1}) \setminus Im(Z_i, S_{i+1})$;
$X_1^{i+1} := Im(X_1^i, S_{i+1})$;
$X_2^{i+1} := Im(X_2^i, S_{i+1}); ...$;
$X_m^{i+1} := Im(X_m^i, S_{i+1})$;
$i := i + 1$;

END

Table 5 Inputs and outputs used in the Zoo classification problem.

Attributes (1 - 17)	Output	Output Classes
animal name	Animal	1 (41 samples of mammals)
hair		2 (20 samples of birds)
feathers		3 (5 samples of reptiles)
eggs		4 (13 samples of fish)
milk		5 (4 samples of frogs)
airborne		6 (8 samples of insects)
aquatic		7 (10 samples of sea shell)
predator		
toothed		
backbone		
breathes		
venomous		
fins		
legs		
tail		
domestic		
catsize		
type		

Table 6 Inputs and outputs used in the Iris classification problem.

Attributes (In cm)	Output	Output Classes
Sepal Length	Iris plant type	Iris Setosa
Sepal Width		Iris Versicolour
Petal Length		Iris Virginica
Petal Width		

Table 7 Inputs and outputs used in the Wisconsin Breast Cancer classification problem.

Attributes (1 - 10)	Output	Output Classes
Clump Thickness	Class	Benign
Uniformity of Cell Size		Malignant
Uniformity of Cell Shape		
Marginal Adhesion		
Single Epithelial Cell Size		
Bare Nuclei		
Bland Chromatin		
Normal Nucleoli		
Mitoses		

The Zoo benchmark data set contains 17 Boolean-valued attributes with a type of animal as output (5). A total of 101 samples are included (mammals, birds, reptiles, fish, frogs, insects and sea shells).

The Iris dataset classifies a plant as being an Iris Setosa, Iris Versicolour or Iris Virginica. The dataset describes every iris plant using four input parameters (Table 6). The dataset contains a total of 150 samples with 50 samples for each of the three classes. All the samples of the Iris Setosa class are linearly separable from the rest of the samples (Iris Versicolour and Iris Virginica). Some of the publications that used this benchmark include: [7] [8] [2] and [4].

The Soybean classification problem contains data for the disease diagnosis of the Soybean crop. The dataset describes the different diseases using symptoms. The original dataset contains 19 diseases and 35 attributes. The attribute list was limited to those attributes that had non trivial values in them (Table 4). Thus there were only 20 out of the 35 attributes included in the tests. Only 15 of the 19 have no missing values. Therefore, only these 15 classes were used for the comparisons.

The Wisconsin Breast Cancer dataset [9, 1, 15] consists of a binary classification problem to distinguish between benign and malignant breast cancer. The data set contains 699 instances and 9 attributes (Table 7). The class distribution is: Benign 458 instances (65.5 %), and Malignant 241 instances (34.5 %).

The technique of cross validation was applied to split the benchmarks into training and testing data sets. The datasets were randomly divided into 'n' equal sized testing sets that were mutually exclusive [14]. The remaining samples were used to train the networks. In this study, the classification benchmark data sets were divided into ten equally sized data sets. On one hand sixty percent of the samples were used for training the networks and the remaining forty percent were used for testing purposes. On the other hand the training dataset consisted of eighty percent of the samples and the remaining twenty percent were used for the testing dataset.

The simplex algorithm was used on this study for testing for linear separability. This algorithm was remarkably faster than the Perceptron one when searching for LS subsets. Other algorithms for testing linear separability include the Class of Linear Separability [5] and the Fisher method (see [6] for a survey on methods for testing linear separability).

These results provide a good basis for further developing this study and comparing the effects of using single or multiple output neurons for multiple class classification problems using the m-class RDP method and Backpropagation and Cascade Correlation. After describing the experimental setup, some conclusions are presented in the next section.

4 Results and Discussion

We now present a comparison of the m-class RDP construction method, Backpropagation and Cascade Correlation based on their level of generalisation on previously

Table 8 Results obtained with the m-class, and backpropagation, using the Glass data set benchmark in terms of the level of generalisation with 60% of the data used for training and 40% for testing.

Data Set	m-RDP	BackProp	BP 1 out	CC Mout
1	48.84	56.98	52.33	53.49
2	52.33	61.63	53.49	60.47
3	53.49	61.63	59.30	56.98
4	52.33	65.12	55.81	54.65
5	51.16	65.12	50.00	62.79
6	66.28	62.79	55.81	58.14
7	63.95	72.09	51.16	59.30
8	52.33	55.81	59.30	52.33
9	59.30	56.98	61.63	48.84
10	55.81	62.79	52.33	55.81
Mean	55.58	62.01	55.12	56.28
Std	5.76	4.84	3.92	4.14

Table 9 Results obtained with the m-class, and backpropagation, using the Glass data set benchmark in terms of the level of generalisation with 80% of the data used for training and 20% for testing.

Data Set	m-RDP	BackProp	BP 1 out	CC Mout
1	67.44	67.44	67.44	51.16
2	65.12	65.12	55.81	62.79
3	53.49	72.09	55.81	58.14
4	60.47	62.79	60.47	48.84
5	44.19	76.74	62.79	65.12
6	58.14	76.74	65.12	67.44
7	67.44	62.79	51.16	58.14
8	53.49	58.14	51.16	55.81
9	58.14	55.81	55.81	65.12
10	46.51	74.42	67.44	51.16
Mean	57.44	67.21	59.30	58.37
Std	8.13	7.55	6.23	6.62

unseen data and the number of neurons needed for each method to solve the classification problems (i.e. the size of the topology).

As specified before, the m-class RDP uses a single output neuron for multiple classes. Backpropagation and Cascade Correlation are tested using two different topologies. The first one uses a unique output neuron and is named *BP 1out* and *CC 1out* in the tables. The second type of topology uses as many neurons in the output layer as the number of classes in the data set (*Backprop* and *CC Mout* in the tables). Only the first type of topology is used when the dataset defines a binary classification problem such as the Wisconsin Breast Cancer dataset.

Table 10 Results obtained with the m-class, and backpropagation, using the Wine data set benchmark in terms of the level of generalisation with 60% of the data used for training and 40% for testing.

Data Set	m-RDP	BackProp	BP 1out	CC Mout
1	93.06	95.83	94.44	90.28
2	93.06	95.83	98.61	93.06
3	88.89	97.22	95.83	90.29
4	87.50	97.22	94.44	90.28
5	94.44	97.22	95.83	86.11
6	91.67	97.22	93.06	93.06
7	90.28	97.22	94.44	91.67
8	91.67	98.61	95.83	91.67
9	95.83	95.83	94.44	90.28
10	95.83	94.44	95.83	91.67
Mean	92.22	96.66	95.27	90.84
Std	2.79	1.17	1.49	1.99

Table 11 Results obtained with the m-class, and backpropagation, using the Wine data set benchmark in terms of the level of generalisation with 80% of the data used for training and 20% for testing.

Data Set	m-RDP	BP 1out	CC Mout
1	88.89	100.00	88.89
2	86.11	97.22	91.67
3	86.11	97.22	91.67
4	94.44	100.00	94.44
5	94.44	94.44	94.44
6	97.22	100.00	94.44
7	97.22	100.00	97.22
8	97.22	100.00	100.0
9	97.22	97.22	94.44
10	94.44	97.22	91.67
Mean	88.57	98.33	93.88
Std	7.84	1.94	3.15

Overall, considering all the results obtained from tables 8 to 19 in terms of generalisation obtained using the m-class RDP, it appears that the method is broadly comparable with CC and BP, but has slightly poorer results. It appears to be more variable in its performance. While it does generally perform better than the other methods when they are used with a single output neuron, it is arguable that the nature of the data makes this an inappropriate choice of topology for a BP or CC network.

Considering the size of the network produced (tables 20 to 31), the number of neurons in an m-class RDP is usually significantly lower than in the corresponding

Table 12 Results obtained with the m-class, and backpropagation, using the Zoo data set benchmark in terms of the level of generalisation with 60% of the data used for training and 40% for testing.

Data Set	m-RDP	BackProp	BP 1out	CC Mout	CC 1out
1	90.24	87.80	90.24	78.05	97.56
2	92.68	87.80	90.24	92.68	97.56
3	95.12	95.12	95.12	90.24	100.00
4	87.80	97.56	95.12	95.12	97.56
5	92.68	100.00	95.12	92.68	100.00
6	90.24	100.00	97.56	92.68	100.00
7	90.24	100.00	97.56	97.56	100.00
8	92.68	100.00	97.56	97.56	100.00
9	90.24	95.12	100.00	90.24	95.12
10	85.37	90.24	97.56	82.93	92.68
Mean	90.73	95.36	95.60	90.97	98.04
Std	2.77	5.07	3.21	6.20	2.52

Table 13 Results obtained with the m-class, and backpropagation, using the Zoo data set benchmark in terms of the level of generalisation with 80% of the data used for training and 20% for testing.

Data Set	m-RDP	BackProp	BP 1out	CC Mout	CC 1out
1	80.95	100.00	100.00	95.24	100.00
2	80.95	100.00	100.00	95.24	100.00
3	95.24	100.00	100.00	95.24	100.00
4	95.24	95.24	90.48	95.24	100.00
5	85.71	100.00	100.00	95.24	90.48
6	90.48	100.00	100.00	100.00	90.48
7	95.24	95.24	100.00	95.24	100.00
8	76.19	95.24	100.00	85.71	100.00
9	85.71	95.24	100.00	85.71	100.00
10	100.00	100.00	100.00	100.00	100.00
Mean	88.57	98.10	99.05	94.28	98.10
Std	7.84	2.46	3.01	4.92	4.01

BP and CC networks with multiple output neurons. The single output neuron BP and CC networks sometimes have fewer neurons but, as discussed above, this is probably an inappropriate architecture for the data. This will lead to future research and exploring a multiple output architecture for the m-class RDP model.

Table 14 Results obtained with the m-class, and backpropagation, using the Iris data set benchmark in terms of the level of generalisation with 60% of the data used for training and 40% for testing.

Data Set	m-RDP	BackProp	BP 1out	CC Mout	CC 1out
1	98.33	98.33	98.33	98.33	98.33
2	98.33	98.33	98.33	96.67	98.33
3	95.00	96.67	96.67	95	96.67
4	95.00	98.33	98.33	96.67	100.00
5	91.67	98.33	98.33	95	98.33
6	91.67	96.67	96.67	91.67	98.3
3 7	96.67	98.33	96.67	93.33	96.67
8	96.67	96.67	98.33	95	100.00
9	95.00	96.67	96.67	93.33	98.33
10	93.33	91.67	96.67	91.67	96.6
Mean	95.167	97	97.5	94.667	98.166
Std	2.41	2.05	0.87	2.19	1.23

Table 15 Results obtained with the m-class, and backpropagation, using the Iris data set benchmark in terms of the level of generalisation with 80% of the data used for training and 20% for testing.

Data Set	m-RDP	BackProp	BP 1out	CC Mout	CC 1out
1	90.00	93.33	96.67	93.33	91.67
2	90.00	96.67	96.67	93.33	96.67
3	93.33	100.00	100.00	100.00	100.00
4	100.00	100.00	100.00	100.00	96.67
5	96.67	100.00	100.00	100.00	93.33
6	93.33	96.67	100.00	100.00	100.00
7	100.00	96.67	100.00	96.67	96.67
8	96.67	100.00	100.00	100.00	100.00
9	96.67	96.67	96.67	96.67	96.67
10	90.00	93.33	90.00	90.00	93.33
Mean	94.67	97.33	98.00	97	96.50
Std	3.91	2.63	3.22	3.67	2.98

Table 16 Results obtained with the m-class, and backpropagation, using the Soybean data set benchmark in terms of the level of generalisation with 60% of the data used for training and 40% for testing.

Data Set	m-RDP	BackProp	BP 1out	CC Mout	CC 1out
1	83.18	87.85	49.53	87.85	54.21
2	73.83	87.85	54.21	89.71	50.47
3	74.77	84.11	45.79	85.98	47.66
4	61.68	80.37	37.38	83.17	54.21
5	75.70	84.11	45.79	84.11	51.40
6	70.09	81.31	39.25	85.05	51.40
7	75.70	84.11	43.93	85.50	52.34
8	74.77	78.50	47.66	82.24	52.34
9	73.83	84.11	45.79	85.98	52.34
10	71.03	85.05	48.60	91.58	57.94
Mean	73.46	83.74	45.79	86.12	52.43
Std	5.42	3.0	4.87	2.89	2.70

Table 17 Results obtained with the m-class, and backpropagation, using the Soybean data set benchmark in terms of the level of generalisation with 80% of the data used for training and 20% for testing.

Data Set	m-RDP	BackProp	BP 1out	CC Mout	CC 1out
1	74.07	88.89	38.89	83.33	51.85
2	62.96	87.04	40.74	81.48	38.89
3	77.78	92.59	44.44	85.19	64.81
4	87.04	90.74	61.11	92.59	61.11
5	74.07	85.19	42.59	81.48	33.33
6	72.22	90.74	50.00	85.19	50.00
7	83.33	90.74	53.70	87.04	50.00
8	77.78	94.44	55.56	88.89	61.11
9	68.52	92.59	48.15	88.89	55.56
10	74.07	87.04	48.15	88.89	50.00
Mean	75.18	90	48.33	86.30	51.66
Std	6.90	2.92	7.01	3.62	9.85

Table 18 Results obtained with the m-class, and backpropagation, using the Wisconsin Breast Cancer data set benchmark in terms of the level of generalisation with 60% of the data used for training and 40% for testing.

Data Set	m-RDP	BackProp	CC
1	94.16	97.08	98.17
2	95.62	95.26	97.08
3	94.53	95.26	96.00
4	95.62	96.72	97.00
5	93.07	98.18	96.35
6	91.61	97.45	97.00
7	94.16	95.62	96.70
8	89.78	97.45	97.00
9	91.61	97.45	97.45
10	94.53	97.45	98.54
Mean	93.47	96.79	97.13
Std	1.92	1.04	0.77

Table 19 Results obtained with the m-class, and backpropagation, using the Wisconsin Breast Cancer data set benchmark in terms of the level of generalisation with 80% of the data used for training and 20% for testing.

Data Set	m-RDP	BackProp	CC
1	97.08	95.62	93.43
2	94.16	97.81	94.89
3	94.89	97.81	97.08
4	94.16	98.54	93.43
5	95.62	97.81	97.08
6	94.89	96.35	94.16
7	94.16	97.08	94.16
8	92.70	95.62	93.20
9	93.43	96.35	94.16
10	93.43	96.35	93.43
Mean	94.45	96.93	94.65
Std	1.25	1.02	1.46

Table 20 Results obtained with the m-class, and backpropagation, using the Glass data set benchmark in terms of the topology size (number of hidden/intermediate neurons) with 60% of the data used for training and 40% for testing.

Data Set	m-RDP	BackProp	BP 1out	CCMout
1	13.00	22.00	7.00	51.00
2	12.00	22.00	7.00	50.00
3	13.00	22.00	7.00	50.00
4	14.00	22.00	7.00	47.00
5	14.00	22.00	7.00	51.00
6	15.00	22.00	7.00	41.00
7	16.00	22.00	7.00	57.00
8	13.00	22.00	7.00	47.00
9	15.00	22.00	7.00	68.00
10	13.00	22.00	7.00	46.00
Mean	13.8	22	7	7.57
Std	1.23	0	0	7.33

Table 21 Results obtained with the m-class, and backpropagation, using the Glass data set benchmark in terms of the topology size (number of hidden/intermediate neurons) with 80% of the data used for training and 20% for testing.

Data Set	m-RDP	BackProp	BP 1out	CC Mout
1	18.00	32.00	16.00	47
2	19.00	32.00	16.00	49
3	18.00	32.00	16.00	55
4	16.00	32.00	16.00	55
5	20.00	32.00	16.00	44
6	18.00	32.00	16.00	53
7	17.00	32.00	16.00	61
8	20.00	32.00	16.00	59
9	17.00	32.00	16.00	53
10	20.00	32.00	16.00	47
Mean	18.3	32	16	52.3
Std	1.42	0	0	5.50

Table 22 Results obtained with the m-class, and backpropagation, using the Wine data set benchmark in terms of the topology size (number of hidden/intermediate neurons) with 60% of the data used for training and 40% for testing.

Data Set	m-RDP	BackProp	BP 1out	CCMout
1	4.00	28.00	31.00	6.00
2	4.00	28.00	31.00	93.00
3	4.00	28.00	31.00	50.00
4	4.00	28.00	31.00	89.00
5	4.00	28.00	31.00	123.00
6	4.00	28.00	31.00	86.00
7	4.00	28.00	31.00	99.00
8	4.00	28.00	31.00	123.00
9	4.00	28.00	31.00	10.00
10	4.00	28.00	31.00	32.00
Mean	4	28	31	71.1
Std	0	0	0	43.63

Table 23 Results obtained with the m-class, and backpropagation, using the Wine data set benchmark in terms of the topology size (number of hidden/intermediate neurons) with 80% of the data used for training and 20% for testing.

Data Set	m-RDP	BP 1out	CC Mout
1	4.00	31.00	123
2	4.00	31.00	123
3	4.00	31.00	123
4	4.00	31.00	107
5	4.00	31.00	59
6	4.00	31.00	123
7	4.00	31.00	123
8	4.00	31.00	60
9	4.00	31.00	76
10	4.00	31.00	123
Mean	4	31	104
Std	0	0	27.73

Table 24 Results obtained with the m-class, and backpropagation, using the Zoo data set benchmark in terms of the topology size (number of hidden/intermediate neurons) with 60% of the data used for training and 40% for testing.

Data Set	m-RDP	BackProp	BP 1out	CCMout	CC out
1	8.00	32.00	7.00	7.00	1.00
2	8.00	32.00	7.00	7.00	1.00
3	8.00	32.00	7.00	7.00	1.00
4	8.00	32.00	7.00	7.00	1.00
5	8.00	32.00	7.00	7.00	1.00
6	8.00	32.00	7.00	7.00	1.00
7	8.00	32.00	7.00	7.00	1.00
8	8.00	32.00	7.00	7.00	1.00
9	8.00	32.00	7.00	7.00	1.00
10	8.00	32.00	7.00	7.00	1.00
Mean	8	32	7	7	0
Std	0	0	0	0	0

Table 25 Results obtained with the m-class, and backpropagation, using the Zoo data set benchmark in terms of the topology size (number of hidden/intermediate neurons) with 80% of the data used for training and 20% for testing.

Data Set	m-RDP	BackProp	BP 1out	CC Mout	CC 1out
1	8.00	27.00	3.00	7	1
2	8.00	27.00	3.00	7	1
3	8.00	27.00	3.00	7	1
4	8.00	27.00	3.00	7	1
5	8.00	27.00	3.00	7	1
6	8.00	27.00	3.00	7	1
7	8.00	27.00	3.00	7	1
8	8.00	27.00	3.00	7	1
9	8.00	27.00	3.00	7	1
10	8.00	27.00	3.00	7	1
Mean	8	27	3	7	0
Std	0	0	0	0	0

Table 26 Results obtained with the m-class, and backpropagation, using the Iris data set benchmark in terms of the topology size (number of hidden/intermediate neurons) with 60% of the data used for training and 40% for testing.

Data Set	m-RDP	BackProp	BP 1out	Cc Mout	CC 1out
1	4.00	9.00	5.00	6.00	12.00
2	4.00	9.00	5.00	7.00	13.00
3	4.00	9.00	5.00	6.00	10.00
4	5.00	9.00	5.00	8.00	13.00
5	5.00	9.00	5.00	6.00	11.00
6	5.00	9.00	5.00	8.00	11.00
7	4.00	9.00	5.00	6.00	13.00
8	4.00	9.00	5.00	6.00	13.00
9	4.00	9.00	5.00	6.00	13.00
10	4.00	9.00	5.00	5.00	10.00
Mean	4.3	9	5	6.4	11.9
Std	0.48	0	0	0.97	

Table 27 Results obtained with the m-class, and backpropagation, using the Iris data set benchmark in terms of the topology size (number of hidden/intermediate neurons) with 80% of the data used for training and 20% for testing.

Data Set	m-RDP	BackProp	BP 1out	CC Mout	CC 1out
1	4.00	13.00	5.00	4	14
2	4.00	13.00	5.00	6	15
3	6.00	13.00	5.00	8	15
4	6.00	13.00	5.00	9	16
5	6.00	13.00	5.00	8	16
6	6.00	13.00	5.00	7	15
7	6.00	13.00	5.00	10	16
8	7.00	13.00	5.00	8	16
9	6.00	13.00	5.00	8	16
10	5.00	13.00	5.00	6	13
Mean	5.6	13	5	7.4	15.2
Std	0.97	0	0	1.71	1.03

Table 28 Results obtained with the m-class, and backpropagation, using the Soybean data set benchmark in terms of the topology size (number of hidden/intermediate neurons) with 60% of the data used for training and 40% for testing.

Data Set	m-RDP	BackProp	BP 1out	CCMout	CC 1out
1	19.00	60.00	51.00	17.00	15.00
2	17.00	60.00	51.00	17.00	15.00
3	18.00	60.00	51.00	17.00	15.00
4	16.00	60.00	51.00	17.00	14.00
5	17.00	60.00	51.00	16.00	14.00
6	16.00	60.00	51.00	16.00	14.00
7	16.00	60.00	51.00	16.00	15.00
8	16.00	60.00	51.00	16.00	14.00
9	16.00	60.00	51.00	16.00	14.00
10	16.00	60.00	51.00	17.00	15.00
Mean	16.7	60	51	16.5	14.5
Std	1.06	0	0	0.53	0.53

Table 29 Results obtained with the m-class, and backpropagation, using the Soybean data set benchmark in terms of the topology size (number of hidden/intermediate neurons) with 80% of the data used for training and 20% for testing.

Data Set	m-RDP	BackProp	BP 1out	CC Mout	CC 1out
1	17.00	50.00	51.00	17	18
2	19.00	50.00	51.00	17	17
3	18.00	50.00	51.00	18	18
4	17.00	50.00	51.00	17	18
5	17.00	50.00	51.00	17	17
6	19.00	50.00	51.00	17	18
7	17.00	50.00	51.00	17	18
8	18.00	50.00	51.00	18	17
9	17.00	50.00	51.00	18	18
10	17.00	50.00	51.00	18	19
Mean	17.6	50	51	17.4	17.8
Std	0.84	0	0	0.52	0.63

Table 30 Results obtained with the m-class, and backpropagation, using the Wisconsin Breast Cancer data set benchmark in terms of the topology size (number of hidden/intermediate neurons) with 60% of the data used for training and 40% for testing.

Data Set	m-RDP	BackProp	CC
1	10.00	16.00	6.00
2	7.00	16.00	5.00
3	9.00	16.00	5.00
4	10.00	16.00	7.00
5	11.00	16.00	8.00
6	12.00	16.00	7.00
7	11.00	16.00	7.00
8	9.00	16.00	7.00
9	10.00	16.00	6.00
10	11.00	16.00	6.00
Mean	10	16	6.4
Std	1.41	0	0.97

Table 31 Results obtained with the m-class, and backpropagation, using the Wisconsin Breast Cancer data set benchmark in terms of the topology size (number of hidden/intermediate neurons) with 80% of the data used for training and 20% for testing.

Data Set	m-RDP	BackProp	CC
1	11.00	16.00	7
2	11.00	16.00	8
3	12.00	16.00	8
4	14.00	16.00	8
5	13.00	16.00	9
6	12.00	16.00	7
7	14.00	16.00	8
8	12.00	16.00	7
9	11.00	16.00	6
10	10.00	16.00	8
Mean	12	16	7.67
Std	1.33	0	0.87

References

1. Bennett, K.P., Mangasarian, O.L.: Robust linear programming discrimination of two linearly inseparable sets. Optimization Methods and Software 1, 23–34 (1992)
2. Dasarathy, B.W.: Nosing around the neighborhood: A new system structure and classification rule for recognition in partially exposed environments. IEEE Transactions on Pattern Analysis and Machine Intelligence 2(1), 67–71 (1980)
3. Newman, D.J., Hettich, S., Merz, C.B.: C.: UCI repository of machine learning databases (1998), `http://www.ics.uci.edu/~mlearn/MLRepository.html`
4. Elizondo, D.: The recursive determinist perceptron (rdp) and topology reduction strategies for neural networks. Ph.D. thesis, Université Louis Pasteur, Strasbourg, France (1997)
5. Elizondo, D.: Searching for linearly separable subsets using the class of linear separability method. In: Proceedings of the IEEE-IJCNN, pp. 955–960 (2004)
6. Elizondo, D.: The linear separability problem: Some testing methods. Accepted for Publication: IEEE TNN 17, 330–344 (2006)
7. Fisher, R.A.: The use of multiple measurements in taxonomic problems. Annual Eugenics 7(II), 179–188 (1936)
8. Gates, G.W.: The reduced nearest neighbor rule. IEEE Transactions on Information Theory, 431–433 (1972)
9. Mangasarian, O.L., Wolberg, W.H.: Cancer diagnosis via linear programming. SIAM News 23(5), 1–18 (1990)
10. Preparata, F.P., Shamos, M.I.: Computational Geometry. An Introduction. Springer, New York (1985)
11. Rosenblatt, F.: Principles of Neurodynamics. Spartan, Washington D.C (1962)
12. Tajine, M., Elizondo, D.: Enhancing the perceptron neural network by using functional composition. Tech. Rep. 96-07, Computer Science Department, Université Louis Pasteur, Strasbourg, France (1996)
13. Tajine, M., Elizondo, D., Fiesler, E., Korczak, J.: Adapting the 2−class recursive deterministic perceptron neural network to $m-$classes. In: The International Conference on Neural Networks (ICNN), IEEE, Los Alamitos (1997)
14. Weiss, S.M., Kulikowski, C.A.: Computer Systems That Learn. Morgan Kaufmann Publishers, San Mateo (1991)
15. Wolberg, W.H., Mangasarian, O.: Multisurface method of pattern separation for medical diagnosis applied to breast cytology. Proceedings of the National Academy of Sciences 87, 9193–9196 (1990)

Active Learning Using a Constructive Neural Network Algorithm

José L. Subirats, Leonardo Franco, Ignacio Molina, and José M. Jerez

Abstract. Constructive neural network algorithms suffer severely from overfitting noisy datasets as, in general, they learn the set of available examples until zero error is achieved. We introduce in this work a method for detect and filter noisy examples using a recently proposed constructive neural network algorithm. The new method works by exploiting the fact that noisy examples are in general harder to be learnt than normal examples, needing a larger number of synaptic weight modifications. Different tests are carried out, both with controlled and real benchmark datasets, showing the effectiveness of the approach. Using different classification algorithms, it is observed an improved generalization ability in most cases when the filtered dataset is used instead of the original one.

1 Introduction

A main issue at the time of implementing feed-forward neural networks in classification or prediction problems is the selection of an adequate architecture [1, 2, 3].

José L. Subirats
Departamento de Lenguajes y Ciencias de la Computación, Universidad de Málaga, Campus de Teatinos S/N, 29071 Málaga, Spain
e-mail: jlsubirats@lcc.uma.es

Leonardo Franco
Departamento de Lenguajes y Ciencias de la Computación, Universidad de Málaga, Campus de Teatinos S/N, 29071 Málaga, Spain
e-mail: lfranco@lcc.uma.es

Ignacio Molina
Departamento de Tecnología Electrónica,, Universidad de Málaga, Campus de Teatinos S/N, 29071 Málaga, Spain
e-mail: aimc@dte.uma.es

José M. Jerez
Departamento de Lenguajes y Ciencias de la Computación, Universidad de Málaga, Campus de Teatinos S/N, 29071 Málaga, Spain
e-mail: jja@lcc.uma.es

L. Franco et al. (Eds.): Constructive Neural Networks, SCI 258, pp. 193–206.
springerlink.com

Feed-forward neural networks trained by back-propagation have been widely used in several problems but still the standard approach for selecting the number of layers and number of hidden units of the neural architecture is the inefficient trial-by-error method. Several constructive methods and pruning techniques [1] have been proposed as an alternative for the architecture selection process but it is a research issue whether these methods can achieve the same level of prediction accuracy. Constructive algorithms start with a very small network, normally comprising a single neuron, and work by adding extra units until some convergence condition is met [4, 5, 6, 7, 8]. A totally opposite approach is the used by pruning techniques, as these methods start with large architectures and work by eliminating unnecessary weights and units [9].

Despite the existence of several constructive algorithms, they have not been extensively applied in real problems. This fact is relatively surprising, given that they offer a systematic and controlled way of obtaining a neural architecture together with the set of weights, and also because in most cases they offer the possibility of an easier knowledge extraction procedure. In a 1993 work, Smieja [10] argued that constructive algorithms might be more efficient in terms of the learning process but cannot achieve a generalization ability comparable to back-propagation neural networks. Smieja arguments were a bit speculative rather than based on clear results, but nevertheless might explain the fact that constructive methods have not been widely applied to real problems. In recent years new constructive algorithms have been proposed and analyzed, and the present picture might have changed [8, 7].

One of the problems that affects predictive methods in general, is the problem of overfitting [11, 12]. The problem of overfitting arises when an algorithm specializes in excess in learning the available training data causing a reduction on the generalization ability, computed on unseen data. In particular, overfitting affects severely neural network constructive algorithms as they, in general, learn towards zero error on the training set. One of the strategies used in constructive algorithms for avoiding overfitting is the search of very compact architectures, as models with fewer number of parameters may suffer less from overfitting. Other standard methods to avoid overfitting, like early stopping using a validation set or weight decay, can also be applied to constructive methods but they tend to be computationally costly and sometimes difficult to adapt to work in conjunction with some constructive algorithms. When the input data is noisy, as it is normally the case of real data, the simple use of compact architectures is not enough to avoid overfitting as it will be shown later in this chapter. A possible solution to this problem might be the implementation of methods that exclude noisy instances from the training dataset [13, 14, 15, 16, 17, 18], in a process that is usually considered a pre-processing stage. In this work, we refer to the whole problem of learning and filtering noisy examples as "Active learning", as we considered both stages together in an on-line procedure in which noisy instances are eliminated during the learning procedure. Nevertheless, we also show in this work that the new introduced filtering process can be applied as a separate stage and the selected instances used later with any available predictive algorithm. The usual name given to the process of selecting or filtering some examples from the available dataset is "Instance selection" and we refer to [19] for previous work on the field. Instance selection can be also used for reducing the size of the dataset in order to speed up the training process and can be also lead to prototype selection when the selected data are very much reduced. The approach taken in this chapter is to use the proposed instance selection method as a pre-processing step , as way of improving the generalization ability of predictive algorithms. The method is developed inside a recently introduced constructive neural network algorithm named C-Mantec [20] (Competitive MAjority Network Trained by Error Correction) and leads to an improvement in the generalization ability of the algorithm, permitting also to obtain more compact neural network architecures. The reduced, filtered, datasets are also tested with other standard classification methods like standard multilayer perceptrons, decision trees and support vector machines, analyzing the generalization ability obtained. This chapter is organized as follows: Next we give details about the C-Mantec constructive neural network algorithm and in Section 3 the method for eliminating noisy instances is introduced, to follow with some experiments, results and conclusions.

2 The C-Mantec Algorithm

The C-Mantec algorithm is a constructive neural network algorithm that creates architectures with a single layer of hidden nodes with threshold activation functions. For the most general case of input data comprising 2 output classes, the constructed networks have a single output neuron computing the majority function of the responses of the hidden nodes (i.e., if more than half of the hidden neurons are activated the output neuron will be active). The case of multiclass classification will be considered separately below. The learning procedure starts with an architecture comprising a single neuron in the hidden layer and as the learning advances more neurons are added every time the present ones are not able to learn the whole set of training examples. The synaptic weight modification rule used at the single neuron level is the thermal perceptron learning rule proposed by Frean [5, 21]. The thermal perceptron rule is a modification of the standard perceptron rule [22] that incorporates a modulation factor that makes the perceptron to learn only inputs that are similar to the already acquired knowledge, as the introduced factor limits the value of the modifications of the synaptic vector. The idea behind the thermal perceptron is to introduce stability to the standard perceptron for the case of non-linearly separable tasks and this is achieved by permitting large changes of the synaptic vector only at the beginning and later on only allow small modifications.

We consider neurons with a threshold activation function receiving input signals ψ_i through synaptic weights w_i that are active if the synaptic potential, ϕ is larger than zero. The synaptic potential is defined as:

$$\phi = (\sum_i \psi_i * w_i) - b. \quad (1)$$

Note that the definition of the synaptic potential includes the value of the threshold or bias, b, as this will be useful because for wrongly classified inputs the absolute value of the synaptic potential, $|\phi|$, quantifies the error committed as it gives the distance to the bordering hyperplane dividing the classification regions.

The neuron model used is a threshold gate where the output of the neuron, S, is given by a step function depending on the value of the synaptic potential.

$$S = f(p) = \begin{cases} 1 \text{ if } p \geq 0 \\ 0 \text{ otherwise} \end{cases} \tag{2}$$

As said above, the synaptic modification rule that is used by the C-Mantec algorithm is the thermal perceptron rule for which the change of the weights, δw_i, is given by the following equation:

$$\delta w_i = (t - S)\ \psi_i\ \frac{T}{T_0}\ \exp\{-\frac{|\phi|}{T}\}\ , \tag{3}$$

where t is the target value of the example being considered, S represents the actual output of the neuron and ψ is the value of the input unit i. T is a parameter introduced in the thermal perceptron definition, named temperature, T_0 the starting temperature value and ϕ, the synaptic potential defined in Eq. 1. For rightly classified examples, the factor $(t - S)$ is equals to 0 and then no synaptic weight changes take place. The thermal perceptron rule can be seen as a modification to the standard perceptron rule where the change of weights is modified by the factor, m, equals to:

$$m = \frac{T}{T_0}\ \exp\{-\frac{|\phi|}{T}\}\ . \tag{4}$$

At the single neuron level the C-Mantec algorithm uses the thermal perceptron rule, but at a global network level the C-Mantec algorithm incorporates competition between the neurons, making the learning procedure more efficient and permitting to obtain more compact architectures [20]. The main novelty introduced in the new C-Mantec algorithm is the fact that once a new neuron is added to the network, the existing synaptic weights are not frozen, as it is the standard procedure in constructive algorithms. Instead, after an input instance is presented to the network all existing neurons can learn the incoming information by modifying its weights in a competitive way, in which only one neuron will learn the incoming information. The norm in standard constructive algorithms is to freeze weights not connected to the last added neurons in order to preserve the stored information, in the C-Mantec algorithm this is not necessary as the thermal perceptron is a quite conservative learning algorithm and also because the C-Mantec algorithm incorporates a parameter $gfac$ that further controls the size of the allowed changes in synaptic weights, in particular when the Temperature is large when this large changes are allowed at the single neuron level by the thermal perceptron.

The C-Mantec algorithm generates architectures with a single hidden layer of neurons. The output neuron of the network computes the majority function of the activation values of the hidden units and thus the set of weights connecting the hidden neurons with the output are fix from the beginning and not modified during

```
1. Start a network with one hidden neuron and one output
neuron.
2. Input a random training example and check the output of
the network.
3. If the input example is not rightly classified then:
  3a. Compute the value of φ for all existing hidden neurons
that wrongly classify the input example.
  3b. Modify the weights of the neuron with the smallest
value of φ, provided that the value of the factor m is larger
than the value of the parameter gfac. Lower the internal
temperature of the modified neuron.
  3c. If there is no neuron with a value of m larger than gfac
then introduce a new neuron that learns the incoming example.
4. Go to instruction 2 until all examples are classified
correctly.
```

Fig. 1 Pseudocode of the C-Mantec algortihm

the learning procedure. As the output neuron computes the majority of the hidden layer activations, a correct functioning of the network is a state in which for every instance in the training set the output of more than half of the hidden units coincides with the respective target value of the instances.

As mentioned before, the algorithm also incorporates a parameter named growing factor, *gfac*, as it adjustment affects the size of the resulting architecture. Once an instance is presented and the output of the network does not coincide with the target value, a neuron in the hidden layer will be selected to learn it if some conditions are met. The selected neuron will be the one with the lowest value of ϕ among those neurons whose output is different from the target one, but only if the value of m (see Eq. 4) is larger than the *gfac* value, set at the beginning of the learning process. Thus, the *gfac* parameter will prevent the learning of misclassified examples that will involve large weight modifications, as for high values of T the thermal perceptron rule would not avoid these large changes, that can cause instability to the algorithm. After a neuron modifies it weights, its internal temperature value is lowered. In the case in which for a wrongly classified instance there are no neurons available for learning, a new neuron is added to the network and this unit will learn the current input, ensuring the convergence of the algorithm. After a new unit is added to the network the temperature, T, of all neurons is reset to the initial value T_0 and the learning process continues until all training examples are correctly classified. In Fig. 1 a pseudocode of the algorithm is shown, summarizing the most important steps of the C-Mantec algorithm and in Fig. 2 a flow diagram of the algorithm is shown.

Regarding the setting of the two parameters of the algorithm, T_0 and *gfac*, several experiments have shown that the C-Mantec algorithm is quite robust against changes of these two parameters and the finding of some optimal values is not difficult. The parameter T_0 (initial temperature) ensures that a certain number of learning iterations will take place, permitting an initial phase of global exploration for the weights values, as for high temperature values larger changes are easier to be

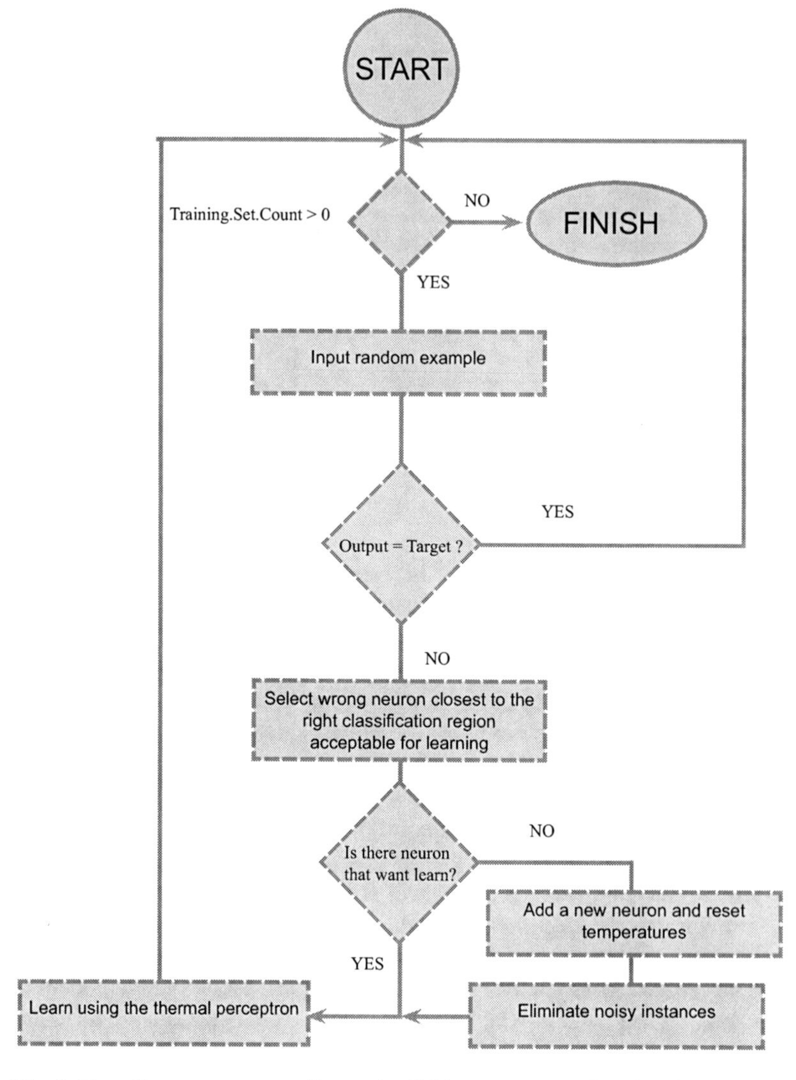

Fig. 2 Flow diagram corresponding to the C-Mantec constructive algorithm.

accepted. The value of the parameter $gfac$ affects the size of the final architecture, and it has been observed that different values are needed in order to optimize the algorithm towards obtaining more compact architectures or a network with a better generalization ability.

The convergence of the algorithm is ensured because the learning rule is very conservative in their changes, preserving the acquired knowledge of the neurons and given by the fact that new introduced units learn at least one input example. Tests performed with noise-free Boolean functions using the C-Mantec algorithm show that it generates very compact architectures with less number of neurons than existing constructive algorithms [20]. However, when the algorithm was tested on real datasets, it was observed that a larger number of neurons was needed because the algorithm overfit noisy examples. To avoid this overfitting problem the method introduced in the next section is developed in this work .

3 The "Resonance Effect" for Detecting Noisy Examples

We introduce in this section a method designed to eliminate instances considered noisy, as a way to increase the classification ability of predictive algorithms. It is worth mentioning that deciding whether an input example is a true input datum or a noise-contaminated one is a difficult issue that can in principle be carried out only if one knows a priori the level of noise present in the system. However, a reasonable approach is to discard suspicious noisy inputs and test the generalization ability obtained, without making claims about whether the eliminated instances are noise or not. The filtering method to be introduced is developed from an effect observed during the application of the C-Mantec algorithm to real datasets. The effect, named "resonance effect" can be exemplified by the picture displayed in Fig. 3, where an schematic drawing shows the effect that is produced when a thermal perceptron tries to learn a set of instances containing a contradictory pair of examples. In Fig. 3, the set of "good" examples is depicted in the left part of the figure, while the contradictory pair is on the right. When a single neuron tries to learn this set, the algorithm will find an hyperplane from a beam of the possible ones (indicated in the figure) that classifies correctly the whole set except for one of the noisy examples. Further

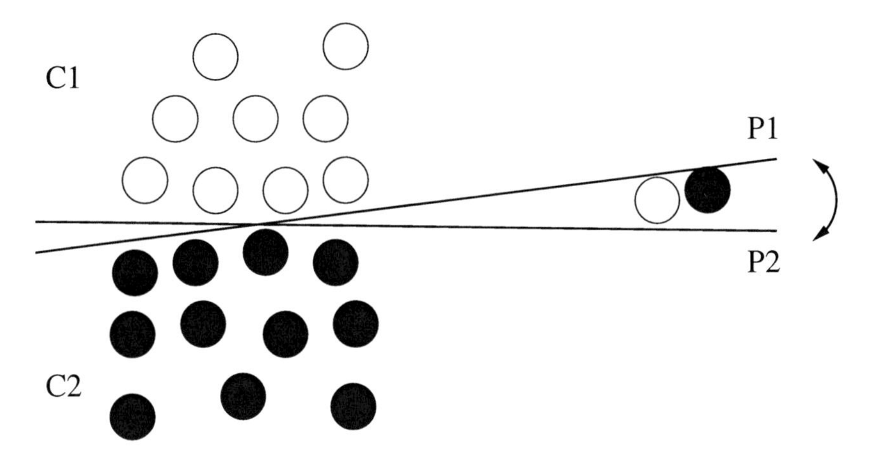

Fig. 3 Schematic drawing of the "Resonance effect" that occurs when noisy examples are present in the training set. A thermal perceptron will learn the "good" examples, represented at the left of the figure, but will classify rightly only one of the noisy samples. Further learning iterations in which the neuron tries to learn the wrongly classified example will produce an oscillation of the separating hyperplane. The number of times the synaptic weights are adjusted upon presentation of an example can be used to detect noisy inputs.

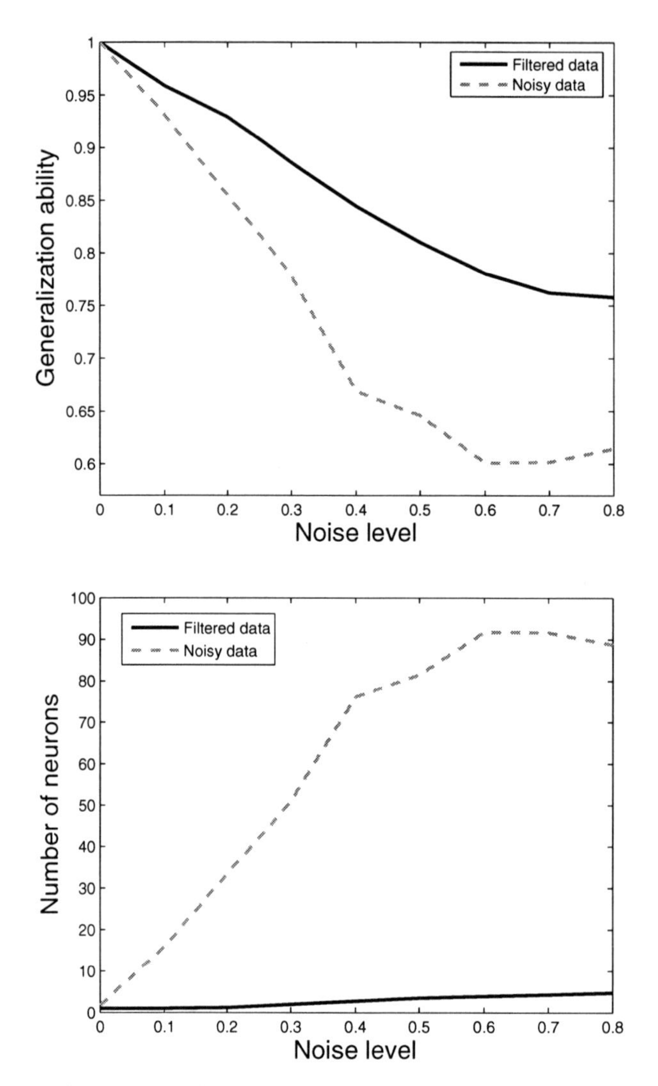

Fig. 4 The effect of adding attribute noise. Top: Generalization ability as a function of the level of attribute noise for the "modified" Pima indians diabetes dataset for the C-Mantec algorithm applied with and without the filtering stage. Bottom: The number of neurons of the generated architectures as a function of the level of noise. The maximum number of neurons was set to 101.

learning iterations produce a resonant behavior, as the dividing hyperplane oscillates trying to classify correctly the wrong example. Eventually, the iterations will end as the whole set cannot be learnt by a simple perceptron and a new neuron will be added to the network. It was observed that these noisy examples make the network to grow in excess, degrading the generalization ability. The filtering method works by counting the number of times each training example is presented to the network,

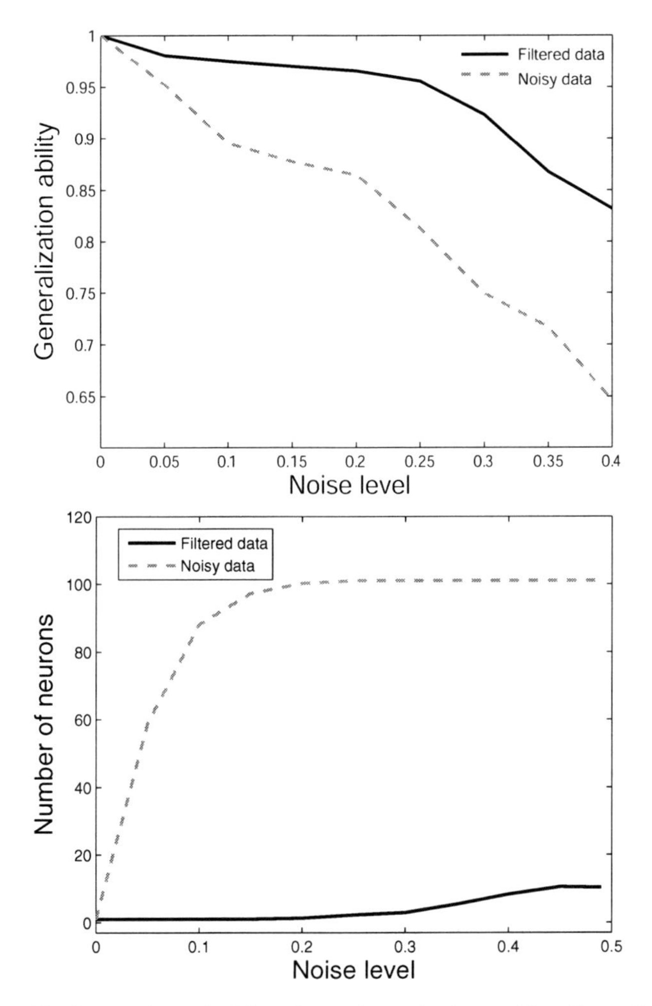

Fig. 5 The effect of adding class noise to the dataset. Top: Generalization ability as a function of the level of class noise for the modified Pima indians diabetes dataset for the cases of implementing the filtering stage and for the case of using the whole raw dataset. Bottom: The number of neurons of the generated architectures for the two mentioned cases of the implementation of the C-Mantec algorithm.

and if the number of presentations for an example is larger by two standard deviations from the mean, it is removed from the training set. The removal of examples is made on-line as the architecture is constructed and a final phase is carried out where no removal of examples is allowed.

To test the new method for removal of noisy examples a "noise-free" dataset is created from a real dataset, and then controlled noise was added to the attributes (input variables) and to the class (output), in separate experiments to analyze whether there is any evident difference between the two cases [23]. Knowing the origin of the noise is an interesting issue with practical applications, as it can help to detect the sources of noise and consequently help to eliminate it. The dataset chosen for this analysis is the Pima Indians Diabetes dataset, selected because it has been widely studied and also because it is considered a difficult set with an average generalization ability around 75%. To generate the "noise-free" dataset, the C-Mantec algorithm was run with a single neuron that classified correctly approximately 70% of the dataset, and then the "noise-free" dataset was constructed by presenting the whole set of inputs through this network to obtain the "noise-free" output. Two different experiments were carried out: in the first one, noise was added to the attributes of the dataset and the performance of the C-Mantec algorithm was analyzed with and without the procedure for noisy examples removal. In Fig. 4 (top) the generalization ability for both cases is shown for a level of noise between 0 and 0.8 and the results are the average over 100 independent runs. For a certain value of added noise, x, the input values were modified by a random uniform value between $-x$ and x. The bottom graph shows the number of neurons in the generated architectures when the filtering process was and was not applied as a function of the added attribute noise. It can be clearly seen that the removal of the noisy examples helps to obtain much more compact architectures while a better generalization ability is observed. The second experiment consisted in adding noise to the output values and the results are shown on Fig. 5. In this case the noise level indicate the probability of modifying the class value to a binary value, chosen randomly between 0 or 1.

From the experiments carried out with the two types of noise introduced to the Diabetes dataset we can observe that the resonance effect helps to detect and eliminate the noisy instances in both cases, helping to increase the generalization ability, even if the change is not enough to recover the generalization ability obtained in the noise-free case. It can also be observed that the size of the neural architectures obtained after the removal of the noisy instances is much lower than the size of the architectures needed for the noisy cases. Also, it has to be said that the experiments did not lead to a way of differentiating the sources of noise, as the results obtained for the two noise-contaminated datasets considered were not particularly different.

4 Experiments and Results on Public Domain Datasets

We tested the noise filtering abilities of the method introduced in this work using the C-Mantec constructive algorithm on a set of 11 well known benchmark functions [24]. The set of analyzed functions contains 6 two-classes functions and 5 multi-class problems with a number of classes up to 19. The C-Mantec algorithm was run with a maximum number of iterations of 50.000 and an initial temperature value (T_0) equals to the number of inputs of the analyzed functions. It is worth noting that different tests showed that the algorithm is quite robust to changes on these parameter values. The results are shown in Table 1, where it is shown the number of neurons of the obtained architectures and the generalization ability obtained, including the standard deviation values, computed over 100 independent runs. The last column of Table 1 shows, as a comparison, the generalization ability values

Table 1 Results for the number of neurons and the generalization ability obtained with the C-Mantec algorithm using the data filtering method introduced in this work. The last column shows the results from [25] (See text for more details).

Function	Inputs	Classes	Neurons	Generalization C-Mantec	Generalization NN [25]
Diab1	8	2	3.34 ± 1.11	76.62 ± 2.69	74.17 ± 0.56
Cancer1	9	2	1 ± 0.0	96.86 ± 1.19	97.07 ± 0.18
Heart1	35	2	2.66 ± 0.74	82.63 ± 2.52	79.35 ± 0.31
Heartc1	35	2	1.28 ± 0.57	82.48 ± 3.3	80.27 ± 0.56
Card1	51	2	1.78 ± 0.87	85.16 ± 2.48	86.63 ± 0.67
Mushroom	125	2	1 ± 0.0	99.98 ± 0.04	100.00 ± 0.0
Thyroid	21	3	3 ± 0.0	91.91 ± 0.59	93.44 ± 0.0
Horse1	58	3	3 ± 0.0	66.56 ± 5.08	73.3 ± 1.87
Gene1	120	3	3.03 ± 0.22	88.75 ± 1.07	86.36 ± 0.1
Glass	9	6	17.84 ± 1.19	63.75 ± 6.38	53.96 ± 2.21
Soybean	82	19	171 ± 0.0	91.63 ± 1.89	90.53 ± 0.51
Average	50.27	4.18	18.99 ± 0.43	84.21 ± 2.03	82.50 ± 0.63

obtained by Prechelt [25] in a work where he analyzed in a systematic way the prediction capabilities of different topologies neural networks. The size of training and test sets were chosen in a similar way in both compared cases: the training set comprises 75% of the total number of instances and the remaining 25% was used for testing the generalization ability. The results obtained with the C-Mantec algorithm outperforms the ones obtained by Prechelt in 6 out of 11 problems and on average the generalization ability is 2.1% larger. Regarding the size of the networks obtained using the new method introduced, the architectures are very small for all problems with 2 or 3 classes, for which the architectures contain less than 4 neurons. For multi-class problems the algorithm generates networks with a larger number of hidden neurons but this is because of the method used to treat multiclass problems that will be reported elsewhere [20].

A further set of experiments was carried out using as classification algorithms other standard methods in machine learning. Three different available algorithms were used and tested on the original datasets and on the filtered dataset where the noisy examples were eliminated. The three algorithms used were standard multilayer perceptrons (MLP) trained by backpropagation, Support Vector Machines (SVM) [27] and the C4.5 algorithm based on decision trees (C4.5) [26], all implemented under the WEKA package ([28]) using the default parameter settings. The results are presented in table 2 where the generalization ability obtained for six different datasets are shown for the two cases considered: filtered and original datasets. It can be observed that when multilayer perceptron are used, in all analyzed cases the generalization ability obtained with the filtered dataset was larger than with the original set and the difference was on average larger by a 1.31%, with values in some cases as large as 2.32%. For the case of the C4.5 algorithm the results with and without filtering instances were similar with an average difference of 0.18% in favor of the filtered case. When the support vector machines were tested with both

Table 2 Results for generalization ability obtained using standard multilayer perceptrons (MLP), decision trees (C4.5) and support vector machines (SVM) algorithms using both the filtered and original datasets (See the text for more details).

	Filtered data			Original data		
	MLP	C4.5	SVM	MLP	C4.5	SVM
Diab1	75.63 ± 4.21	75.21 ± 2.85	76.36 ± 3.43	74.58 ± 2.36	73.54 ± 2.78	77.81 ± 1.57
Cancer1	95.06 ± 0.78	93.91 ± 2.05	95.63 ± 1.00	94.84 ± 1.28	93.93 ± 1.83	95.90 ± 0.95
Heart1	82.49 ± 2.87	79.44 ± 1.15	81.80 ± 2.70	80.17 ± 2.23	78.70 ± 2.94	81.74 ± 2.38
Heartc1	83.20 ± 4.43	78.13 ± 3.22	82.13 ± 4.58	81.05 ± 2.70	79.47 ± 5.23	84.74 ± 4.21
Card1	85.81 ± 2.50	85.81 ± 3.68	86.74 ± 2.71	83.72 ± 1.81	85.93 ± 2.55	86.05 ± 1.90
Mushroom	100.00 ± 0.00	99.95 ± 0.05	100.00 ± 0.00	100.00 ± 0.00	99.86 ± 0.07	100.00 ± 0.00
Average	87.04 ± 3.65	85.41 ± 3.99	87.11 ± 3.69	85.73 ± 3.95	85.23 ± 4.09	87.71 ± 3.48

datasets, the generalization ability observed decreases with the filtered instances in average by approximately 0.61%, but noting that in 2 out of the 6 cases considered the prediction improved.

Regarding the generalization ability obtained by the different methods, we first note that the average generalization ability for the 6 functions shown in table 2 is of 87.29 ± 3.72 for the C-Mantec algorithm with the active learning procedure incorporated. Thus, the best method with these limited set of 6 functions turns out to be the SVM approach, close followed by the constructive C-Mantec algorithm and by the MLP; while the C4.5 came last with a lower generalization ability.

The number of instances in the filtered datasets was on average 2.73% smaller than the original sets, being the smaller ones found in those for which the generalization ability was lower, as for Diabetes dataset. The standard deviation of the results shown in tables 1 and 2 is computed over 5 randomly selected datasets, using 75% of the examples for training the models and the remaining 25% for testing the generalization ability.

5 Discussion

We introduced in this chapter a new method for filtering noisy examples using a recently developed constructive neural network algorithm. The new C-Mantec algorithm generalizes very well on free-noise dataset but have shown to overfit with noisy datasets and thus, a filtering scheme for noisy instances have been implemented. The filtering method devised is based on the observation that noisy examples needs more number of weights updates than regular ones. This "resonant effect" observed, permits to distinguish these instances and eliminate them in an on-line procedure. Simulations performed show that the generalization ability and size of the resulting networks are very much improved after the removal of the noisy examples. A comparison of results was done against previous reported values obtained using standard feed-forward neural networks [25] and showed that the generalization ability was on average a 2.1% larger, indicating the effectiveness of the C-Mantec algorithm implemented with the new filtering stage. The introduced method of data selection can also be used as a pre-processing stage for other prediction algorithms, and for this reason a second comparison was carried out using three well known predictive algorithms: MLP, C4.5 decision trees and SVM. The results obtained and shown in table 2 indicate that the instance selection procedure appears to work quite well with MLP and less with the other two algorithms. It might be possible, given the neural nature of the C-Mantec algorithm, that the filtering stage developed works better with neural-based algorithms but further studies might be needed to extract a final conclusion. Overall we have observed that the active learning procedure implemented using the new C-Mantec algorithm is working very efficiently in the task of avoiding overfitting problems and that comparable results to those obtained using MLP's and SVM's can be obtained with a constructive neural network algorithm.

Acknowledgements. The authors acknowledge support from CICYT (Spain) through grants TIN2005-02984 and TIN2008-04985 (including FEDER funds) and from Junta de Andalucía through grants P06-TIC-01615 and P08-TIC-04026. Leonardo Franco acknowledges support from the Spanish Ministry of Science and Innovation (MICIIN) through a Ramón y Cajal fellowship.

References

1. Haykin, S.: Neural Networks: A Comprehensive Foundation. Macmillan/IEEE Press (1994)
2. Lawrence, S., Giles, C.L., Tsoi, A.C.: What Size Neural Network Gives Optimal Generalization? Convergence Properties of Backpropagation. In: Technical Report UMIACS-TR-96-22 and CS-TR-3617, Institute for Advanced Computer Studies, Univ. of Maryland (1996)
3. Gómez, I., Franco, L., Subirats, J.L., Jerez, J.M.: Neural Networks Architecture Selection: Size Depends on Function Complexity. In: Kollias, S.D., Stafylopatis, A., Duch, W., Oja, E. (eds.) ICANN 2006. LNCS, vol. 4131, pp. 122–129. Springer, Heidelberg (2006)
4. Mezard, M., Nadal, J.P.: Learning in feedforward layered networks: The tiling algorithm, J. Physics A 22, 2191–2204 (1989)
5. Frean, M.: The upstart algorithm: A method for constructing and training feedforward neural networks. Neural Computation 2, 198–209 (1990)
6. Parekh, R., Yang, J., Honavar, V.: Constructive Neural-Network Learning Algorithms for Pattern Classification. IEEE Transactions on Neural Networks 11, 436–451 (2000)
7. Subirats, J.L., Jerez, J.M., Franco, L.: A New Decomposition Algorithm for Threshold Synthesis and Generalization of Boolean Functions. IEEE Transactions on Circuits and Systems I 55, 3188–3196 (2008)
8. Nicoletti, M.C., Bertini, J.R.: An empirical evaluation of constructive neural network algorithms in classification tasks. International Journal of Innovative Computing and Applications 1, 2–13 (2007)
9. Reed, R.: Pruning algorithms - a survey. IEEE Transactions on Neural Networks 4, 740–747 (1993)

10. Smieja, F.J.: Neural network constructive algorithms: trading generalization for learning efficiency? Circuits, systems, and signal processing 12, 331–374 (1993)
11. Bramer, M.A.: Pre-pruning classification trees to reduce overfitting in noisy domains. In: Yin, H., Allinson, N.M., Freeman, R., Keane, J.A., Hubbard, S. (eds.) IDEAL 2002. LNCS, vol. 2412, pp. 7–12. Springer, Heidelberg (2002)
12. Hawkins, D.M.: The problem of Overfitting. Journal of Chemical Information and Computer Sciences 44, 1–12 (2004)
13. Angelova, A., Abu-Mostafa, Y., Perona, P.: Pruning training sets for learning of object categories. In: IEEE Computer Society Conference on Computer Vision and Pattern Recognition, CVPR 2005, vol. 1, pp. 494–501 (2005)
14. Cohn, D., Atlas, L., Ladner, R.: Improving Generalization with Active Learning. Mach. Learn. 15, 201–221 (1994)
15. Cachin, C.: Pedagogical pattern selection strategies. Neural Networks 7, 175–181 (1994)
16. Kinzel, W., Rujan, P.: Improving a network generalization ability by selecting examples. Europhys. Lett. 13, 473–477 (1990)
17. Franco, L., Cannas, S.A.: Generalization and Selection of Examples in Feedforward Neural Networks. Neural Computation 12(10), 2405–2426 (2000)
18. Sánchez, J.S., Barandela, R., Marqués, A.I., Alejo, R., Badenas, J.: Analysis of new techniques to obtain quality training sets. Pattern Recognition Letters 24, 1015–1022 (2003)
19. Jankowski, N., Grochowski, M.: Comparison of Instances Seletion Algorithms I. Algorithms Survey. In: Rutkowski, L., Siekmann, J.H., Tadeusiewicz, R., Zadeh, L.A. (eds.) ICAISC 2004. LNCS (LNAI), vol. 3070, pp. 598–603. Springer, Heidelberg (2004)
20. Subirats, J.L., Franco, L., Jerez, J.M.: Competition and Stable Learning for Growing Compact Neural Architectures with Good Generalization Abilities: The C-Mantec Algorithm (2009) (in preparation)
21. Frean, M.: Thermal Perceptron Learning Rule. Neural Computation 4, 946–957 (1992)
22. Rosenhlatt, F.: The perceptron: A probabilistic model for information storage and organization in the brain. Psychological Review 65, 386–408 (1959)
23. Zhu, X., Wu, X.: Class noise vs. attribute noise: a quantitative study of their impacts. Artif. Intell. Rev. 22, 177–210 (2004)
24. Merz, C.J., Murphy, P.M.: UCI Repository of Machine Learning Databases. Department of Information and Computer Science. University of California, Irvine (1998)
25. Prechelt, L.: Proben 1 – A Set of Benchmarks and Benchmarking Rules for Neural Network Training Algorithms. Technical Report (1994)
26. Quinlan, J.R.: C4.5: Programs for Machine Learning. Morgan Kauffman, CA (1992)
27. Shawe-Taylor, J., Cristianini, N.: Support Vector Machines and other kernel-based learning methods. Cambridge University Press, Cambridge (2000)
28. Witten, I.H., Frank, E.: Data Mining: Practical Machine Learning Tools and Techniques with Java Implementations. Morgan Kaufmann Publishers, San Francisco (2000), http://www.cs.waikato.ac.nz/ml/weka

Incorporating Expert Advice into Reinforcement Learning Using Constructive Neural Networks

Robert Ollington, Peter Vamplew, and John Swanson

Abstract. This paper presents and investigates a novel approach to using expert advice to speed up the learning performance of an agent operating within a reinforcement learning framework. This is accomplished through the use of a constructive neural network based on radial basis functions. It is demonstrated that incorporating advice from a human teacher can substantially improve the performance of a reinforcement learning agent, and that the constructive algorithm proposed is particularly effective at aiding the early performance of the agent, whilst reducing the amount of feedback required from the teacher. The use of constructive networks within a reinforcement learning context is a relatively new area of research in itself, and so this paper also provides a review of the previous work in this area, as a guide for future researchers.

1 Introduction

Reinforcement learning is a learning paradigm in which an autonomous agent learns to execute actions within an environment in such a way as to maximise the reinforcement which it receives from the environment. Scaling reinforcement learning to large, complex problems is an ongoing area of research, and this paper deals with the combination of two approaches which have been applied to this scaling issue – the use of constructive neural networks, and the incorporation of human guidance into the learning process. Specifically we propose and empirically test a novel algorithm for utilising human advice within a reinforcement learning system, which exploits the properties of a particular constructive neural network known as the Resource Allocating Network.

Robert Ollington and John Swanson
School of Computing and Information Systems, University of Tasmania, Tasmania, Australia

Peter Vamplew
Center for Informatics and Applied Optimisation, School of Information Technology and Mathematical Sciences, University of Ballarat, Victoria, Australia

L. Franco et al. (Eds.): Constructive Neural Networks, SCI 258, pp. 207–224.
springerlink.com

As many readers of this volume may not previously be familiar with reinforcement learning, Section 2 will present a brief introduction to this learning paradigm – for a more thorough review the reader is directed to the excellent textbook by Sutton and Barto (1998). This section will also discuss the relationship between reinforcement learning and function approximation methods such as neural networks, whilst Section 3 will review the previous research on utilising constructive neural networks within a reinforcement learning context. The following sections examine how a reinforcement learning agent can be aided in learning a difficult task. Section 4 reviews existing approaches to guiding an agent during learning, whilst Section 5 proposes a new algorithm for incorporating human advice into the learning process, based on the use of a constructive radial basis function network. Section 6 documents experiments carried out to assess the effectiveness of this new algorithm relative to alternative approaches. Section 7 offers conclusions and suggestions for future work.

2 Reinforcement Learning and Function Approximation

Reinforcement learning addresses the problem of an autonomous agent learning to perform a particular task whilst interacting with an environment. The agent is required to carry out a series of actions which will lead to the overall task being achieved. Unlike other forms of machine learning such as supervised learning, the agent is given no direct instructions on the decisions it should make and the actions which it should select to perform. Instead it receives indirect feedback in the form of a reward signal, which is highest when the desired results are achieved. At each time-step the agent observes the current state and selects an action. The action is executed, which may change the environment, and the agent receives a scalar reward[1] and observes the new state of the environment.

The agent attempts over time to learn a mapping from state-to-action to maximise the long-term reward. This mapping, which determines the behaviour of the agent, is known as its *policy*. Policies which achieve higher long-term rewards will be favoured over those achieving lesser rewards, and in this way the reward signal guides the agent towards behaviour which is desirable, without explicitly instructing the agent on which actions to take in a specific state. This leaves the agent free to discover effective policies which may previously have been unknown. In contrast a system trained using supervised learning methods will be limited to reproducing the policy embedded in the training data provided to it. Perhaps the best-known example illustrating this difference between reinforcement learning and supervised learning is the work on backgammon carried out by Tesauro (1995). An agent trained to play backgammon using reinforcement learning methods over a large number of games played against itself (TDGammon) was shown to significantly outperform another agent trained using supervised learning on a

[1] In fact, whilst the majority of reinforcement learning signals use a scalar reward signal, tasks which require the agent to balance multiple conflicting objectives may be dealt with by using a vector reward, with an element for each distinct objective – see for example (Vamplew et al. 2008).

large data-set of game states labelled by a human expert. In fact, TDGammon achieved an extremely high level of play, comparable to that of human experts. Of course, the benefits of reinforcement learning do not come without cost – the reinforcement learning task is considerably harder and slower than supervised learning.

One common approach to reinforcement learning is to learn the expected return associated with each state and action. Once these state-action values have been learnt, an optimal policy can be found simply by following the greedy policy of selecting at each state the action which has the highest expected return for that state. Many of the algorithms for learning these values are based on the use of the method of temporal differences (TD) where the value of the current state at each step is used to update the estimated value of previous states (Sutton, 1988).

In order for TD-based methods to work successfully, it is important that all states and actions are sampled a large number of times – in fact formal proofs of the convergence of TD algorithms generally require the assumption that all states and actions are visited an infinite number of times. Whilst clearly this can not be achieved in practice, experience has shown that excellent results can still be obtained on many tasks as long as a degree of exploration is implemented by the agent. That is to say, the agent must not simply follow the greedy policy dictated by its current estimate of the state-action values, as this could easily lead to it become trapped in a sub-optimal policy. Instead on some time-steps exploratory, non-greedy actions must be selected. Two broad categories of TD methods exist – on-policy methods such as Q-learning (Watkins and Dayan, 1992) in which the values learnt are those corresponding to the greedy policy (i.e. ignoring the exploratory actions), and off-policy methods such as SARSA (Rummery and Niranjan, 1994) in which the values learnt are for the policy actually being followed by the agent (i.e. including the exploratory actions).

The earliest work on TD methods examined simple problems with relatively small, discrete state-spaces, which allowed the state-action values to be stored in a lookup table. However the storage requirements of these tabular methods rapidly become impractical as the dimensionality or resolution of the state associated with the problem environment increases. In addition learning can be extremely slow, as tabular algorithms can only learn about states and actions which the agent has experienced. Therefore in order to scale TD methods to larger, more complex tasks function approximation must be used to estimate the values. The number of parameters required by the function approximator is generally far fewer than the number of discrete states, thereby reducing storage requirements. In addition function approximators can generalise from states which have been experienced to similar states that are yet to be visited, which can substantially increase the rate of learning.

Two main approaches to function approximation have been explored in the reinforcement learning literature. One involves the use of global approximators such as neural networks. These have been applied successfully to a range of problems, such as elevator control (Crites and Barto, 1996) and the previously mentioned work on backgammon (Tesauro, 1996). However this success has failed to be replicated on other, seemingly similar tasks. The second approach is to use locally sensitive approximators such as CMACs (Sutton, 1996) or radial-basis functions (Kretchmar and Anderson, 1997). The local approach has been shown to be both more stable and more amenable to formal analysis, but may scale less well to higher-dimensional input spaces (Coulom, 2002).

Function approximation is, of course, not specific to the field of reinforcement learning, and many different approaches to function approximation have previously been investigated within the context of supervised learning. One of the most promising approaches (as evidenced by the existence of this volume!) is the use of constructive neural networks. Unlike many other forms of function approximation (including fixed-architecture neural networks), constructive networks do not have a fixed structure. They generally start with a minimal architecture and add processing units (neurons) during the training process, thereby tailoring their structure to the demands of the task being learnt. A wide variety of constructive algorithms have been proposed so, rather than replicate material which will be covered in more detail elsewhere in this volume, the next section of this paper will focus on the particular constructive network algorithms which have so far been applied within the context of reinforcement learning.

3 Reinforcement Learning Using Constructive Neural Networks

3.1 Motivation for Using Constructive Networks in Reinforcement Learning

Various studies have previously identified that constructive neural networks can be an extremely effective tool for function approximation, both in terms of the speed of learning and the accuracy of their approximation (see for example, Prechelt (1997). This in itself would make them of interest for application in reinforcement learning tasks. However there are also a number of other reasons why constructive approaches may be particularly beneficial for reinforcement learning.

Thrun and Schwartz (1993) provides both a theoretical argument and empirical evidence that function approximators are prone to systematic overestimation of values when used with temporal difference algorithms (particularly for off-policy algorithms). Whilst not directly advocating the use of constructive networks, this work argues that function approximators with a bounded memory (such as a fixed-architecture neural network) are less likely to overcome this systematic bias than those with an unbounded memory (such as a constructive network). Therefore it may be that constructive algorithms can overcome some of the failures which have been observed when fixed forms of function-approximation have been applied to large, complex reinforcement learning tasks.

Whilst Tesauro (1996) achieved exceptional results in backgammon from the combination of temporal difference learning and fixed multi-layer perceptron neural networks, he suggests the use of cascade-correlation networks as a future development of TD-Gammon, to avoid the need to completely retrain the network when new input features are added. More generally, constructive algorithms are well-suited to tasks which require building on previously acquired knowledge. This aspect of constructive networks forms the rationale for the experimental work reported later in this paper, and so we will explore this issue in greater detail in Sections 4 and 5.

Finally we observe that the fundamental characteristic of reinforcement learning is that it aims to produce autonomous learning agents that can adapt to carry out tasks in an unknown environment with relatively little guidance. Using a constructive neural network as the function approximator allows the agent to determine its own internal structure, thereby further increasing its level of autonomy.

3.2 Previous Work on Reinforcement Learning Using Constructive Networks

3.2.1 Locally Responsive Constructive Reinforcement Learning

The first application of a constructive algorithm within a reinforcement learning context was probably by Anderson (1993) who adapted the constructive Resource Allocating Network (RAN) (Platt 1991) for use within the Q-learning algorithm. In a strict sense Anderson's RAN was no longer a constructive algorithm, as it used a fixed number of hidden nodes. Rather than adding a new node when an unexpectedly large error was detected, the algorithm identified the least useful current hidden node, and reinitialised its weights to correct for the current error.

Since Anderson's pioneering work, constructive networks for reinforcement learning based on radial basis functions have been further developed and applied by a number of other authors. For example, Jun and Duckett (2005) applied a combination of Q-learning and a Resource Allocating Network to train a robot to carry out a wall-following task. An interesting observation of this work is that the temporal difference error fails to converge to zero during learning, and in fact oscillates over the entire training period. This illustrates an unusual aspect of function approximation within TD algorithms - unlike most regression tasks, in this case the absolute accuracy of the function approximator is less relevant than its relative accuracy. As long as the optimal action for each state is valued higher than the other actions then the optimal policy will be followed.

Santos and Touzet (1999) also applied a constructive radial basis function network to a robotic navigation task (wall following). Their work differs from that of Li and Duckett in that it is not based on temporal difference methods for learning state-action values – instead the network is trained to directly perform a mapping from state to actions. Hence in this work the constructive network is being used as a classifier rather than as a function approximation system.

An issue which arises in many systems based on radial basis functions is the setting of appropriate widths for the basis functions. A constructive reinforcement learning system which neatly addresses this problem was proposed by Šter and Dobnikar (2003). Their Adaptive Radial Basis Decomposition algorithm implements two different methods of adding new hidden neurons. If a large error is detected in a currently unoccupied region of state space, a new neuron is added much as in a conventional RAN. In addition the local TD error is monitored for each hidden neuron, and if this is found to be high the neuron is decomposed, and replaced by two neurons with narrower widths. This algorithm is particularly effective in problems where the value function is smooth in some areas but with large local variations in other regions of the state space.

3.3.2 Globally-Responsive Constructive Reinforcement Learning

All of the systems described above use neurons with radial basis functions, where each neuron responds only to inputs within localised regions of the state-space. The application in reinforcement learning of constructive networks based on neurons with non-local response functions has been minimal, despite this style of system being widely and successfully applied within the supervised learning field. Many constructive algorithms have been proposed in the supervised learning literature, but amongst the most widely adopted has been Cascade-Correlation (Cascor) (Fahlman and Lebiere, 1990). This constructive algorithm, based on non-localised neurons, has been shown to equal or outperform fixed-architecture networks on a wide range of supervised learning tasks (Fahlman and Lebiere, 1990; Waugh 1995) and its possible utility for reinforcement learning was first identified by Tesauro (1993). Despite this, the first work using a cascade constructive network for reinforcement learning appears to be by (Rivest and Precup (2003); Bellemare, Precup and Rivest, (2004)). Possibly this is explained by the added difficulty in incorporating this style of network into a reinforcement learning environment. Whereas the RAN was initially designed for on-line learning where the network is updated after each input, Cascade-Correlation is usually used in conjunction with batch training algorithms such as Quickprop. The approach used by this group was to modify the reinforcement learning process so as to allow direct application of the Cascor algorithm. They propose a learning algorithm with two alternating stages. In the first stage the agent selects and executes actions, and stores the input state and the target value generated via TD in a cache. Once the cache is full, a network is trained on the cached examples, using the standard Cascor algorithm. Once the network has been trained, the cache is cleared and the algorithm returns to the cache-filling phase.

In contrast Vamplew and Ollington (2005) adapted the training algorithm for the cascade network to use simple gradient descent backpropagation, thereby allowing it to be trained in a fully on-line fashion within the temporal difference algorithm with no need for caching. They propose the use of parallel candidate training whereby the weights for both the output and candidate neurons in the cascade network are updated after each interaction with the environment. This eliminates the need to maintain a cache, and ensures that the policy is updated immediately after the results of each action are known. However a possible disadvantage of parallel candidate training is the issue of moving targets. The temporal difference errors used as targets for the candidate nodes are themselves changing during training as the weights of the output neurons are adapted. Therefore the task facing the candidate neurons is more complex than it would be if these values were static.

Nissen (2007) provides the most extensive examination so far of the use of cascade networks for reinforcement learning, and introduces algorithms which attempt to find a middle-ground between the two previous approaches. Nissen's algorithms allow batched training algorithms such as Quickprop to be applied in an on-line fashion by dynamically maintaining a '*sliding window*' cache of the most recent states, actions and temporal difference errors.

Most recently Girgin and Preux (2008) have explored the use of Cascade-Correlation networks in conjunction with the Least Squares Policy Iteration reinforcement learning algorithm. In this case the candidate neurons trained using the cascade-correlation algorithm form the basis features for the LSPI algorithm, which forms a linear weighting of these basis features.

3.3 Summary of Previous Work

At this point in time very little evidence exists on which to judge the relative merits of the different approaches to constructive reinforcement learning. No large-scale comparative studies have been performed, and there is little overlap in the sample problems used by different authors.

Vamplew and Ollington provide results comparing their cascade algorithm against a RAN-based reinforcement learning algorithm on a small set of simple test problems. These results show that in some cases the cascade approach can perform comparably or even slightly better than the RAN system, whilst producing a much more compact network. However the cascade network is less stable, and has much poorer worst-case performance than the RAN. Similarly Precup, Rivest and Bellemare report promising results for tic-tac-toe and car-rental tasks, but their system did not perform well on the more complex backgammon task. Nissen's system demonstrates improved performance on the backgammon task, but is still very sensitive to some of the parameter settings.

The ability of the globally-responsive networks to learn policies based on a far smaller number of neurons would appear to hold potential for scaling to more complex and higher-dimensional state spaces. However in practice the instability of the learning algorithms based on these networks means this potential has yet to be fulfilled.

A possible clue to the cause of this instability may lie in the observations made by Baird (1995) regarding the potential for divergence when even simple linear function approximators are combined with the TD algorithm. The 'bootstrapping' nature of TD, in which the estimated value of the current state-action pair is updated based on an error derived from the estimate of the next state-action can lead to instability when those estimates are based on common weights. As the weights are modified to improve the estimate for the current state and action, this may in fact also affect the estimated value of the next state and action. If these states are revisited in the future, this can easily lead to an endless cycle driving both estimates towards infinite values. Clearly this problem is more likely to arise if temporally adjacent states share similar features – this is more likely to occur in a network in which the hidden neurons have globally responsive behaviour than in a network with locally-responsive hidden neurons. Baird (1995) has proposed the residual gradient algorithm as a means of addressing this issue of diverging values. The application of residual gradient learning in combination with cascade networks remains an interesting possibility for future research.

4 Guiding Reinforcement Learning

One of the major tasks facing reinforcement learning researchers is to develop techniques which effectively and efficiently scale to large, complex problems. Whilst there have been some spectacular successes such as TDGammon, often algorithms which been effective on simple test problems have failed when applied to more difficult, real-world applications.

Learning to perform a task based only on a simple reinforcement signal with no other guidance may simply be too difficult to be practical for complex problems. Therefore a variety of methods have been proposed to aid the agent in this learning process. Transfer of knowledge between tasks (Perkins and Precup, 1999) and shaping (Randlov and Alstrom, 1998) aim to allow the agent to apply learning from previous, possibly simpler, tasks to the task currently being learnt. Another way to aid the agent is to allow it to initially learn the values of policies being executed by a human expert or another program, before entering a reinforcement learning phase during which it may improve on these initial policies (Nechyba and Bagnell, 1999). All of these approaches require the agent to be able to build on previously acquired knowledge. Constructive networks are well suited to this style of learning, as they have the ability to extend their structure to support new knowledge, whilst retaining the learnt knowledge encapsulated in their existing structure and weights, and so some attempts have been made to use them in this manner within reinforcement learning tasks.

Both Jun and Duckett (2005) and Großmann and Poli (1998) have utilised constructive networks within robot learning applications in which the robot first learns to mimic actions selected by a human expert, before using reinforcement learning methods to find an improved policy. Jun and Duckett use a RAN as their choice of network, whilst Großmann and Poli's system is based on a specialised high-order constructive network known as the Temporal Transition Hierarchy. Nissen (2007) discusses the possible extension of the NFQ-SARSA(λ) algorithm to this combination of reinforcement learning and '*learning-by-example*', but does not actually implement or test this approach.

One of the difficulties faced when using reinforcement learning algorithms is the almost random nature of exploration in the early stages of training. This is particularly problematic when the agent has a manifestation in the real world or when there is some real cost associated with poor action selection. In many such cases, a human (or non-human) expert may have already learned a reasonable, if not optimal, solution to the problem. It would be nice to be able to incorporate that knowledge into the reinforcement learning agent.

One approach to achieving this is to allow an external expert to provide advice to the learning agent (Clouse and Utgoff, 1991, 1992; Maclin and Shavlik, 1996; Papudesi and Huber, 2003; Maclin et al., 2005). There are two questions that need to be answered when designing a RL agent that receives advice - "how should the advice be represented?" and "how should the advice be utilized?".

Ideally advice would be represented in a manner that is easy to describe and in a format that is easily understood by the expert. The simplest method that satisfies these criteria is for the expert to observe a situation and suggest the action that they would perform in that situation. Such a system has been used successfully by Clouse and Utgoff (1992). However, incorporating this form of advice directly into the agent is not as straightforward. Clouse and Utgoff (1992) incorporated the advice by forcing the agent to perform the suggested action, and providing an additional bonus reward to encourage similar behaviour in the future. This method has two main disadvantages. Firstly, there is no guarantee that the bonus reward provided will be enough to make this action more favourable than alternative actions. Secondly, it is likely that such a method will result in incorrect action values being learnt by the agent – and unlearnt when the expert stops providing advice.

Therefore we have investigated an alternative approach to providing advice based on directly adding neurons to a constructive network when advice is provided to the agent. Section 5 describes this new algorithm, whilst Section 6 provides experimental results comparing it against an agent learning without advice, and an agent learning via the 'bonus reward' approach to advice-giving.

5 Providing Expert Advice via a Resource Allocating Network

Using a RAN (or other locally constructive algorithm) for function approximation, as described in the previous sections, allows a more direct method of incorporating advice that results in faster and more robust learning. The method, which we will refer to as the *difference advice method*, is implemented as follows.

If advice is presented to the agent, and that advice contradicts the agent's own action choice for the current state, the error novelty criterion for the RAN is considered to be satisfied automatically, and the 'error' for that action is set to a value that will make the action more favourable than the agent's own action choice. The error for actions other than the advised action is set to zero. This has the effect of immediately making this action the preferred choice for states that are locally similar to the current state, without affecting action values for dissimilar states. In addition, assuming the expert's action choice is better than the agent's preferred action, the action value learnt is at least no more incorrect than the learnt value for the agent's own action choice. Figure 1 provides a formal description of the integration of this difference advice method into the Q-learning framework.

6 Experimental Methodology and Results

To test this new algorithm for providing expert advice to a reinforcement learning agent, experiments were conducted for a simple car-racing game, where a human expert could clearly visualise the current state and provide on-line advice using keyboard input. The simulation environment was configured to ensure the learning agent was able to learn the problem domain sufficiently enough to draw strong

$$\begin{array}{l}
\textit{initialise network, initialise current state } \vec{s} \\
\textit{repeat} \\
\quad \textit{compute } Q(\vec{s},a): \quad \forall j: h_j = \exp\left[-(\vec{s}-\vec{c}_j)^2 / \omega_j^{\,2}\right] \\
\quad \qquad \forall a: Q(\vec{s},a) = \sum_j w_{ja} h_j \\
\quad \textit{choose next action } a \textit{ via } \varepsilon-\textit{greedy action selection} \\
\quad \textit{if advice received} (a^* = \textit{advised action}): \\
\qquad \textit{if } Q(\vec{s},a^*) < Q(\vec{s},a): \\
\qquad\quad \textit{if } \min_j(|\vec{s}-\vec{c}_j|) > \textit{dist threshold} \\
\qquad\qquad \textit{add new node } j' \textit{ at } \vec{s}, \textit{with } \omega_{j'} = \rho \min_j(|\vec{s}-\vec{c}_j|) \\
\qquad\qquad \textit{setting } w_{j'k} = Q(\vec{s},a) - Q(\vec{s},a^*), \textit{ for } k = a^*; \\
\qquad\qquad\qquad w_{j'k} = 0, \textit{ for } k \neq a^* \\
\qquad a = a^*,\ Q(\vec{s},a) = Q(\vec{s},a) + q \\
\quad \textit{take action } a, \textit{observe } r, s' \\
\quad \textit{calculate error}: \delta = r + \gamma \max_a Q(\vec{s}',a) - Q(\vec{s},a) \\
\quad \textit{update trace}: \forall j,k: \quad \textit{if } k = a, \textit{and } h_j > e_{jk} : e_{jk} = h_j \\
\qquad\qquad\qquad \textit{else}: e_{jk} = \lambda \gamma e_{jk} \\
\quad \textit{if } \delta > \textit{error threshold}, \textit{and } \min_j(|\vec{s}-\vec{c}_j|) > \textit{dist threshold} \\
\qquad \textit{add new node } j' \textit{ at } \vec{s}, \textit{with } \omega_{j'} = \rho \min_j(|\vec{s}-\vec{c}_j|) \\
\qquad \textit{setting } w_{j'k} = \delta, \textit{ for } k = a;\ w_{j'k} = 0, \textit{ for } k \neq a \\
\quad \textit{else update weights}: \\
\qquad \forall j,k: \quad w_{jk} = w_{jk} + \alpha w_{jk} e_{jk} \delta \\
\quad \vec{s} = \vec{s}'
\end{array}$$

Fig. 1 The Q-Learning algorithm using a Resource Allocating Network, incorporating the difference advice method (α is the learning rate, ρ is the overlap parameter, h indicates a hidden node weight, w indicates an output weight).

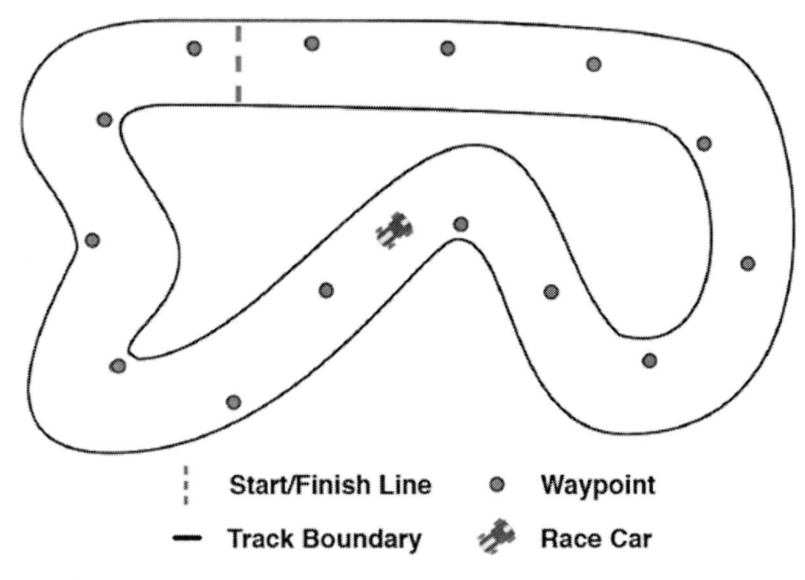

Fig. 2 The test environment, showing the track boundaries, and the waypoint configuration.

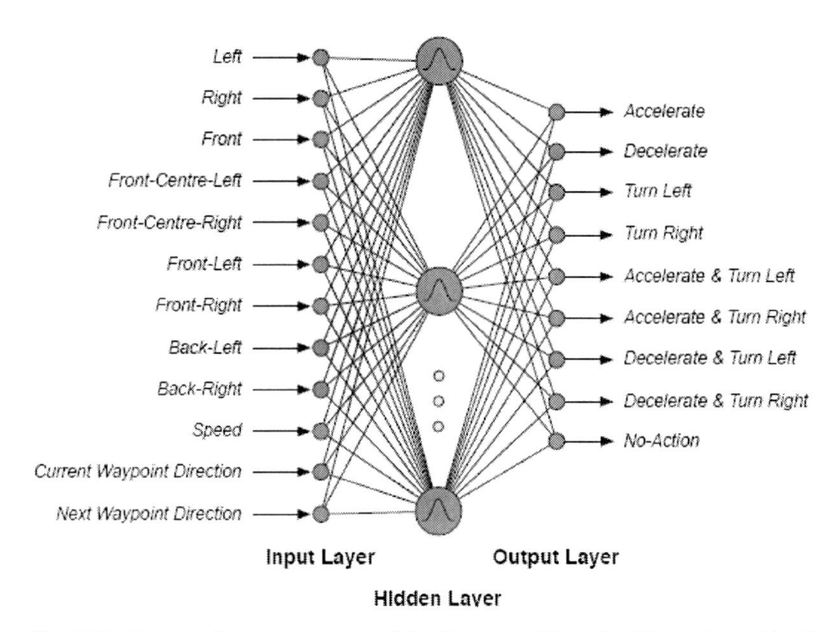

Fig. 3 The input and output structure of the Resource Allocating Network used in all trials.

conclusions on the effectiveness between an RL agent that receives advice and one that does not.

The simulation problem domain is a car navigation task loosely based on the Race Track game (Gardner 1973). Adaptions of this problem domain have also been used in other advice systems (Clouse and Utgoff 1992, Clouse 1995). The track can be described as being of any length and shape with a left boundary, a right boundary, a start line, a finish line and way-points in between. Figure 2 depicts an outline of the track employed in the simulation environment. In our adaption the objective is to drive the car from the start line to the finish line, while avoiding colliding with the track walls and minimising lap time. The reward signal combines a positive reward based on progress around the track and a negative term for collisions.

Figure 3 shows the input and output configuration of an agent RAN for this task. There are 12 inputs for state information and 9 outputs for action Q-values. The first nine inputs correspond to a set of sensors that record the relative placement of the car between track boundaries. The sensor configuration can be seen in Figure 4. The tenth input value corresponds to the car's speed, where a positive value means that the car is travelling forward and a negative value means the car is travelling in reverse. The eleventh input is the direction to the nearest way-point. The final input corresponds to the direction to the second nearest way-point. These directional values were provided to encourage smoother movement between way-points.

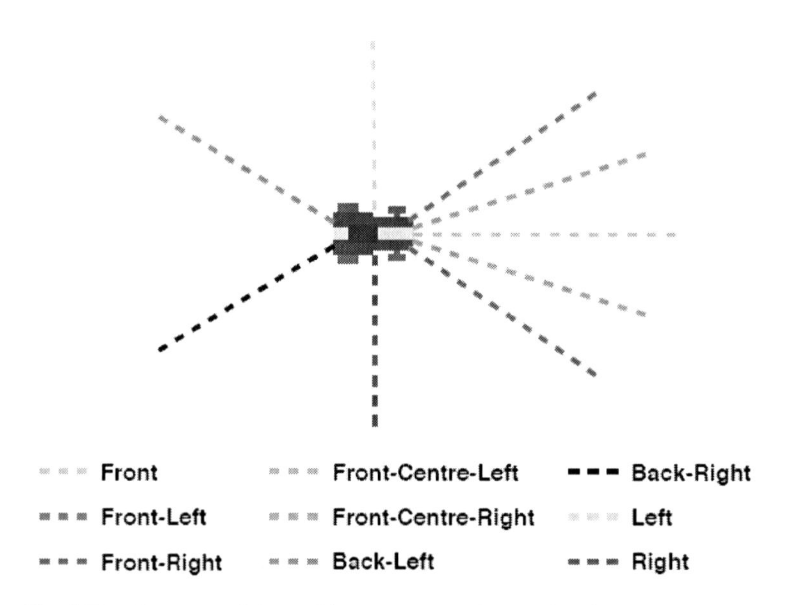

Fig. 4 The placement of the position sensors on the simulated car.

5 trials were run for each of three different learning agents – one using the new method of providing advice (referred to as the *difference* method), an agent receiving *bonus* rewards for advised actions, and an agent receiving no advice. For each of five trials, the agent received advice for the first 250000 steps. Performance was then monitored for a further 250000 steps during which standard Q-learning

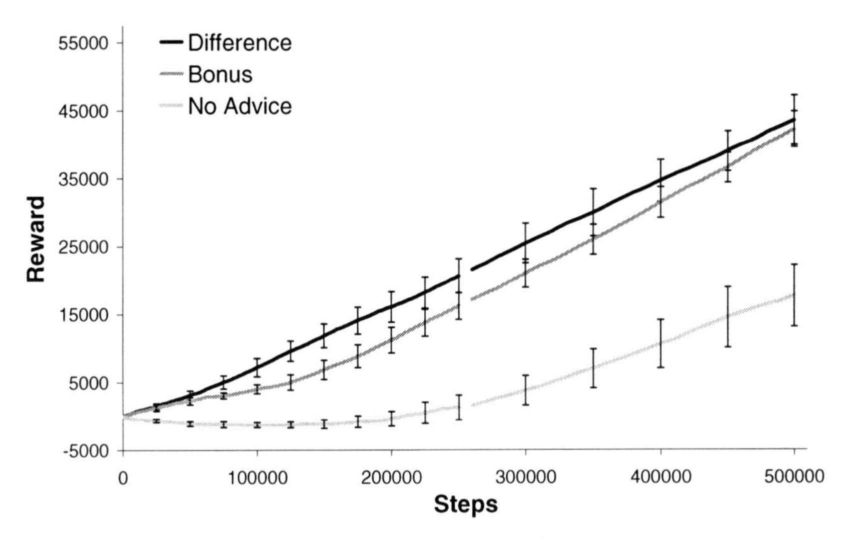

Fig. 5 Cumulative reward received by networks trained using the *difference* and *bonus* advice methods, and by a network trained with no advice. Advice was provided for the first 250000 steps only. (Error bars show 95% confidence intervals)

was performed with no advice being provided. The same system parameters were used for all trials.

Figure 5 compares the performance of these three agents based on cumulative reward. The results show that the new method produces more rapid early learning than either an agent receiving no advice or the bonus reward agent. However, the bonus reward agent does learn a better final solution to the problem, as indicated by the upward trend of its curve in Figure 5. This may be due to the difference agent exploring less and in particular exploring fewer difficult situations and therefore, when the teacher is no longer available, not knowing how to deal with these difficult situations when they arise. Another possible reason for this is overfitting. As will be discussed below, the difference method tends to produce a much larger network, which may be resulting in overfitting.

Figure 6 illustrates the number of advice actions provided by the human teacher during the first 250,000 training steps. Clearly the difference method places a much lower load on the teacher than the bonus method. This is due to the teacher being more satisfied with the performance of the agent and therefore not feeling the need to provide advice as frequently.

As noted above, the difference method results in a much larger network than the bonus method or an agent trained without advice, as shown in 7. This is not surprising as this method does not require the error novelty criterion of the RAN to be satisfied when advice is given to the network.

Finally, Figure 8 shows the number of collisions experienced over time. This is important since, while the teacher did not have an accurate idea of the reward received at each time step, the concept of avoiding collisions is intuitive to a human teacher in this context. As with the cumulative reward indicator, the difference

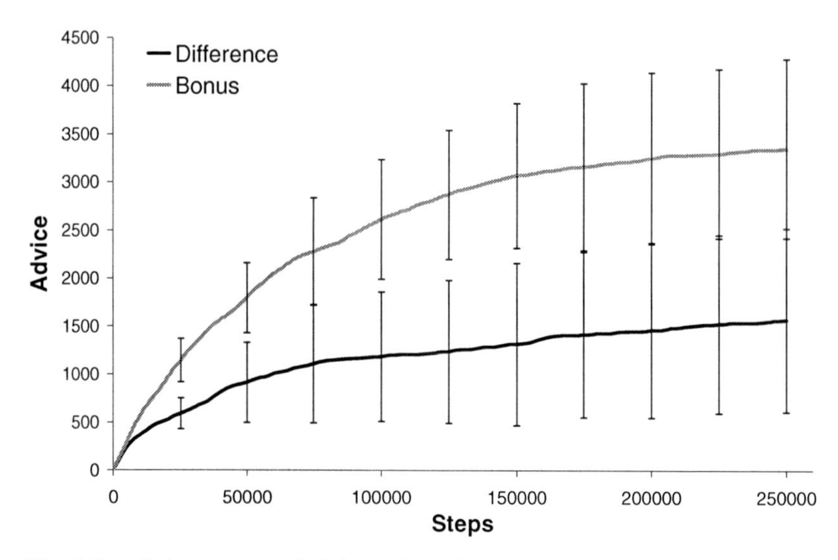

Fig. 6 Cumulative average of advice actions given to networks trained using the *difference* and *bonus* methods. (Error bars show 95% confidence intervals).

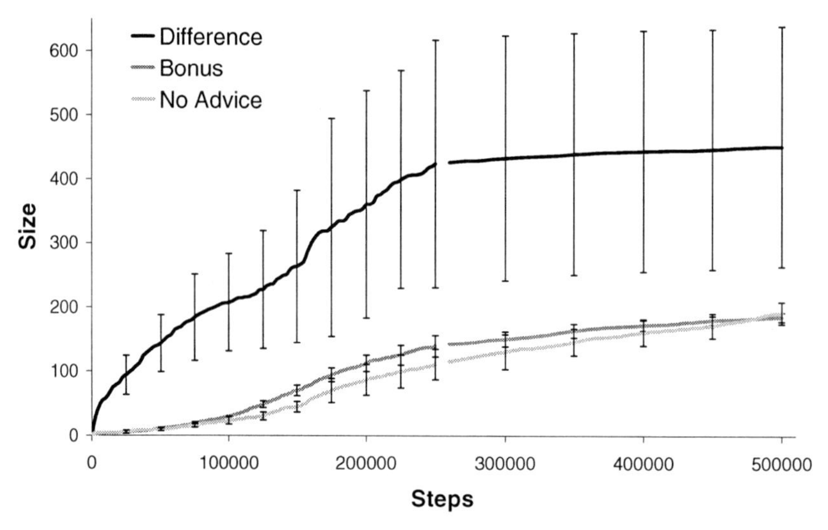

Fig. 7 Network growth for the *difference* and *bonus* advice methods, and for a network trained with no advice. Advice was provided for the first 250000 steps only. (Error bars show 95% confidence intervals).

agent performs considerably better than the bonus reward agent in terms of number of collisions while advice is being provided. In contract to cumulative reward however, the difference agent appears to perform equally as well as the bonus agent, and better than the control agent, even after advice from the teacher has

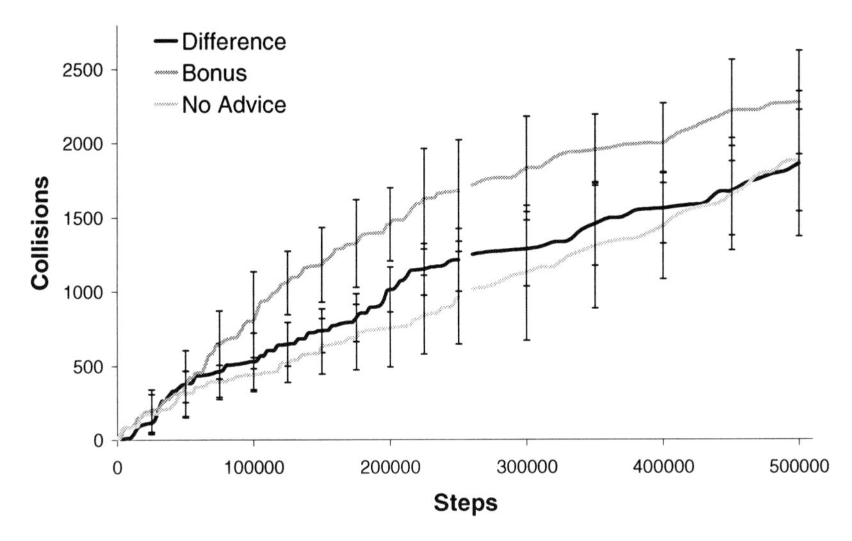

Fig. 8 Cumulative average of track collisions for networks trained using the *difference* and *bonus* advice methods, and by a network trained with no advice. Advice was provided for the first 250000 steps only. (Error bars show 95% confidence intervals).

ceased. The strong performance of the 'no advice' agent on this metric is accounted for by its tendency to drive slowly, which reduces the size of the penalty imposed when a collision occurs.

It should be noted that, for one of the five difference method trials, almost double the number of hidden nodes were added during the advice period (compared to the next highest for the other trials), as a result of the teacher providing more than double the amount of advice. This agent also performed the most poorly in terms of cumulative reward, and number of collisions. If this trial was not included in the results, we would expect to see even better results for the difference agent.

7 Conclusion

The results presented here have showed significant benefits for using human advice to guide the learning of a reinforcement learning agent for this particular learning problem. Both agents which were provided with advice during training performed significantly better than the pure reinforcement-learning agent. The use of the constructive Resource Allocation Network for the function approximation component of the Q-learning algorithm enabled the implementation of a new method for directly incorporating expert advice into the training procedure. This new *difference* method of advice-giving resulted in better performance early in training than for competing methods, while at the same time placing a lesser load on the teacher.

Once the teacher stopped providing advice however, the results were mixed. While the agent was able to maintain good performance in terms of catastrophic failures (collisions), it performed worse in terms of overall reward. The agent taught with the new method also tended to result in a much larger network size. Closer examination of the results show the large average network size and the subsequent poor average performance of the network, were in part due to a single trial where the teacher provided much more advice than for the other trials. This may be an indicator that the proposed method is more influenced by the quality of the teacher, and suggests that with some coaching of the teacher even better performance could be achieved.

The results presented here clearly warrant further investigation of methods of using constructive networks to incorporate advice into the learning of an agent in a reinforcement-learning task. The proposed algorithm needs to be tested within a wider variety of learning tasks to ensure that the results observed in this case are representative of the more general situation. In particular it needs to be established whether the larger networks produced by our new algorithm are a significant hindrance to the performance of the agent, and if so, how the algorithm can be modified to address this issue. One possible modification would be to periodically 'prune' the network by removing those hidden neurons which are contributing least to the agent's performance (Yingwei et al, 1998).

References

Anderson, C.W.: Q-learning with hidden unit restarting. Advances in Neural Information Processing Systems 5, 81–88 (1993)

Baird, L.: Residual algorithms: Reinforcement learning with function approximation. In: Proceedings of the Twelfth International Conference on Machine Learning, pp. 30–37 (1995)

Bellemare, M., Precup, D., Rivest, F.: Reinforcement Learning Using Cascade-Correlation Neural Networks, Technical Report RL-3.04, McGill University, Canada (2004)

Clouse, J.: Learning from an automated training agent. In: Working Notes of the ICML 1995 Workshop on Agents that Learn from Other Agents (1995)

Clouse, J., Utgoff, P.: Two kinds of training information for evaluation function learning. In: Proceedings of the Ninth Annual Conference on Artificial Intelligence, pp. 596–600 (1991)

Clouse, J., Utgoff, P.: A teaching method for reinforcement learning. In: Proceedings of the Ninth International Workshop on Machine Learning, pp. 92–101 (1992)

Coulom, R.: Feedforward Neural Networks in Reinforcement Learning Applied to High-dimensional Motor Control. In: International Conference on Algorithmic Learning Theory, pp. 402–413. Springer, Heidelberg (2002)

Crites, R.H., Barto, A.G.: Improving Elevator Performance Using Reinforcement Learning. Advances in Neural Information Processing Systems 8, 1017–1023 (1996)

Fahlman, S.E., Lebiere, C.: The Cascade-Correlation Learning Architecture. In: Touretzky, D.S. (ed.) Advances in Neural Information Processing II, pp. 524–532. Morgan Kauffman, San Francisco (1990)

Gardner, M.: Mathematical games: Fantastic patterns traced by programmed worms. Scientific American 229(5), 116–123 (1973)

Girgen, S., Preux, P.: Incremental Basis Function Expansion in Reinforcement Learning using Cascade-Correlation Networks, Research Report No. 6505, Institut National de Recherche en Informatique et en Automatique, Lille, France (2008)
Großmann, A., Poli, R.: Continual Robot Learning with Constructive Neural Networks. In: Birk, A., Demiris, J. (eds.) EWLR 1997. LNCS (LNAI), vol. 1545, pp. 95–108. Springer, Heidelberg (1998)
Jun, L., Duckett, T.: Q-Learning with a Growing RBF Network for Behavior Learning in Mobile Robotics. In: Proc. IASTED International Conference on Robotics and Applications, Cambridge, USA (2005)
Kretchmar, R.M., Anderson, C.W.: Comparison of CMACs and RBFs for local function approximators in reinforcement learning. In: IEEE International Conference on Neural Networks, pp. 834–837 (1997)
Maclin, R., Shavlik, J.: Creating advice-taking reinforcement learners. Machine Learning 22(1), 251–281 (1996)
Maclin, R., Shavlik, J., Torrey, L., Walker, T., Wild, E.: Giving advice about preferred actions to reinforcement learners via knowledge-based kernel regression. In: Proceedings of the 20th National Conference on Artificial Intelligence, pp. 819–824 (2005)
Nechyba, M.C., Bagnell, J.A.: Stabilizing Human Control Strategies Through Reinforcement Learning. In: Proceedings of the IEEE Hong Kong Symp. on Robotics and Control (1999)
Nissen, S.: Large Scale Reinforcement Learning using Q-SARSA(λ) and Cascading Neural Networks, M.Sc. Thesis, Department of Computer Science, University of Copenhagen, Denmark (2007)
Papudesi, V., Huber, M.: Learning from reinforcement and advice using composite reward functions. In: Proceedings of the 16th International FLAIRS Conference, pp. 361–365 (2003)
Perkins, T.J., Precup, D.: Using Options for Knowledge Transfer in Reinforcement Learning, Technical Report 99-34, Department of Computer Science, University of Massachusetts (1999)
Platt, J.: A Resource-Allocating Network for Function Interpolation. Neural Computation 3, 213–225 (1991)
Randlov, J., Alstrom, P.: Learning to Drive a Bicycle Using Reinforcement Learning and Shaping. In: International Conference on Machine Learning, pp. 463–471 (1998)
Rivest, F., Precup, D.: Combining TD-learning with Cascade-correlation Networks. In: Twentieth International Conference on Machine Learning, Washington DC, pp. 632–639 (2003)
Rummery, G., Niranjan, M.: On-line Q-Learning Using Connectionist Systems, Technical report, Cambridge University Engineering Department (1994)
Santos, J.M., Touzet, C.: Exploration tuned reinforcement function. Neurocomputing 28(1-3), 93–105 (1999)
Šter, B., Dobnikar, A.: Adaptive Radial Basis Decomposition by Learning Vector Quantisation. Neural Processing Letters 18, 17–27 (2003)
Sutton, R.S.: Learning to predict by the methods of temporal differences. Machine Learning 3, 9–44 (1988)
Sutton, R.S.: Generalisation in reinforcement learning: Successful examples using sparse coarse coding. In: Touretzky, D.S., Mozer, M.C., Hasselmo, M.E. (eds.) Proceedings of the, Conference:Advances in Neural Information Processing Systems, pp. 1038–1044. The MIT Press, Cambridge (1996)

Sutton, R., Barto, S.: Reinforcement Learning. MIT Press, Cambridge (1998)
Tesauro, G.J.: Temporal difference learning and TD-Gammon. Communications of the ACM 38(3), 58–68 (1995)
Thrun, S., Schwartz, A.: Issues in Using Function Approximation for Reinforcement Learning. In: Proceedings of the Fourth Connectionist Models Summer School, Hillsdale, NJ (December 1993)
Vamplew, P., Ollington, R.: Global Versus Local Constructive Function Approximation for On-Line Reinforcement Learning. In: Zhang, S., Jarvis, R.A. (eds.) AI 2005. LNCS (LNAI), vol. 3809, pp. 113–122. Springer, Heidelberg (2005)
Vamplew, P., Yearwood, J., Dazeley, R., Berry, A.: On the Limitations of Scalarisation for Multi-objective Reinforcement Learning of Pareto Fronts. In: Wobcke, W., Zhang, M. (eds.) AI 2008. LNCS (LNAI), vol. 5360, pp. 372–378. Springer, Heidelberg (2008)
Watkins, C., Dayan, P.: Q-learning. Machine Learning 8(3), 279–292 (1992)
Waugh, S.G.: Extending and benchmarking Cascade-Correlation, PhD thesis, Department of Computer Science, University of Tasmania, Australia (1995)
Yingwei, L., Sundararajan, N., Saratchandran, P.: Performance evaluation of a sequential minimal radial basis function (rbf) neural network learning algorithm. IEEE Transactions on Neural Networks 9(2), 308–318 (1998)

A Constructive Neural Network for Evolving a Machine Controller in Real-Time

Andreas Huemer, David Elizondo, and Mario Gongora

Abstract. A novel method is presented to allow a machine controller to evolve while the machine is acting in its environment. The method uses a single spiking neural network with a minimum number of neurons and no initial connections. New connections and neurons are grown by evaluating reward values which can represent either the internal state of the machine or the rating of its task performance. This way the topology and the level of connectivity of the network are kept to a minimum. The method will be applied to a controller for an autonomous mobile robot.

Keywords: Constructive Neural Network, Spiking Neural Network, Reinforcement Learning, Growing Machine Controller.

1 Introduction

Typically constructive neural networks are used to solve classification problems. It has been shown that using this type of network results in less computation requirement, smaller topologies, faster learning and better classification [5, 9]. Additionally, [4] shows that certain constructive neural networks can always be evolved to a stage in which it can classify 100% of the training data correctly.

For machine controllers, classification is only a secondary issue, but still an important one, as will be discussed later. The main machine task is to select a suitable action in a certain situation. It need not be the best possible action but must certainly

Andreas Huemer
Institute Of Creative Technologies, De Montfort University, Leicester, UK
e-mail: ahuemer@dmu.ac.uk

David Elizondo
Centre for Computational Intelligence, De Montfort University, Leicester, UK
e-mail: elizondo@dmu.ac.uk

Mario Gongora
Centre for Computational Intelligence, De Montfort University, Leicester, UK
e-mail: mgongora@dmu.ac.uk

L. Franco et al. (Eds.): Constructive Neural Networks, SCI 258, pp. 225–242.
springerlink.com

be suitable. Much more important is that the action is selected in the required time: machines, especially mobile robots, must act in real-time. In [10] real-time systems are discussed and a definition is provided.

One approach to enable a machine acting in real-time is to create the controller in advance and then use it on the machine without further intervention. This is the traditional way, but it works only for simple applications or it needs much development time for more complex applications.

As machines are required to fulfil increasingly complex tasks, people have looked for various possible solutions to this problem. On the one hand, methods were sought to speed up the development of the controller, which is then used on the machine. The other idea was to improve the controller when already in use.

Both approaches are useful and can also be combined, and have resulted in methods inspired by nature. Evolutionary algorithms are inspired by the evolution of life an can be used to improve the controller. For example [23] presents a very effective method for evolving neural networks that can be used to control robots.

However, the power of this approach is limited, because it does not work with a single controller but with a whole population of controllers. More controllers require more computational power, which either may not be available or if available, may slow down the embedded computer thereby making this option too expensive. Of course there could also be a population of machines and not only a population of controllers, but the increasing expense in material and energy is obvious in this case.

Alternatively, machines or their controllers could be simulated on a computer, but this only transfers several problems from the robot to the computer and involves the additional problem of it being hard to simulate the complex environments that robots should often act in.

On-line learning, which means adapting the machine controller while it is running, can help with overcoming the problems of evolutionary methods, but it does not substitute them. The remainder of this chapter is dedicated to on-line learning. In Sect. 2 we summarise the history of on-line learning up to the current state-of-the-art methods and we identify remaining problems tackled in this chapter.

Section 3 shows the basic characteristics of a novel constructive machine controller. The crucial issue of how feedback can be defined, how it is fed into the controller and how it is used for basic learning strategies is discussed in Sect. 4. Section 5 shows how the feedback is used for the more sophisticated methods of growing new neurons and connections.

A simulation of a mobile robot was used to test the novel methodology. The robot learns to wander around in a simulated environment avoiding obstacles, such as walls. The results are presented in Sect. 6, and Sect. 7 draws some conclusions and outlines future lines of work.

2 History of Machine Controllers

This section will provide a brief overview of the history of machine controllers. A machine controller is needed as soon as a machine is required to behave differently in different situations. The controller decides which parts of the machine to adjust (output) in which situation (input). The complexity of the controller increases with the number of different situations that can be sensed and the number of different actions that can be chosen.

We do not cover the history of control technologies (e.g. mechanical or electronic control) here but we discuss control concepts with respect to control methods. Also remote control is not given explicit consideration, because we interpret the remote signals as an input to the actual controller.

2.1 Fixed Behaviour

Early controllers were developed so that they behaved in a strictly predefined way as long as no error occurred. These controllers are still needed and it is hard to imagine why they should not be needed in the future.

However, each decision of the machine is planned and programmed manually and the number of situations that have to be considered grows exponentially:

$$N_I = \prod N_v \tag{1}$$

where the number of all input combinations N_I is the product of all states that have to be considered for the input variables N_v. Because of this, the development time or the number of errors increases dramatically with the number of states that have to be differentiated for an input variable and especially with the number of input variables. This is without even considering the output side.

Consequently, there have been different approaches to tackle this problem. One idea was to adapt the environment in which the machine is situated. For example by situating a mobile robot in a laboratory where the ground is level and there is constant standardized lighting the number of situations the controller has to consider is minimised.

The problem is that mobile robots and other machines are usually not needed in a quasi-perfect environment. Many robots are however required to fulfil increasingly complex tasks in our real and complex world. The good news is that many of those robots need not act in a perfect way but just in a way that is good enough. There is some room for error and it is possible to concentrate on minimising the time required to develop a controller.

Controllers that have a fixed behaviour for regular operations have one important advantage and one important drawback by definition. The advantage is that the machines do not change their behaviour in known situations: their behaviour is predictable. The drawback is that the machines do not change their behaviour in unknown situations either, which may result in intolerable errors. An increasing complexity of the environment increases the probability for intolerable errors, if the time for developing a controller that should work in it is not increased significantly.

2.2 Adaptive Behaviour

This problem has led to research in on-line adaptation processes for machines, especially for mobile robots used in real-world applications. Traditional methods work with controller parts which all have a certain predefined task. Adaptation processes can change the reactions to certain input, but the machine is still limited to the behaviour of preprogrammed modules of different complexity. An interesting example of these adaptation methods was presented in [20].

Increased flexibility can be achieved by keeping the modules as simple as possible and as similar as possible. This way it is not only possible to change the interactions between existing preprogrammed modules but it is also possible to change those modules themselves. Additionally, it is possible to create completely new modules.

Artificial neural networks can achieve a high level of flexibility and therefore are a useful tool for reducing development time to a minimum. For example Alnajjar and Murase [1] have shown that a controller consisting of a spiking neural network (SNN) enables a robot to learn certain tasks successfully.

Spiking neurons, unlike traditional artificial neurons, encode the information they send to other neurons, called postsynaptic neurons, in a sequence of spikes (Boolean signals) instead of a single value. Spiking neurons can have a lot of features that are impossible or at least difficult to implement with traditional neurons, which encode their information in a single value and send them to other neurons at a particular moment. In [12] some interesting features that can be implemented in different types of spiking neural networks are discussed.

Some advantages of spiking neural networks compared to traditional neural networks are:

- Single spikes are not very important to the overall behaviour of the network. This network behaviour is probabilistic and fuzzy and is therefore more robust to errors. For example, if a neuron sends spikes at a slightly higher frequency, it obtains slightly more influence; if a neuron sends a few additional spikes, its influence lasts a little longer; if a spike is sent a little late, it may already be compensated for by the next spike. There are no drastic effects as there would be with wrong single values that transport the whole information.
- Spiking neurons can react faster to certain time-dependent input patterns (e.g. a sound). Those input patterns can be easily implemented in spiking patterns because of their time-dependent nature. If a time-dependent input pattern has to be encoded in a single value, the controller has to wait some time before being able to calculate a statistic value.
- Probably the most interesting advantage regarding reinforcement learning is that feedback assignment is much easier with spiking neural networks. An action is not initiated at a certain moment and then left alone, but rather it is encoded in a continuous spike train. This way the neurons that are responsible for a certain action can still be active when the machine starts to receive feedback for that action. This issue is discussed in more detail in Sect. 4.3.

2.3 *Constructive Machine Controllers*

For full flexibility, to create completely new behaviour that reacts to completely new situations, the spiking neural network must be constructive. This means that new connections and new neurons need to be created in certain circumstances.

Alternative approaches that try to generate neurons at the beginning and later only change the connections of the network (e.g. "Echo State Networks" [14] and "Liquid State Machines" [19]) have two main disadvantages for adaptive control systems:

- Usually, not all parts of the network are needed for the control task. The unnecessary parts are processed anyway and decrease the performance of the system.
- It cannot be guaranteed that the network can learn all tasks successfully, because its general structure is fixed after its random initialisation.

A very flexible and effective approach for classification problems is the "Growing Neural Gas" algorithm [7, 8], which is an improvement of the "Neural Gas" algorithm [21]. Drawbacks of "Neural Gases" that are similar to the problems mentioned previously with "Echo State Networks" and "Liquid State Machines", could be solved successfully with the "Growing Neural Gas" algorithm.

Examples for constructive neural networks for classification problems without the need for recurrent connections are "Support Vector Machines" [22] and the "Recursive Deterministic Perceptron" [24].

In recent years some ideas for constructive methods have been published for control problems. A typical test application for the adaptive controllers is to enable a two-wheeled mobile robot to learn to run around while avoiding crashing into obstacles.

In [17, 18] a robot controller is presented that creates neurons to map a certain input pattern to an output using "pleasure cells". In [2] a two-level control system has been introduced. A low-level unit contains a number of spiking neural networks which are all trained to perform the robot's task in a local environment. The high-level unit switches between the SNNs depending on the current environment or it can create a new SNN, if it cannot find an SNN for the current environment.

The constructive neural network discussed in this chapter integrates the complete control task into a single neural network. The construction method enables the network to evolve from an initial stage with a minimum number of neurons and no connections. Even basic movements are learnt autonomously (in [18] "random cells" are used for random initial behaviour, in [2] the robot initially goes straight forward). The main goal of the work presented here is to minimise the time for developing a highly adaptive control system by making it as flexible and autonomous as possible.

3 A Constructive Machine Controller

Based on the issues we have discussed in the previous sections we present a constructive neural network that will enable a machine to learn to fulfil its tasks autonomously. In addition to the main task, the action selection, we also discuss classification. In a machine controller, classification can be used to reduce the size of the network.

A novel method is presented which not only includes the classification task and the action selection task in a single neural network, but it is also capable of defining the classes autonomously and it can grow new neurons and connections to learn to select the correct actions from the controller's experience based on reward values. Positive and negative reward values build the feedback, which can be generated internally, without human intervention. Alternatively it is possible for a human to feed the controller with feedback during runtime acting as a teacher.

The initial neural network consists of a minimum number of neurons and no connections. The designer of the controller need not concern him/herself with the topology of the network. Only the interface from the controller to the machine has to be defined, which reduces the design effort considerably.

The interface includes:

- *Input neurons*, which convert sensory input into values that can be sent to other neurons. For our experiments we used spiking neurons, so the sensory values are converted into spikes by the input neurons.
- *Output neurons*, which convert the output of the neural network into values that can be interpreted by the actuators.
- *A reward function*, which feeds feedback into the machine controller.

3.1 Controller Characteristics

Our "growing" machine controller consists of a layered spiking neural network as illustrated in Fig. 1.a, in which neurons send Boolean signals that transport the information depending on whether or not the presynaptic neuron (the first of two connected neurons) was activated and fired (a spike). The hidden layer "B" contains two neurons that store input combinations. One of them excites dendrite "a" via axon "d", which itself excites an output neuron. The neuron in layer "C" inhibits the same output neuron by axon "c". Also axon "b" inhibits that neuron but implements local inhibition. A neuron is activated when a certain threshold potential is exceeded. The neuron potential is increased by spikes arriving at excitatory connections and decreased by spikes arriving at inhibitory connections. A basic explanation of spiking neural networks can be found in [25].

For a controller system it is evident, that the action selection task needs to be fulfilled. Classification of input signals and recursive classification of classes are important issues, because this reduces the necessary number of neurons and connections.

The use of sparse neural networks has been discussed in [5] and [9]. These models result in fewer computational requirements and better development of the network as well as smaller topologies. The main reason for obtaining smaller topologies when using classification is obvious:

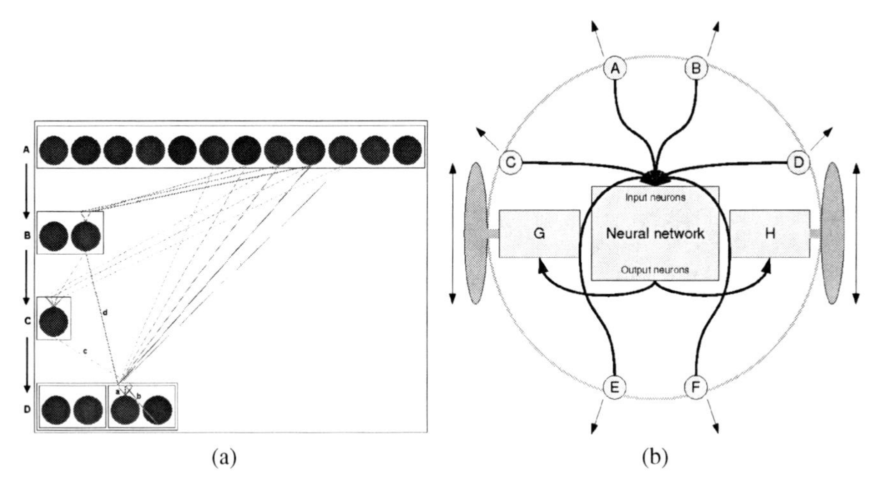

Fig. 1 (a) Neural network topology after a short simulation period. For clarity not all connections are shown. Layer A consists of the input neurons which are connected to sensors A to F from the robot shown in (b). Layer D contains the motor neurons, connected to G and H in (b).

If there are no specialised neurons that represent a certain class in the subsequent layers in the network, all neurons of this class have to connect to the next layer separately and not with a single connection from the specialised "class neuron". There is no problem if one neuron connects only to one other neuron. In this case no additional connections are required. However, when a neuron connects to a second neuron, only one additional connection is made instead of more connections from all the neurons of a class. Also the total number of neurons is reduced if the neurons represent the possible combinations of neurons in the previous layer, because in a subsequent layer only the class neurons have to be considered.

To achieve an efficient topology along with action selection and classification, in a single neural network, we separate the connections into two parts: artificial dendrites and axons. An axon of the presynaptic neuron is connected to a dendrite of the postsynaptic neuron. Excitatory connections have to be used for operations that are similar to the logical AND and OR operations. For inhibitory connections this separation is not necessary, because they represent an operation similar to the logical NOT operation.

A dendrite has a low threshold potential and is activated when only one presynaptic neuron (or a few neurons) have fired a spike via their axons. All presynaptic neurons are handled equally at this point (logical OR operation) and represent neurons which are combined into one class. An axon weight defines to what degree one presynaptic neuron belongs to the class.

The neuron finally fires only if a certain combination of dendrites has fired (logical AND operation). For this operation, the threshold of the neuron potential, which is modified by the dendrites, is very high. This causes a neuron to select an action for a certain combination of input signals. The dendrite weight defines the importance of a class in the action selection process. In our model a single neuron can fulfil both tasks, classification and action selection, and it learns both tasks automatically.

In the following we show the computations performed when signals travel from one neuron to another.

Input of a dendrite:

$$I_d = \sum O_{a+} \cdot w_{a+} \tag{2}$$

where I_d is the dendrite input. O_{a+} is the output of an excitatory axon, which is 1 if the presynaptic neuron has fired and 0 otherwise. w_{a+} is the weight of the same excitatory axon, which must be a value between 0 and 1.

Output of a dendrite:

$$O_d = \frac{1}{1 + e^{-b \cdot (I_d - \theta_d)}} \tag{3}$$

where O_d is the dendrite output and I_d its input. θ_d is a threshold value for the dendrite. b is an activation constant and defines the abruptness of activation.

Input of a neuron:

$$I_j = \sum O_d \cdot w_d - \sum O_{a-} \cdot w_{a-} \tag{4}$$

where I_j is the input of the postsynaptic neuron j, O_d is the output of a dendrite, w_d is the weight of this dendrite, O_{a-} is the output of an inhibitory axon and w_{a-} is the weight of this inhibitory axon. Dendrite weights and axon weights are in the range $[0,1]$ and all dendrite weights add up to 1.

Change of neuron potential:

$$P_j(t+1) = \delta \cdot P_j(t) + I_j \tag{5}$$

where the new neuron potential $P_j(t+1)$ is calculated from the potential of the last time step t, $P_j(t)$, and the current contribution by the neuron input I_j. δ is a constant between 0 and 1 for recovering the resting potential with time (which is 0 in this case).

The postsynaptic neuron is activated when its potential reaches the threshold θ_j and becomes a presynaptic neuron for neurons which its own axons are connected to. After firing, the neuron resets its potential to the resting state. In contrast to similar neuron models that are for example summarised by [15], a refractory period is not implemented here.

4 Feedback and Reward Values

4.1 Calculation of Reward Values

The main challenge for the designer of a machine controller that uses the methods described in this chapter, is to define an appropriate reward function. Positive and negative reward values are fed into the neural network as explained below and are used for all adaptation processes, like basic learning (adaptation of connection weights), creating new connections and creating new neurons (growing mechanisms).

In our experiments we have used a single global reward value, which represents a positive "rewarding" or negative "aversive" feedback depending on the machine's performance in the task that it has been assigned. The objective is to create a network which maximises positive feedback.

Depending on current measurements like fast movement, crashes, recovering from crashes, or the energy level, the reward value is set from -1 (very bad), to 1 (very good). The challenge is to provide a useful value in all situations. For example, as experiments have shown (see Sect. 6), a reward function that is not capable of providing enough positive feedback may result in a machine malfunction, because despite all of its effort to find a good action, it is not evaluated properly. Also uniform positive feedback may result in a similar situation because of a lack of contrast.

The reward value $\rho(t)$, which is the result of the reward function at time t, has to be back-propagated to all neurons, where it is analysed and used for the adaptation mechanisms. To do so, the neurons of each layer, starting with the output layer and going back to the input layer, calculate their own reward value. The value of the output neurons is equivalent to the global reward value. All other neurons calculate their reward as follows:

$$\rho_i(t) = \frac{\Sigma \rho_{j+}(t) - \Sigma \rho_{j-}(t)}{N_+ + N_-} \tag{6}$$

where $\rho_i(t)$ is the reward value of the presynaptic neuron i at time step t. $\rho_{j+}(t)$ is the reward value of a postsynaptic neuron that has an excitatory connection from neuron i, while $j-$ refers to a postsynaptic neuron that has an inhibitory connection from neuron i. N_+ is the number of postsynaptic neurons of the first kind and N_- is the number of the other postsynaptic neurons.

4.2 Adaptation of Connection Weights

First the reward value of a neuron is used to adapt the connection weights. This is done after the basic processes of Sect. 3 for each neuron. All calculations for all neurons are done once at each time step t.

The reward value can be added to a learning rule as an additional factor. Different authors, all of them using different neuron functions and learning functions, have shown that this surprisingly simple method can be used successfully to implement reinforcement learning in a neural network [3, 6, 13]. Networks no longer need an external module that evaluates and changes the connections after each processing step.

Activation Dependent Plasticity is used to adapt connection weights in the experiments. Activation Dependent Plasticity (ADP) is based on Hebb's ideas of strengthening connections that fire together [11]. As shown by [6, 13] reward can also be integrated into the more sophisticated Spike Time Dependent Plasticity (STDP) learning model.

Adaptation of an excitatory axon weight (axon connected to dendrite):

$$w_{a+}(t+1) = w_{a+}(t) + \eta_a \cdot \rho_j(t) \cdot \phi_i \cdot \phi_j \tag{7}$$

where $w_{a+}(t)$ and $w_{a+}(t+1)$ are the axon weights before and after the adaptation. η_a is the learning factor for axons and $\rho_j(t)$ is the current reward value of the postsynaptic neuron. ϕ_i and ϕ_j represent the recent activity of the presynaptic and the postsynaptic neuron. In our experiments ϕ_i was kept between -1 and 0 for very little activity and from 0 to 1 for more activity, ϕ_j is kept between 0 and 1. For positive reward much activity in the presynaptic neuron strengthens the axon weight if also the postsynaptic neuron was active but little presynaptic activity weakens the axon weight. A negative reward value reverses the direction of change.

Adaptation of a dendrite weight (always excitatory):

$$w_d(t+1) = w_d(t) + \eta_d \cdot \rho_j(t) \cdot \phi_d \cdot \phi_j \tag{8}$$

where $w_d(t)$ and $w_d(t+1)$ are the dendrite weights before and after the adaptation. η_d is the learning factor for dendrites. ϕ_d represents the recent dendrite activity (which joins the activity of the connected axons) and is kept between 0 and 1. $\rho_j(t)$ and ϕ_j are discussed with Equ. 7. When all dendrite weights of a neuron are adapted they are normalised to add up to 1 again because of the dependencies between the weights (see Sect. 3).

Adaptation of an inhibitory axon weight (axon connected to neuron):

$$w_{a-}(t+1) = w_{a-}(t) - \eta_a \cdot \rho_j(t) \cdot \phi_i \cdot \phi_j \tag{9}$$

where $w_{a-}(t)$ and $w_{a-}(t+1)$ are the axon weights before and after the adaptation. η_a, $\rho_j(t)$, ϕ_i and ϕ_j are discussed in Equ. 7. Axons that are part of local inhibition, were not changed in our experiments. Also if ϕ_i is negative, the weight was kept equal. An inhibitory axon is strengthened, if it was not able to prevent bad feedback, and it is weakened, if it tried to prevent good feedback.

4.3 Delayed Feedback

An important issue to consider when dealing with feedback from the environment and the resulting rewards is delayed feedback. When weights are adapted and, as discussed later, neurons are created based on the current reward, it may at first seem to be the wrong time to do so. Typically, the feedback is received *after* the action responsible is executed. In fact, the time difference can vary significantly.

However, because in spiking neural networks there is no single event that is responsible for an action, but a continuous flow of spikes, the discussed methods can be efficient anyway. The input pattern, and hence the spiking pattern, usually does not change rapidly if a certain feedback is received. Figure 2 shows an example situation for this issue.

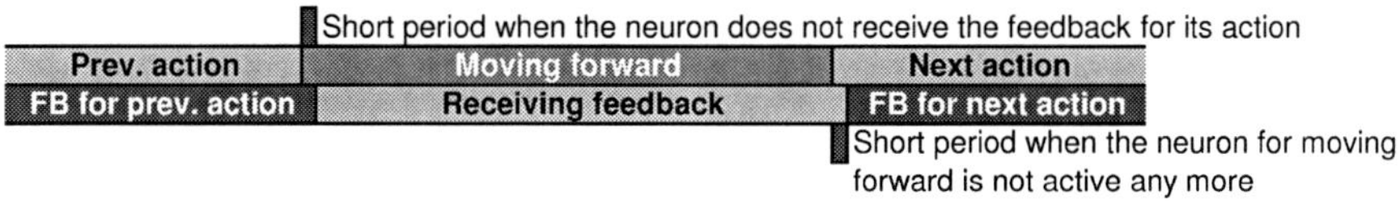

Short period when the neuron does not receive the feedback for its action
Prev. action | Moving forward | Next action
FB for prev. action | Receiving feedback | FB for next action
Short period when the neuron for moving forward is not active any more

Fig. 2 When a spiking neural network controls a machine, there are two short periods where the assignment of the feedback to the corresponding action is problematic. Whenever an action is active for a longer period the feedback can be assigned correctly to the active neurons that are responsible for the current action.

Of course there remain situations in which it is difficult to assign the feedback correctly, for example if there is a big time difference between action and feedback, or if there are many competing actions or feedback values at the same time. However, even humans do not always arrive at the correct conclusions. They can deal with very complex relations but not with all of them.

5 Autonomous Creation of the Neural Network

The methods that were discussed in Sect. 4 tune a neural network. They are necessary to reinforce neural paths that were responsible for good actions and to weaken and finally remove connections that made a neuron output classify incorrectly.

There are different methods that result in new connections or even new neurons.

If a neuron has no input connections, it can connect to a random new predecessor in the previous layer. In our experiments the randomness was reduced by looking for neurons that have similar relative positions. Such a new connection is always excitatory and consists of a single axon and a single dendrite. This makes the postsynaptic neuron (or better: its single dendrite) represent a new class and it has the potential to carry out a new action when activated. Other neurons can be added to the class by creating new axons occasionally. An axon that does not fit into the class will be weakened by the mechanisms of Sect. 4.

New neurons are created to remember activation patterns that were responsible for good or bad actions. Liu, Buller and Joachimczak have already shown that correlations between certain input patterns and a certain reward can be stored by creating new neurons [17, 18]. For this task we add a new potential value to each neuron. The common "neuron potential" function defines when a neuron fires a spike. Our new "reward potential" function defines when a neuron has enough feedback to create a new neuron:

$$R_j(t+1) = \delta_R \cdot R_j(t) + \phi_j \cdot \rho_j(t) \tag{10}$$

where $R_j(t+1)$ is the new reward potential of neuron j while $R_j(t)$ is the old one. ϕ_j is the value for the recent activity that was also used in Sect. 4 and $\rho_j(t)$ is the current reward value of neuron j.

When $|R_j|$ reaches a certain threshold θ_R (different thresholds for positive and negative feedback are possible) all dendrites (excitatory connections) that were active recently are evaluated. Young dendrites are ignored, because they have not yet proved themselves. For a dendrite with more than one axon it may be worth remembering an activation combination. A single axon may still be interesting if there was bad feedback, because the axon should have been inhibitory in this case. A list of axons, that had an influence on the activation of the neuron, is kept. If this combination is not yet stored in an existing preceding neuron, a new neuron is created and each axon of the list is copied. However, each of the new axons is connected to its own dendrite to record the combination. The new neuron will only be activated when the same combination is activated again. When the reward potential is positive, the new neuron is connected to the existing one by a new axon to the currently evaluated dendrite (neurons in layer B in Fig. 1.a and neurons in Fig. 3). A negative reward potential results in the addition of a new inhibitory axon to the existing neuron (neuron in layer C in Fig. 1.a).

Generally, the new neuron is then inserted into the layer previous to the existing postsynaptic neuron as shown in Fig. 3.a. The relative position of the new neuron will be similar to the relative position of the existing one.

However, if one of the axons to the new neuron has its source in the layer the new neuron should be inserted into, a new layer is created in front of the layer of the existing postsynaptic neuron as shown in Fig. 3.b. This way the feed-forward structure which our methods are based on can be preserved.

Once the new neuron is inserted, local inhibition will be generated. Experiments have shown that local inhibition makes learning much faster and much more reliable (see Sect. 6). Hence, new inhibitory axons are created to and from each new neuron. This inhibitory web has been automatically created within one layer in our experiments. In more sophisticated future developments this web should perhaps be limited to certain areas within a layer. Nested layers with neuron containers, which

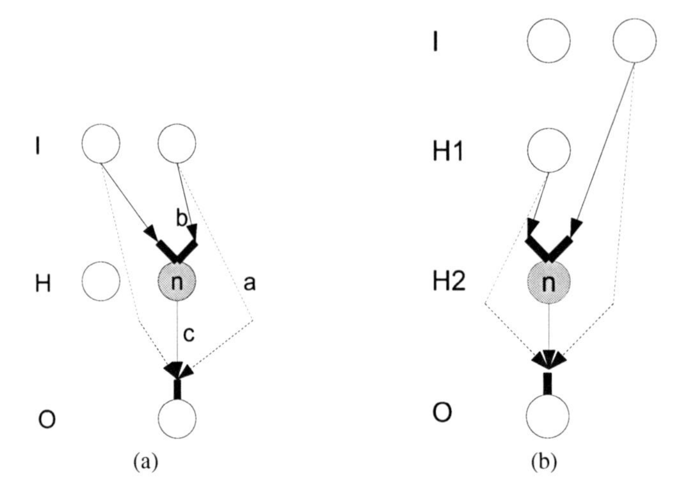

Fig. 3 (a) The new neuron n is created and active axons (a) are used to create new connections (b) to it. Then the new neuron is connected to the existing one (c). In (b) the new layer H2 is created before neuron n is inserted, because the output neuron already has a connection to a neuron in layer H1.

are basically supported by our implementation but are not used for the growing mechanisms yet, could help with this task.

6 Experiments

6.1 Setup

Our novel method for the autonomous creation of machine controllers was tested in a simulation of a mobile robot which moves using differential steering, as shown in figure Fig. 1.b. The initial neural network consists of 12 input neurons (2 for each sensor) and 4 output neurons (2 for each motor, see Fig. 1.a).

The input neurons are fed by values from 6 sonar sensors as shown in Fig. 1.b, each sensor feeds the input of 2 neurons. The sonar sensors are arranged so that 4 scan the front of the robot and 2 scan the rear as shown in the figure. The distance value is processed so that one input neuron fires more frequently as the measured distance increases and the other neuron connected to the same sensor fires more frequently as the distance decreases.

For the actuator control, the output connections are configured so that the more frequently one of the output neurons connected to each motor fires, the faster this motor will try to drive forward. The more frequently the other output neuron connected to the same motor fires, the faster that motor will try to turn backwards. The final speed at which each motor will drive is calculated by the difference between both neurons.

Fig. 4 A simulated Peoplebot is situated in this simulated office provided by MobileRobots/ActivMedia.

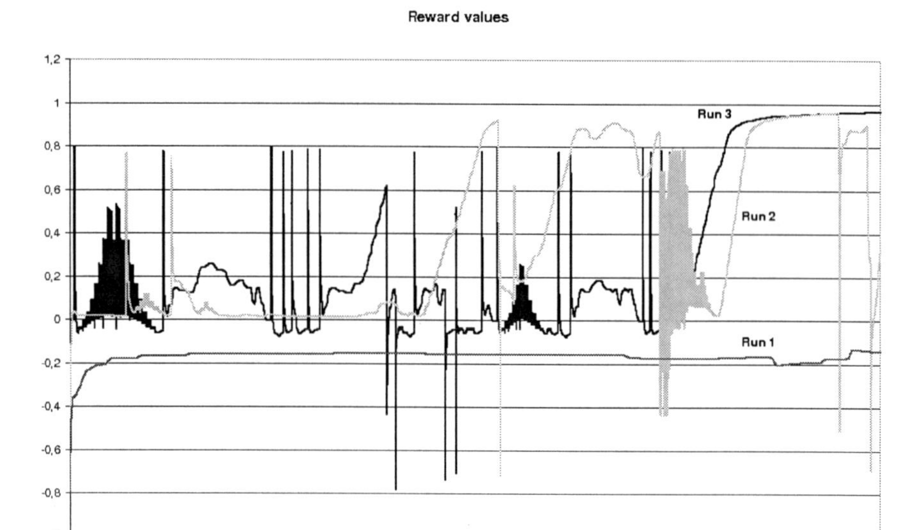

Fig. 5 The three curves show the reward values the robot received in simulation runs of the same duration and starting from the same position. In "Run 1" only forward movement is rewarded. The robot learns to hold its position to minimise negative feedback. In "Run 2" "activity reward" was introduced. In "Run 3" the same reward algorithm is used as in "Run 2", but in "Run 3" no or only small forward movement is punished. The robot learns to receive positive feedback with time which makes it more stable.

With this experimental setup the robot should learn to wander around in the simulated environment shown in Fig. 4 while avoiding obstacles.

The original robot's bumpers are included in the simulation and are used to detect collisions with obstacles, and are used to penalise significantly the reward values when such a collision occurs. The reward is increased continuously as the robot travels farther during its wandering behaviour. Backward movement is only acceptable when recovering from a collision, therefore it will only be used to increase the robot's reward value in that case, while it is used to decrease this value for all other cases. As time increases, linear forward movement will receive higher positive reward and this will discourage circular movement.

6.2 Results

This section discusses the challenges when finding an appropriate reward function. Additionally, we show the importance of local inhibition for the reliability of the learning methods. Finally, we present some results considering the performance and the topology of the constructive neural network.

The reward function that we have used in our experiments delivers -1 in the case where the robot crashes into an obstacle. Backward movement is punished

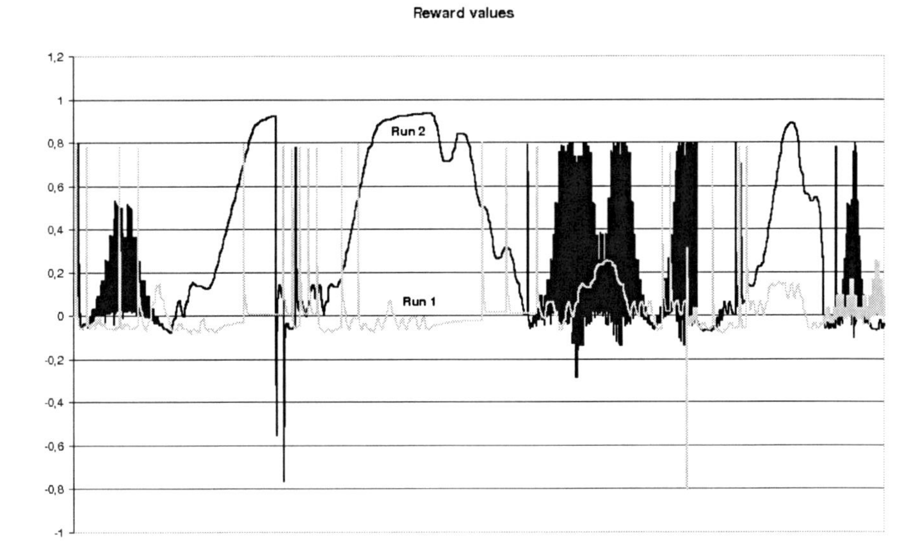

Fig. 6 "Run 1" shows the development of the reward value without local inhibition. This contrasting method increases the produced reward values significantly in "Run 2".

(negative value). There are two features that are not implemented in all of the three simulation runs of Fig. 5: First, no or only small forward movement is punished; second, backward movement is rewarded (positive value), if the robot received bad feedback for a while, to keep the robot active.

Figure 6 shows the importance of local inhibition. Without local inhibition the simulation run did not produce a single phase in which significant positive feedback was received. Only short periods of positive reward can be identified where the robot acted appropriately by chance. Local inhibition increases the contrast of spiking patterns, which makes single neurons and hence single motor actions more powerful and the assignment of reward to a certain spiking pattern more reliable.

Table 1 shows the results of a test of 50 simulation runs. In many cases the robot was able to learn the wandering task with the ability to avoid obstacles. Each run of the test sample was stopped after 20000 control cycles (processing the whole neural network in each cycle).

Table 1 Results of 50 simulation runs.

Performance of the controller			
	Min.	Max.	Avg.
Total reward	664.82	7486.24	4014.02
Average reward	0.03	0.37	0.20
Maximum speed	657.00	1056.00	1015.92
Average speed	102.74	971.63	260.37
Crashes	0.00	13.00	4.50

Topology of the controller			
	Min.	Max.	Avg.
Neurons	16.00	38.00	22.46
Excitatory axons	14.00	178.00	45.28
Excitatory dendrites	4.00	139.00	27.88
Inhibitory axons	4.00	7.00	4.36

The neural network contained no hidden neurons and no connections at the beginning. Connections for local inhibition were created when the controller was started. The speed values of the table are given in internal units of the simulation.

7 Conclusions and Further Work

We have shown that a neural network can be grown based on the reward measured by a feedback function which analyses the performance of a task in real-time. The neural network can control a machine such as a mobile robot in an unpredictable or unstructured environment.

Since the controller constructs itself, only the input layer, the output layer and a feedback function that measures the task performance of the machine and rewards the controller have to be defined. This means that the task of the designer involves only the definition of these elements, no effort is required for the actual design of the network.

Because controlling the machine and learning from experience continuously when running is integrated into a single and robust stage, the system can adapt to completely new situations without changing any part of the control structure manually. This involves three advantages in addition to the possibility of evolving a control system from scratch:

1. The machine can learn to react appropriately in situations that it has encountered previously but can use the experience it has gathered so far.
2. The machine can learn to handle changes to its hardware, for example if a sensor breaks.
3. When new machines are developed, it may be possible to use the neural network of established machines to reduce the time necessary for training them by starting from a semi-optimal state rather than from scratch. This could not only help to save time but also material, because a machine that has no knowledge at the beginning may easily damage itself.

Further analysis and improvements of the growing methodology are necessary to gain even better results from the growing methods for the neural network. For example it will be necessary to investigate the behaviour of the system with concurrent tasks and conflicting or noisy sensory data.

Also time dependent situations like action sequences and timed input (speech, objects moving through a visual field, ...) will be important issues for future research. Combining our results with other work may be very helpful for these issues, for example using Spike Time Dependent Plasticity (STDP) methods and recurrent connections. Another essential issue of further research is to investigate different reward functions for different tasks, because this is the key to effective evolutions of neural networks. Sophisticated reward functions will probably make a machine learn its tasks sooner and fulfil its tasks better (faster, more precisely, ...).

References

1. Alnajjar, F., Murase, K.: Self-organization of Spiking Neural Network Generating Autonomous Behavior in a Real Mobile Robot. In: Proceedings of the International Conference on Computational Intelligence for Modelling, Control and Automation, vol. 1, pp. 1134–1139 (2005)
2. Alnajjar, F., Bin Mohd Zin, I., Murase, K.: A Spiking Neural Network with Dynamic Memory for a Real Autonomous Mobile Robot in Dynamic Environment. In: 2008 International Joint Conference on Neural Networks (2008)
3. Dauc, E., Henry, F.: Hebbian Learning in Large Recurrent Neural Networks. Movement and Perception Lab, Marseille (2006)
4. Elizondo, D., Birkenhead, R., Taillard, E.: Generalisation and the Recursive Deterministic Perceptron. In: International Joint Conference on Neural Networks, pp. 1776–1783 (2006)
5. Elizondo, D., Fiesler, E., Korczak, J.: Non-ontogenetic Sparse Neural Networks. In: International Conference on Neural Networks, vol. 26, pp. 290–295. IEEE, Los Alamitos (1995)
6. Florian, R.V.: Reinforcement Learning Through Modulation of Spike-timing-dependent Synaptic Plasticity. Neural Computation 19(6), 1468–1502 (2007)
7. Fritzke, B.: Fast Learning with Incremental RBF Networks. Neural Processing Letters 1(1), 2–5 (1994)
8. Fritzke, B.: A Growing Neural Gas Network Learns Topologies. In: Advances in Neural Information Processing Systems, vol. 7, pp. 625–632 (1995)
9. Gómez, G., Lungarella, M., Hotz, P.E., Matsushita, K., Pfeifer, R.: Simulating Development in a Real Robot: On the Concurrent Increase of Sensory, Motor, and Neural Complexity. In: Proceedings of the Fourth International Workshop on Epigenetic Robotics, pp. 119–122 (2004)
10. Greenwood, G.W.: Attaining Fault Tolerance through Self-adaption: The Strengths and Weaknesses of Evolvable Hardware Approaches. In: Zurada, J.M., Yen, G.G., Wang, J. (eds.) WCCI 2008. LNCS, vol. 5050, pp. 368–387. Springer, Heidelberg (2008)
11. Hebb, D.O.: The Organization of Behaviour: A Neuropsychological Approach. John Wiley & Sons, New York (1949)
12. Izhikevich, E.M.: Which Model to Use for Cortical Spiking Neurons? IEEE Transactions on Neural Networks 15(5), 1063–1070 (2004)
13. Izhikevich, E.M.: Solving the Distal Reward Problem through Linkage of STDP and Dopamine Signaling. Cerebral Cortex 10, 1093–1102 (2007)
14. Jaeger, H.: The "Echo State" Approach to Analysing and Training Recurrent Neural Networks. GMD Report 148, German National Research Institute for Computer Science (2001)
15. Katic, D.: Leaky-integrate-and-fire und Spike Response Modell. Institut für Technische Informatik, Universität Karlsruhe (2006)
16. Kohonen, T.: Self-organization and Associative Memory. 3rd printing. Springer, Heidelberg (1989)
17. Liu, J., Buller, A.: Self-development of Motor Abilities Resulting from the Growth of a Neural Network Reinforced by Pleasure and Tension. In: Proceedings of the 4th International Conference on Development and Learning, pp. 121–125 (2005)
18. Liu, J., Buller, A., Joachimczak, M.: Self-motivated Learning Agent: Skill-development in a Growing Network Mediated by Pleasure and Tensions. Transactions of the Institute of Systems, Control and Information Engineers 19(5), 169–176 (2006)

19. Maass, W., Natschlaeger, T., Markram, H.: Real-time Computing without Stable States: A New Framework for Neural Computation Based on Perturbations. Neural Computation 14(11), 2531–2560 (2002)
20. Maes, P., Brooks, R.A.: Learning to Coordinate Behaviors. In: AAAI, pp. 796–802 (1990)
21. Martinetz, T.M., Schulten, K.J.: A "Neural-Gas" Network Learns Topologies. In: Kohonen, T., Mäkisara, K., Simula, O., Kangas, J. (eds.) Artificial Neural Networks, pp. 397–402 (1991)
22. Schölkopf, B., Smola, A.J.: Learning with Kernels: Support Vector Machines, Regularization, Optimization, and Beyond. In: Adaptive Computation and Machine Learning. The MIT Press, Cambridge (2001)
23. Stanley, K.O., D'Ambrosio, D., Gauci, J.: A Hypercube-Based Indirect Encoding for Evolving Large-Scale Neural Networks. Accepted to appear in Artificial Life journal (2009)
24. Tajine, M., Elizondo, D.: The Recursive Deterministic Perceptron Neural Network. Neural Networks 11, 1571–1588 (1998)
25. Vreeken, J.: Spiking Neural Networks, an Introduction. Intelligent Systems Group, Institute for Information and Computing Sciences, Utrecht University (2003)

Avoiding Prototype Proliferation in Incremental Vector Quantization of Large Heterogeneous Datasets

Héctor F. Satizábal, Andres Pérez-Uribe, and Marco Tomassini

Abstract. Vector quantization of large datasets can be carried out by means of an incremental modelling approach where the modelling task is transformed into an incremental task by partitioning or sampling the data, and the resulting datasets are processed by means of an incremental learner. Growing Neural Gas is an incremental vector quantization algorithm with the capabilities of topology-preserving and distribution-matching. Distribution matching can produce overpopulation of prototypes in zones with high density of data. In order to tackle this drawback, we introduce some modifications to the original Growing Neural Gas algorithm by adding three new parameters, one of them controlling the distribution of the codebook and the other two controlling the quantization error and the amount of units in the network. The resulting learning algorithm is capable of efficiently quantizing large datasets presenting high and low density regions while solving the prototype proliferation problem.

Keywords: Large Datasets, Vector Quantization, Topology-Preserving Networks, Distribution-matching, Prototype Proliferation, Growing Neural Gas Algorithm.

Héctor F. Satizábal
Institut des Systèmes d'Information (ISI), Hautes Etudes Commerciales (HEC), Université de Lausanne, Switzerland
e-mail: `Hector.SatizabalMejia@unil.ch`

Andres Pérez-Uribe
Reconfigurable and Embeded Digital Systems (REDS), University of Applied Sciences of Western Switzerland (HES-SO) (HEIG-VD)
e-mail: `andres.perez-uribe@heig-vd.ch`

Marco Tomassini
Institut des Systèmes d'Information (ISI), Hautes Etudes Commerciales (HEC), Université de Lausanne, Switzerland
e-mail: `Marco.Tomassini@unil.ch`

L. Franco et al. (Eds.): Constructive Neural Networks, SCI 258, pp. 243–260.
springerlink.com

1 Introduction

Processing information from large databases has become an important issue since the emergence of the new large scale and complex information systems (e.g., satellite images, bank transaction databases, marketing databases, internet). Extracting knowledge from such databases is not an easy task due to the execution time and memory constraints of actual systems. Nonetheless, the need for using this information to guide decision-making processes is imperative.

Classical *data mining* algorithms exploit several approaches in order to deal with this kind of dataset [8, 3]. Sampling, partitioning or hashing the dataset drives the process to a *split and merge*, *hierarchical* or *constructive* framework, giving the possibility of building large models by assembling (or adding) smaller individual parts. Another possibility to deal with large datasets is incremental learning [9]. In this case, the main idea is to transform the modelling task into an *incremental task*[1] by means of a sampling or partitioning procedure, and the use of an incremental learner that builds a model from the single samples of data (one at a time).

Moreover, large databases contain a lot of redundant information. Thus, having the complete set of observations is not mandatory. Instead, selecting a small set of prototypes containing as much information as possible would give a more feasible approach to tackle the knowledge extraction problem. One well known approach to do so is *Vector Quantization* (VQ). VQ is a classical quantization technique that allows the modelling of a distribution of points by the distribution of prototypes or reference vectors. Using this approach, data points are represented by the index of their closest prototype. The codebook, i.e. the collection of prototypes, typically has many entries in high density regions, and discards regions where there is no data [1].

A widely used algorithm implementing VQ in an incremental manner is Growing Neural Gas (GNG) [7]. This neural network is part of the group of *topology-representing networks* which are unsupervised neural network models intended to reflect the topology (i.e. dimensionality, distribution) of an input dataset [12]. GNG generates a graph structure that reflects the topology of the input data manifold (topology learning). This data structure has a dimensionality that varies with the dimensionality of the input data. The generated graph can be used to identify clusters in the input data, and the nodes by themselves could serve as a codebook for vector quantization [5].

In summary, building a model from a large dataset could be done by splitting the dataset in order to make the problem an *incremental task*, then applying an *incremental learning* algorithm performing *vector quantization* in order to obtain a reduced set of *prototypes* representing the whole set of data, and then using the resulting codebook to build the desired model.

Growing Neural Gas suffers from prototype proliferation in regions with high density due to the absence of a parameter stopping the insertion of units in sufficiently-represented[2] areas. This stopping criterion could be based on a local

[1] A learning task is incremental if the training examples used to solve it become available over time, usually one at a time [9].

[2] Areas with low quantization error.

measure of performance. One alternative that exploits this approach to overcome the aforementioned drawback was proposed by Cselenyi [4]. In this case, the proposed method introduces eight new parameters to the GNG algorithm proposed by Fritzke [7]. In our case, we propose a modification that adds three new parameters to the original GNG algorithm in order to restrict the insertion of new units due to points belonging to already covered areas. This approach promotes the insertion of new units in areas with higher quantization error in order to produce network structures covering a higher volume of data using the same number of units.

The rest of the article is structured as follows. In section 2 we make a brief description of the original GNG algorithm. Section 3 describes the proposed modifications made to the algorithm. Section 4 describes some of the capabilities of the resulting method using some "toy" datasets. Section 5 shows two tests of the modified algorithm with a real large size dataset and finally, in section 6 we give some conclusions and insights about prototype proliferation and the exploitation of large datasets.

2 Growing Neural Gas

Growing Neural Gas (GNG) [7] is an incremental *point-based network* [2] which performs vector quantization and topology learning. The algorithm builds a neural network by incrementally adding units using a competitive hebbian learning strategy. The resulting structure is a graph of neurons that reproduces the topology of the dataset by keeping the distribution and the dimensionality of the training data [5].

The classification performance of GNG is comparable to conventional approaches [10] but has the advantage of being incremental. This gives the possibility of training the network even if the dataset is not completely available all the time while avoiding the risk of catastrophic interference. Moreover, this feature makes GNG also suitable for incremental modelling taks where processing a large dataset is not possible due to memory constraints. In such cases, one should proceed in two steps. First, the dataset of the process to be modelled should be split in smaller parts having a size that the system can manage. Second, the resulting parts should be used to train a model like GNG by feeding incrementally each one of the individual datasets resulting from the partitioning procedure. In summary, the methodology consists of transforming the modelling task into an incremental-modelling task, and training an incremental learner in order to build a model of the complete dataset [9].

The algorithm proposed by Fritzke is shown in table 1. In such an approach, every λ iterations (step 8), one unit is inserted halfway between the unit q having the highest error and its neighbour f having also the highest error. Carrying out this insertion without any other consideration makes the network converge to a structure where each cell is the prototype for approximately the same number of data points and hence, keeping the original data distribution.

As an example, a GNG network was trained using the dataset shown in figure 1, and the training parameters shown in table 2. These values were selected after several runs of the algorithm.

Table 1 Original growing neural gas algorithm proposed by Fritzke.

Step 0: Start with two units a and b at random positions w_a and w_b in $\Re^n$

Step 1: Generate an input signal ξ according to a (unknown) probability density function $P(\xi)$

Step 2: Find the nearest unit s_1 and the second-nearest unit s_2

Step 3: Increment the age of all edges emanating from s_1

Step 4: Add the squared distance between the input signal and the nearest unit in input space to a local counter variable:

$$\Delta error(s_1) = \|w_{s_1} - \xi\|^2 \tag{1}$$

Step 5: Move s_1 and its direct topological neighbours towards ξ by fractions ε_b and ε_n, respectively, of the total distance:

$$\Delta w_{s_1} = \varepsilon_b(\xi - w_{s_1}) \tag{2}$$

$$\Delta w_n = \varepsilon_n(\xi - w_n) \quad \text{for all direct neighbours } n \text{ of } s_1 \tag{3}$$

Step 6: If s_1 and s_2 are connected by an edge, set the age of this edge to zero. If such an edge does not exist, create it

Step 7: Remove edges with an age larger than a_{max}. If the remaining units have no emanating edges, remove them as well

Step 8: If the number of input signals generated so far is an integer multiple of a parameter λ, insert a new unit as follows:

- Determine the unit q with the maximum accumulated error.
- Insert a new unit r halfway between q and its neighbour f with the largest error variable:

$$w_r = 0.5(w_q + w_f) \tag{4}$$

- Insert edges connecting the new unit r with units q and f, and remove the original edge between q and f.
- Decrease the error variables of q and f by multiplying them with a constant α. Initialize the error variable of r with the new value of the error variable of q.

Step 9: Decrease all error variables by multiplying them with a constant d

Step 10: If a stopping criterion (e.g., net size or some performance measure) is not yet fulfilled go to step 1

Figure 2 shows the position and distribution of the 200 cells of the resulting structure. As we can see, the distribution of each one of the variables is reproduced by the group of prototypes required.

The GNG algorithm is a vector quantizer which places prototypes by performing entropy maximization [6]. This approach, while allowing the vector prototypes

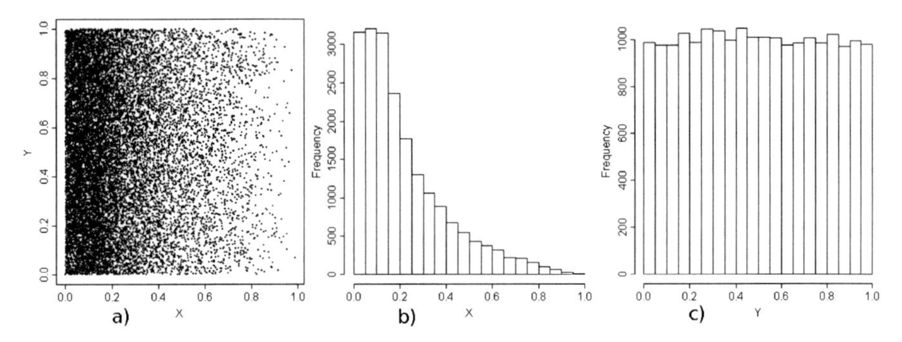

Fig. 1 a) Two dimensional non-uniform data distribution. b) Histogram of variable X. c) Histogram of variable Y.

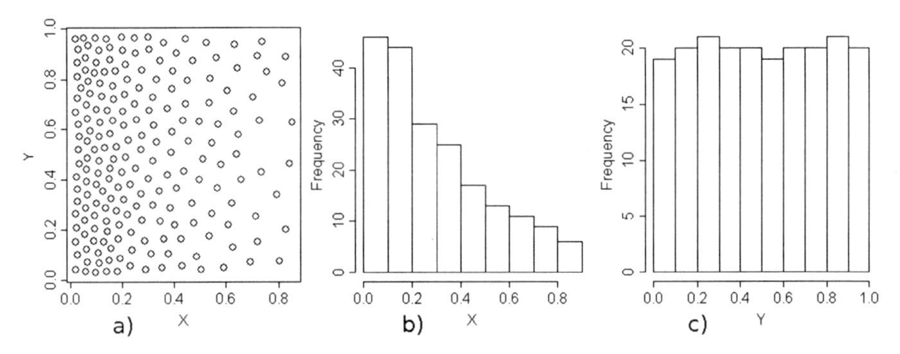

Fig. 2 Positions of neurons of the GNG model. a) Position of the neuron units. b) Distribution of X. c) Distribution of Y.

Table 2 Parameters for the Growing Neural Gas algorithm.

Parameter	ε_b	ε_n	λ	a_{max}	α	d
value	0.05	0.005	100	100	0.5	0.9

to preserve the distribution of the original data, promotes the overpopulation of vector prototypes in high density regions, and therefore, this situation is desirable if the goal is a distribution-matching codebook. Conversely, if we have a huge dataset where there is a lot of redundant information, and we want to keep only a relatively small number of prototypes describing the data with some distortion, then we are not interested in reproducing the data distribution. Instead, we would want to distribute the prototypes over the whole volume of data without exceeding a maximum quantization error or distortion. In such a case, some modifications to the original algorithm are needed in order to modulate the distribution matching property of the algorithm, and therefore, replacing the entropy maximization behaviour in high densitiy regions by an error minimization policy being capable of stopping the insertion of prototypes.

3 Proposed Modifications to the Algorithm

This section describes the three main modifications we propose. As already mentioned, the main goal of modifying the original algorithm is avoiding prototype proliferation in high density regions of data, or in other words, to modulate the distribution matching property of the algorithm. The modified version adds three new parameters to the original algorithm, whose operation is explained in the following subsections.

3.1 *Modulating the Local Measure of Error*

The criterion driving the insertion of units in the GNG algorithm is the accumulated error of each unit (equation 1). This local error measure grows each time a cell becomes the winner unit (i.e. the closest unit to the current data point), producing the insertion of more cells in zones with higher densities of data.

In order to attenuate that effect, we propose the error signal $\Delta error$ when it is produced by a data point having a quantization error smaller than a threshold qE by multiplying it by a factor h, as show in equation 5.

$$accumulatedError = \begin{cases} accumulatedError + \Delta error & \text{if } \Delta error \geq qE \\ accumulatedError + (h * \Delta error) & \text{if } \Delta error < qE \end{cases} \quad (5)$$
$$\text{where } 0 \leq h \leq 1$$

3.2 *Modulating the "Speed" of Winner Units*

The proliferation of prototypes could also be due to neuron movement. Neuron units located in zones with higher densities are chosen as winners with higher probability, attracting their neighbours belonging to less populated zones. As in the previous case, parameter h can be used to modulate the change of position, or speed, of the units in each iteration, replacing equations 2 and 3, by equations 6 and 7.

$$\Delta w_{s_1} = \begin{cases} \varepsilon_b (\xi - w_{s_1}) & \text{if } \Delta error \geq qE \\ h * \varepsilon_b (\xi - w_{s_1}) & \text{if } \Delta error < qE \end{cases} \quad (6)$$

$$\Delta w_n = \begin{cases} \varepsilon_n (\xi - w_n) & \text{if } \Delta error \geq qE \\ h * \varepsilon_n (\xi - w_n) & \text{if } \Delta error < qE \end{cases} \quad \text{for all direct neighbours } n \text{ of } s_1 \quad (7)$$

3.3 *Modulating the Overlapping of Units*

Parameter qE is the radius of a hypersphere determining the region of influence of each prototype. These regions of influence could overlap to a certain extent. In our approach this amount of superposition can be controlled by a parameter sp. Every λ iterations (step 8), one unit is inserted halfway between the unit q having the highest

Table 3 Proposed modification to the original algorithm.

Step 8: If the number of input signals generated so far is an integer multiple of a parameter λ, insert a new unit as follows:

- Determine the unit q with the maximum accumulated error.
- Determine the unit f in the neighbourhood of q with the maximum accumulated error.
- Calculate the distance $dist$ between units q and f.

$$dist = \|q - f\| \tag{8}$$

- Calculate the available space between the two units as follows:

$$available = dist - qE \tag{9}$$

If equation 10 yields true, goto step 9.

$$(available > (sp \times qE)) \quad \text{where } 0 \leq sp \leq 1 \tag{10}$$

Else,

- Insert a new unit r halfway between q and its neighbour f with the largest error variable:

$$w_r = 0.5\left(w_q + w_f\right) \tag{11}$$

- Insert edges connecting the new unit r with units q and f, and remove the original edge between q and f.
- Decrease the error variables of q and f by multiplying them with a constant α. Initialize the error variable of r with the new value of the error variable of q.

---

error and its neighbour f having also the highest error (see equation 4). Knowing the distance between unit q and unit f (equation 8), and taking the quantization error qE as the radius of each unit, one could change the step 8 of the original algorithm proposed by Fritzke as shown in table 3. Hence, we propose to insert a new unit as in the original version (equation 11), but only if there is enough place between unit q and unit f (equation 10).

3.4 Insights about the Proposed Modification

In summary, the proposed modifications add three parameters, qE, sp and h, to the original GNG algorithm. Parameters *quantization error* qE and *superposition percent* sp depend on the application and are strongly related. Both control the amount of units in the resulting neural network structure, the former by controlling the region of influence of each unit, and the latter by controlling the superposition of units. These natural meanings allow them to be tuned according to the requirements of each specific application.

In a less obvious sense, parameter h controls the distribution of units between high and low density areas, modulating the distribution-matching property of the algorithm. In order to do so, parameter h modulates the signal that drives the insertion of new units in the network only if the best matching neuron fulfills a given quantization error condition. In this way, the algorithm does not insert unnecessary units in well represented zones, even if the local error measure increases due to high data density. Some examples exploring this parameters are given in section 4.

4 Toyset Experiments

In section 2, a non-uniform distribution of points in two dimensions was used to train a GNG network. Figure 2 shows a high concentration of prototypes in the zone with higher density due to the property of density matching of the model. This is an excellent result if we do not have any constraint on the amount of prototypes. In fact, having more prototypes increases the execution time of the algorithm since

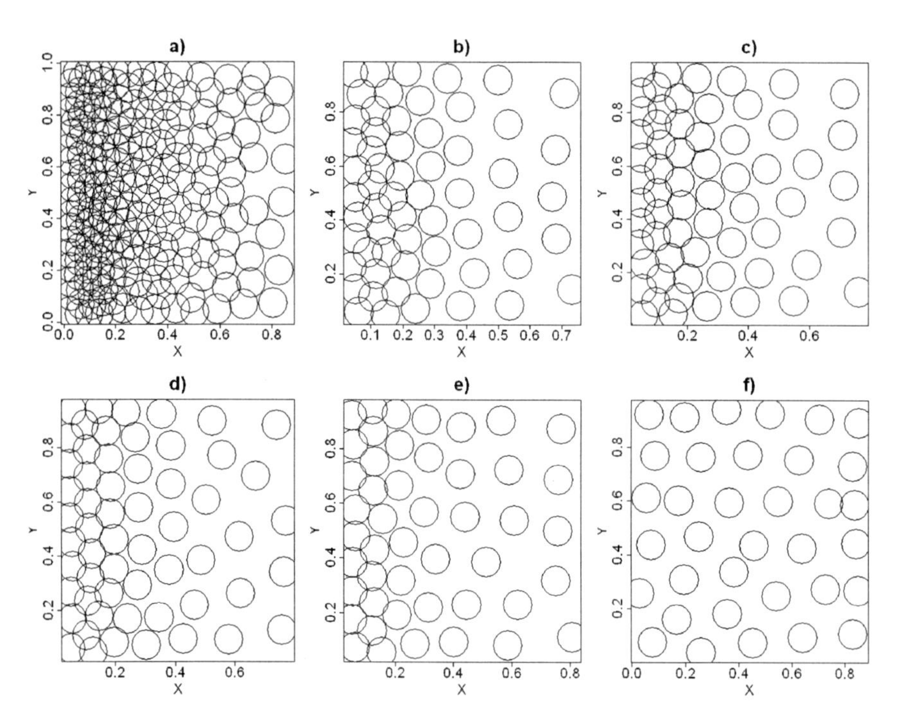

Fig. 3 Results of the modified algorithm varying parameter h ($qE = 0.1$ and $sp = 0.75$). a) Using the original GNG algorithm b) Using $h = 1.00$ c) Using $h = 0.75$ d) Using $h = 0.50$ e) Using $h = 0.25$ f) Using $h = 0.00$.

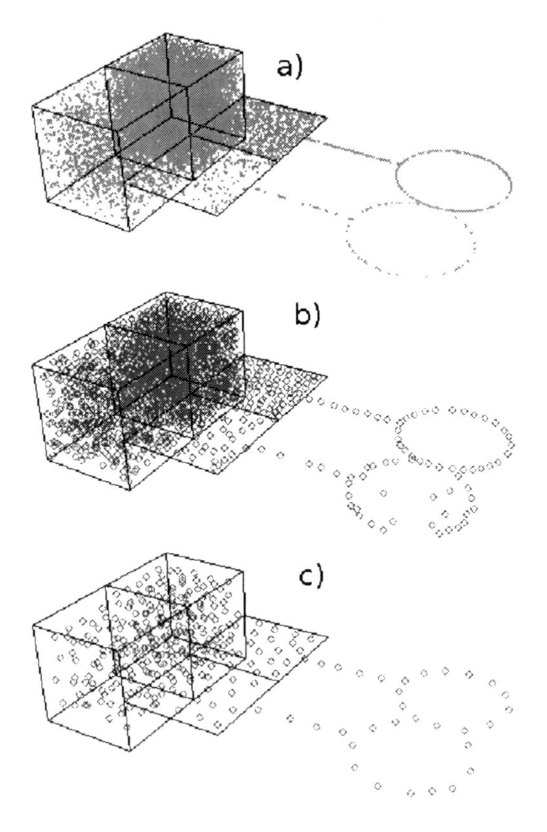

Fig. 4 The original and the modified version of GNG trained with a dataset like the one used by Cselenyi [4]. a) Training data. b) Algorithm of GNG by Fritzke. c) Modified GNG algorithm, $qE = 0.1$, $h = 0.1$, $sp = 0.5$.

there are more units to evaluate each time a new point is evaluated, and this is not desirable if we have a very large dataset. Moreover, we apply vector quantization in order to reduce the number of points to process by choosing a suitable codebok, and therefore, redundant prototypes are not desirable. This section shows how the proposed modification to controlling prototype proliferation allows us to overcome this situation. Two experiments with controlled toysets should help in the testing and understanding of the modified algorithm.

Figure 3 shows the results of the modified algorithm when trained with the data distribution showed in figure 1. Figure 3 shows how parameter h effectively controls the proportion of units assigned to regions with high and low densities. In this case the parameters qE and sp were kept constant ($qE = 0.1$ and $sp = 0.75$) since their effects are more global and depend less on the data distribution. The rest of the parameters were set as shown in table 2.

Another interesting test consists of using a dataset similar to the one proposed by Martinetz [13] in the early implementations of this kind of network (neural gas, growing cell structures, growing neural gas) [5]. This distribution of data has been used by several researchers [4, 5, 7, 12, 13] in order to show the ability of the *topology-preserving* networks in modelling the distribution and dimensionality of data. The generated dataset shown in figure 4 a) presents two different levels of densities for points situated in three, two and one dimension, and has points describing a circle.

When this dataset is used, the model has to deal with data having different dimensionalities, different densities and different topologies. Figures 4 b) and 4 c) show the position of the units of two GNG networks, one of them using the original algorithm and the other one using the modified version. Both structures preserve the topology of the data in terms of dimensionality by placing and connecting units depending on local conditions. Conversely, the two models behave differently in terms of the distribution of the data. The codebook of the original GNG algorithm reproduces the distribution of the training data by assigning almost the same quantity of data points to each vector prototype. In the case of the modified version, the parameter h set to 0.1 makes the distribution of prototypes more uniform due to the fact that the insertion of new units is conditioned with the quantization error. Other parameters were set as shown in table 2.

5 Obtaining a Codebook from a Large Dataset

This section summarizes a series of experiments using a large database of climate. The database contains information of the temperature (minimum, average, and maximum) and the precipitation over approximately the last fifty years in Colombia, with a spatial resolution of 30 seconds (≈900m) (WORLDCLIM) [11]. This database is part of a cooperative project between Colombia and Switzerland named "Precision Agriculture and the Construction of Field-Crop Models for Tropical Fruits", where one of the objectives is finding geographical zones with similar environmental conditions, in order to facilitate the implementation or migration of some crops. There are 1,336.025 data points corresponding to the amount of pixels covering the region, and each one has twelve dimensions corresponding to the months of the year (i.e. one vector of twelve dimensions per pixel), and each month has four dimensions corresponding to the aforementioned variables.

Processing the whole dataset[3] implies the use of a lot of memory resources and takes hours of calculation. Moreover, the situation could get even worse if we consider the use of the whole set of variables at the same time. Therefore, instead of processing every pixel in the dataset, we could use vector quantization to extract a codebook representing the data, and then to process this set of prototypes finding the zones that have similar properties.

[3] Finding zones with similar environmental conditions (i.e, temperature) by means of some measure of distance.

5.1 Incremental Learning of a Non-incremental Task

The first test was done by taking only the data corresponding to the average temperature from the dataset described in section 5. The resulting dataset has 1,336.025 observations corresponding to the amount of pixels on the map, and each one has twelve dimensions corresponding to the months of the year. Figure 5 shows the resulting quantization errors using both algorithms.

Both neural networks have only 89 neuron units, which means having a codebook with only 0.007% of the original size of the dataset. Nonetheless, the quantization error is astonishingly low. This reduction is possible due to the low local dimensionality of the data, and the low range of the variables. Figure 5 shows that the modified algorithm presents quantization error values that are comparable to those from the original version, but with a slightly different distribution.

Having a dataset which allows a representation over two dimensions has some advantages. In this case, we can draw some information from the geographic distribution of the prototypes. Figure 6 shows the geographic representation of the boundaries (white lines) of the Voronoi region of each prototype. The region delimited with a circle is a wide plain at low altitudes which presents homogeneous conditions in terms of temperature. Therefore, this large amount of pixels belongs to a high density zone in the space of twelve dimensions of our data. In this case, this high density zone does not mean more information to quantize. However, the original GNG algorithm is "forced" to proliferate prototypes due to its property of distribution matching. Thus incurring in a high computational cost when one uses

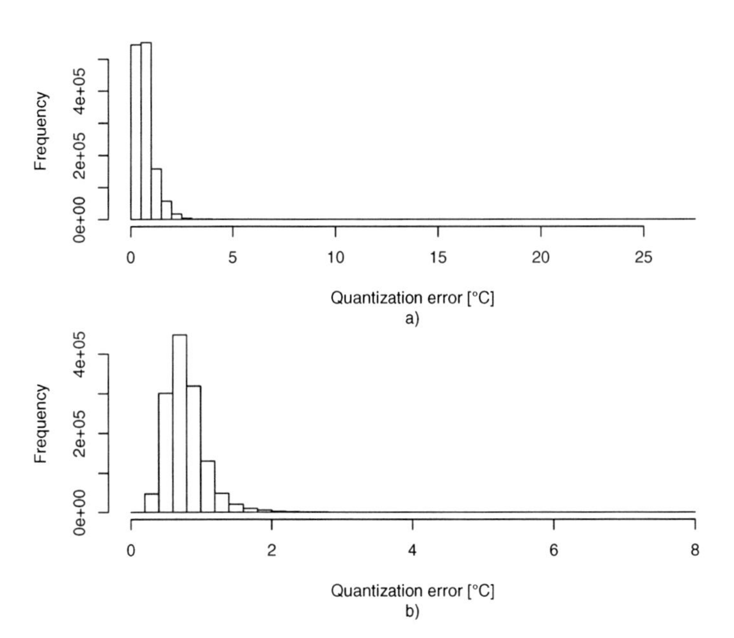

Fig. 5 Histogram of the quantization error for a large dataset. a) Fritzke's original GNG algorithm. b) Modified GNG algorithm, $qE = 1\,°\text{C}$, $h = 0.1$, $sp = 0.5$.

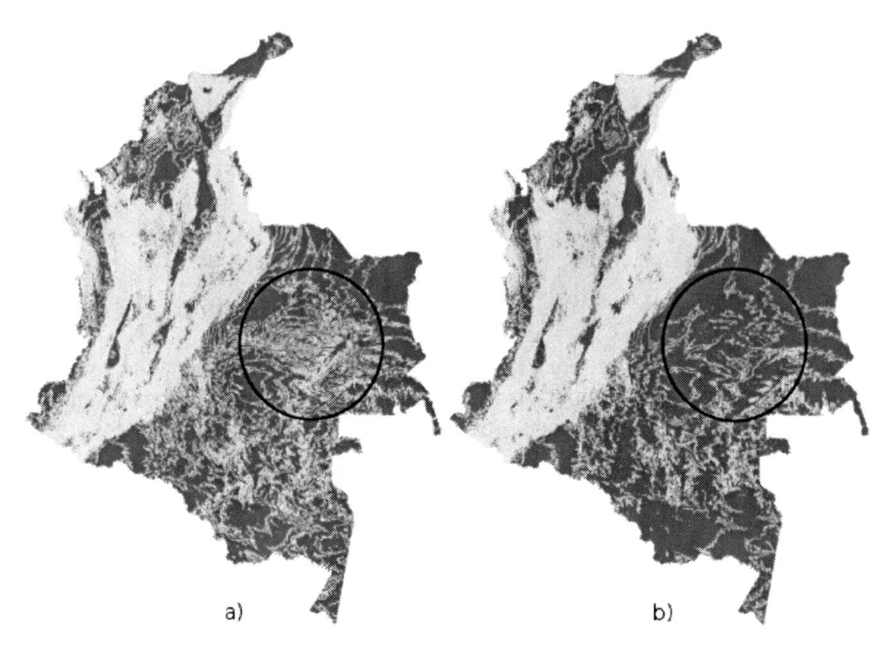

Fig. 6 Prototype boundaries. a) Original algorithm. b) Modified algorithm.

the resulting model. Instead of representing better these areas, our approach is to avoid prototype proliferation in regions with regular conditions in order to better represent heterogeneous zones (e.g., mountains). Figure 6.b) shows that the modified version of GNG places less prototypes in flat areas (i.e., high density regions) than the original version (Figure 6.a), and assigns more prototypes (i.e., cluster centres) to the lower density points belonging to mountain areas (i.e., low density regions).

5.2 Incremental Learning of an Incremental Task

Even if the dataset used in section 5.1 was a large one, and its codebook was extracted by using an incremental algorithm, it was not processed by using the incremental approach. In this case, the whole dataset was used for training, and therefore, the algorithm had access to every point in the dataset at every iteration. Incremental modelling, instead, proposes dividing the dataset, and training on an incremental

Table 4 Parameters for the modified Growing Neural Gas algorithm.

Parameter	ε_b	ε_n	λ	a_{max}	α	d	$qError$	h	sp
value	0.05	0.005	250	1000	0.5	0.9	0.1	0.1	0.5

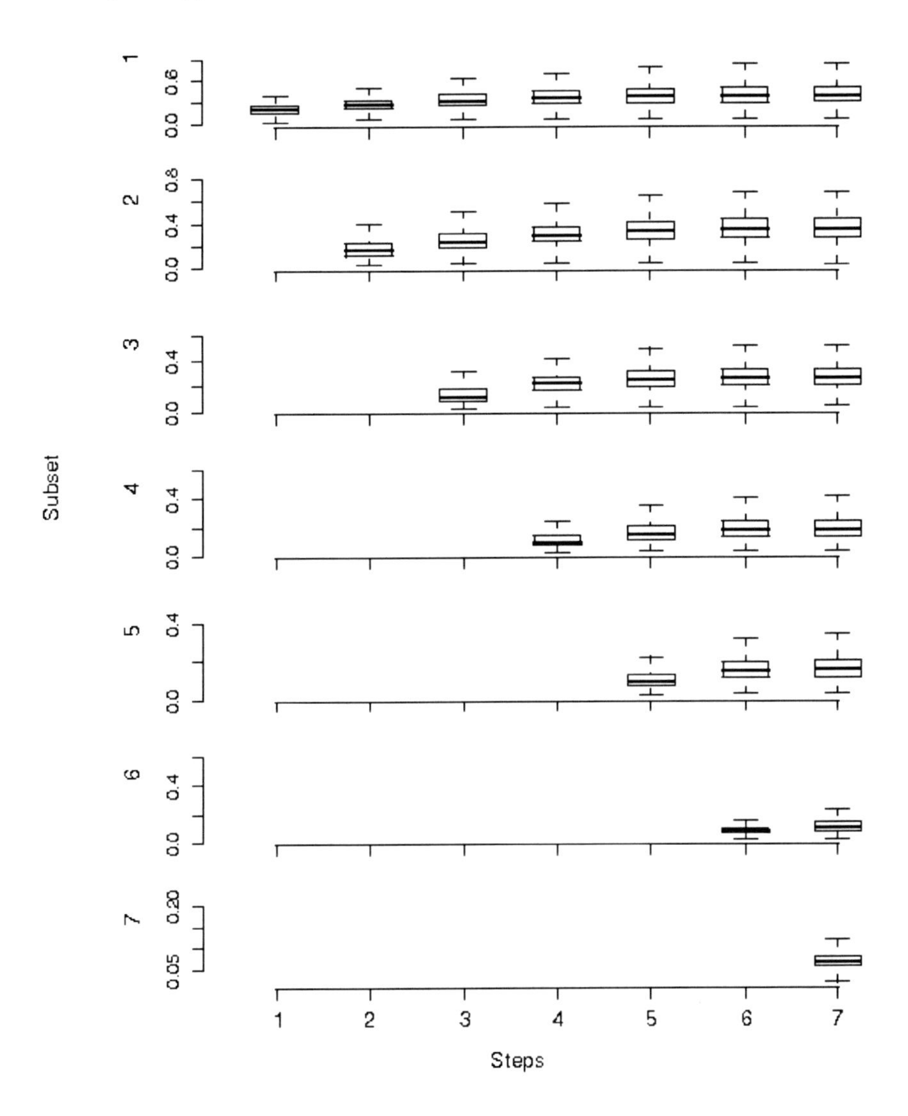

Fig. 7 Error distributions after each incremental step. Each line shows the evolution of the quantization error for a given subset when adding new knowledge to the model.

learner by using the individual parts. This scenario was tested by using the complete dataset mentioned in section 5.

The complete dataset of climate with 1,336.025 data points and forty-eight dimensions was divided in an arbitrary way (i.e. from north to south) into 7 parts; six parts of 200.000 observations, and one final part having 136.025 data points. These individual subsets were used in order to incrementally train a model using our modified version of the GNG algorithm. The parameters used are shown in table 4.

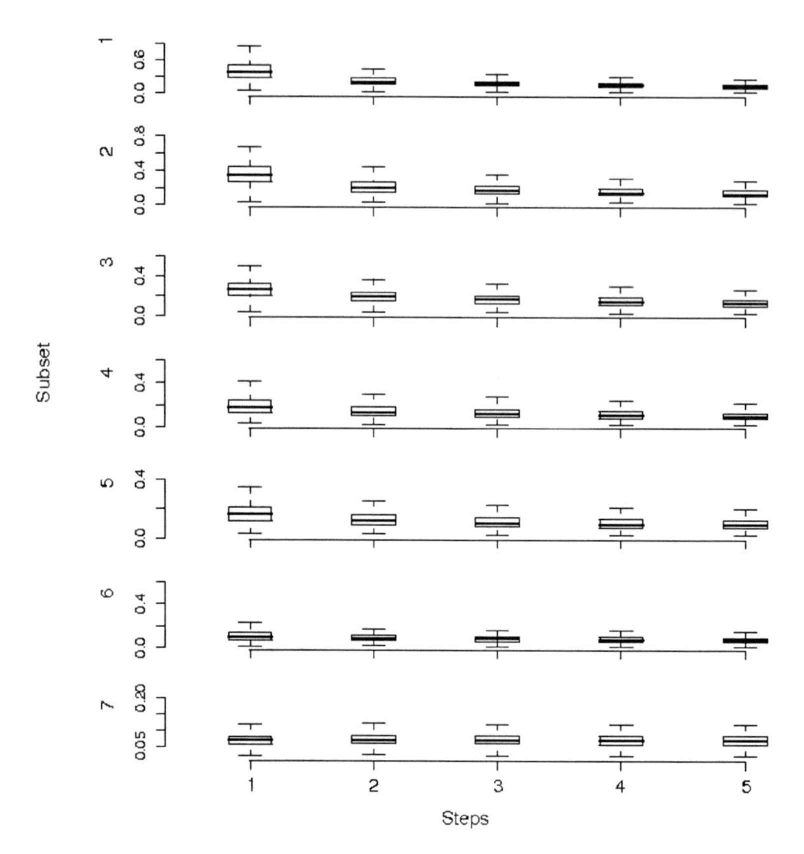

Fig. 8 Error distributions after each training-repetition step. Each line shows the evolution of the quantization error for a given subset.

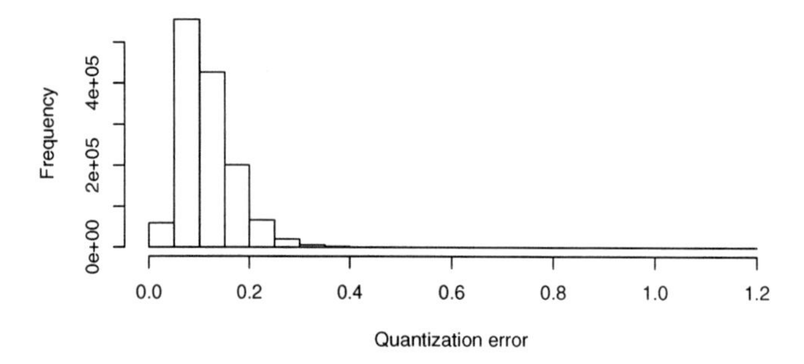

Fig. 9 Histogram of the quantization error for codebook obtained with the modified version of GNG, after feeding in the whole climate dataset

Figure 7 shows the evolution of the quantization error after adding each one of the seven subsets. Each horizontal line of boxplots represents the error for a given subset, and each step means the fact of training the network with a new subset. As

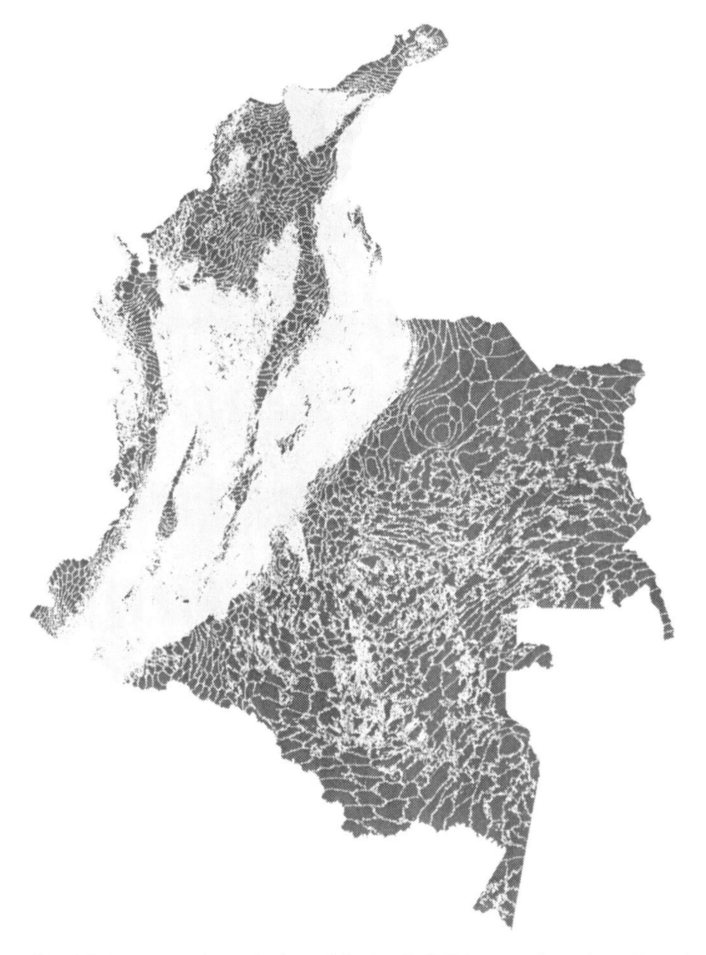

Fig. 10 Prototype boundaries of the final GNG network projected on the geographic map.

can be seen in figure 7, the quantization error for a subset presented in previous steps increases in further steps with the addition of new knowledge. Such behaviour suggests that, even if the GNG algorithm performs incrementally, the addition of new data can produce interference with the knowledge already stored within the network. This undesirable situation happens if the new data are close enough to the existant prototypes to make them move, forgetting the previous information.

In order to overcome this weakness of the algorithm, a sort of repetition policy was added to the training procedure, as follows. The GNG network was trained several times with the same sequence of subsets, from the first to the seventh subset. At each iteration[4] the algorithm had two possibilities: choosing a new observation from the input dataset, or taking one of the prototype vectors from the current codebook as input. The former option, which we named *exploration*, could be taken with a

[4] Each time a new point is going to be presented to the network.

pre-defined probability. Figure 8 shows the evolution of the quantization error for five training-repetition steps of the algorithm, using an exploration level of 50%.

As can be seen in figure 8, the quantization error for each one of the seven subsets decreases with each iteration of the algorithm. Such behaviour means that the GNG network is capable of avoiding catastrophic forgetting of knowledge when using repetition. Moreover, after five iterations the network reaches a low quantization error over the whole dataset. This result is shown in figure 9.

The resulting network has 14.756 prototype vectors in its codebook, which represents 1.1% of the total amount of pixels in the database. The number of prototypes is larger than in the case of the temperature because of the larger dimensionality of the observations (i.e. forty-eight dimensions instead of twelve). Moreover, precipitation data have a wider range than temparature, increasing the area where prototype vectors should be placed.

Figure 10 shows the Voronoi region of each prototype projected over the two-dimensional geographic map. As in the previous case, one can see that prototypes are distributed over the whole map, and they are more concentrated in mountain zones, as desired.

Finally, after quantizing the dataset of 1,336.025 observations, the sets of similar climate zones could be found by analyzing the 14.756 prototypes obtained from the GNG network. This compression in the amount of data to be analyzed is possible due to the existance of redundancy in the original dataset. In other words, pixels with similar characteristics are represented by a reduced number of vector prototypes, even if they are located in regions which are not geographically adjacent.

6 Conclusions

Nowadays, there is an increasing need for dealing with large datasets. A large dataset can be split or sampled in order to divide the modelling task into smaller subtasks that can be merged in a single model by means of an incremental learning technique performing vector quantization. In our case, we chose the Growing Neural Gas (GNG) algorithm as the vector quantization technique. GNG allows us to get a reduced codebook to analyse, instead of analysing the whole dataset. Growing Neural Gas is an excellent incremental vector quantization technique, allowing us to preserve the topology and the distribution of a set of data.

However, in our specific application, we found it necessary to modulate the topology matching property of the GNG algorithm in order to control the distribution of units between zones with high and low density. To achieve this, we modified the original algorithm proposed by Fritzke by adding three new parameters, two controlling the quantization error and the amount of neuron units in the network, and one controlling the distribution of these units. The modified version still has the property of topology-preservation, but contrary to the original version it permits the modulation of the distribution matching capabilities of the original algorithm. These changes allow the quantization of datasets having high contrasts in density while keeping the information of low density areas, and using a limited number of prototypes.

Moreover, we tested the modified algorithm on the task of quantizing a heterogeneous set of real data. First, the difference in the distribution of prototypes between the original and the modified version was tested by using the classical modelling approach where the whole set of data is available during the training process. By doing so, we verified that the modified version modulates the insertion of prototypes in high density regions of data. Finally, we used the modified version of the algorithm to perform an incremental modelling task over a larger version of the former dataset. A repetition policy had to be added to the incremental training procedure in order to carry out this test. This repetition strategy allowed the GNG network to remember previous information, preventing catastrophic forgetting caused by new data interfering with the already stored knowledge.

Acknowledgements. This work is part of a joint project between BIOTEC, CIAT, CENICAÑA (Colombia) and HEIG-VD (Switzerland) named *"Precision agriculture and the construction of field-crop models for tropical fruits"*. The financial support is given by several institutions in Colombia (MADR, COLCIENCIAS, ACCI) and the State Secretariat for Education and Research (SER) in Switzerland.

References

1. Alpaydin, E.: Introduction to Machine Learning. MIT Press, Cambridge (2004)
2. Bouchachia, A., Gabrys, B., Sahel, Z.: Overview of Some Incremental Learning Algorithms. In: Fuzzy Systems Conference. FUZZ-IEEE 2007, vol. 23-26, pp. 1–6 (2007)
3. Bradley, P., Gehrke, J., Ramakrishnan, R., Srikant, R.: Scaling mining algorithms to large databases. Commun. ACM 45, 38–43 (2002)
4. Cselenyi, Z.: Mapping the dimensionality, density and topology of data: The growing adaptive neural gas. Computer Methods and Programs in Biomedicine 78, 141–156 (2005)
5. Fritzke, B.: Unsupervised ontogenic networks. In: Handbook of Neural Computation, ch. C 2.4, Institute of Physics, Oxford University Press (1997)
6. Fritzke, B.: Goals of Competitive Learning. In: Some Competitive Learning Methods (1997), `http://www.neuroinformatik.rub.de/VDM/research/gsn/JavaPaper/` (Cited October 26, 2008)
7. Fritzke, B.: A Growing Neural Gas Learns Topologies. In: Advances in Neural Information Processing Systems, vol. 7. MIT Press, Cambridge (1995)
8. Ganti, V., Gehrke, J., Ramakrishnan, R.: Mining very large databases. Computer 32, 38–45 (1999)
9. Giraud-Carrier, C.: A note on the utility of incremental learning. AI Commun. 13, 215–223 (2000)
10. Heinke, D., Hamker, F.H.: Comparing neural networks: a benchmark on growing neural gas, growing cell structures, and fuzzy ARTMAP. IEEE Transactions on Neural Networks 9, 1279–1291 (1998)

11. Hijmans, R., Cameron, S., Parra, J., Jones, P., Jarvis, A.: Very High Resolution Interpolated Climate Surfaces for Global Land Areas. Int. J. Climatol 25, 1965–1978 (2005)
12. Martinetz, T., Schulten, K.: Topology representing networks. Neural Networks 7, 507–522 (1994)
13. Martinetz, T., Schulten, K.: A neural gas network learns topologies. Artificial Neural Networks, 397–402 (1991)

Tuning Parameters in Fuzzy Growing Hierarchical Self-Organizing Networks

Miguel Arturo Barreto-Sanz[1,2], Andrés Pérez-Uribe[2], Carlos-Andres Peña-Reyes[2], and Marco Tomassini[1]

Abstract. Hierarchical Self-Organizing Networks are used to reveal the topology and structure of datasets. These methodologies create crisp partitions of the dataset producing tree structures composed of prototype vectors, permitting the extraction of a simple and compact representation of a dataset. However, in many cases observations could be represented by several prototypes with certain degree of membership. Nevertheless, crisp partitions are forced to classify observations in just one group, losing information about the real dataset structure. To deal with this challenge we propose Fuzzy Growing Hierarchical Self-Organizing Networks (FGHSON). FGHSON are adaptive networks which are able to reflect the underlying structure of the dataset in a hierarchical fuzzy way. These networks grow by using three parameters which govern the membership degree of data observations to the prototype vectors and the quality of the hierarchical representation. However, different combinations of values of these parameters can generate diverse networks. This chapter explores how these combinations affect the topology of the

Miguel Arturo Barreto-Sanz
Université de Lausanne, Hautes Etudes Commerciales (HEC), Institut des Systèmes d'Information (ISI)
e-mail: Miguel-Arturo.Barreto-Sanz@heig-vd.ch

Andrés Pérez-Uribe
University of Applied Sciences of Western Switzerland (HEIG-VD)(REDS)
e-mail: andres.perez-uribe@heig-vd.ch

Carlos-Andres Peña-Reyes
University of Applied Sciences of Western Switzerland (HEIG-VD)(REDS)
e-mail: Carlos.Pena@heig-vd.ch

Marco Tomassini
Université de Lausanne, Hautes Etudes Commerciales (HEC), Institut des Systèmes d'Information (ISI)
e-mail: Marco.Tomassini@unil.ch

L. Franco et al. (Eds.): Constructive Neural Networks, SCI 258, pp. 261–279.
springerlink.com

network and the quality of the prototypes; in addition the motivation and the theoretical basis of the algorithm are presented.

1 Introduction

We live in a world full of data. Every day we are confronted with the handling of large amounts of information. This information is stored and represented as data, for further analysis and management. One of the essential means in dealing with data is to classify or group it into categories or clusters. In fact, as one of the most ancient activities of human beings [1], classification plays a very important role in the history of human development. In order to learn a new object or distinguish a new phenomenon, people always try to look for the features that can describe it and further compare it with other known objects or phenomena, based on the similarity or dissimilarity, generalized as proximity, according to some standards or rules.

In many cases classification must be done without a priori knowledge of the classes in which the dataset is divided (unlabeled pattern). This kind of classification is called clustering (unsupervised classification). On the contrary, discriminant analysis (supervised classification) is made by providing a collection of labeled patterns; so the problem is to label a newly encountered, unlabeled pattern. Typically, the given labeled patterns are used to learn descriptions of classes which in turn are used to label a new pattern. In the case of clustering, the problem is to group a given collection of unlabeled patterns into meaningful clusters. In a sense, labels are associated with clusters also, but these category labels are data driven; that is, they are obtained solely from the data [15, 23].

Even though the unsupervised classification presents many advantages over supervised classification[1], it is a subjective process in nature. As pointed out by Backer and Jain [2], "in cluster analysis a group of objects is split up into a number of more or less homogeneous subgroups on the basis of an often subjectively chosen measure of similarity (i.e., chosen subjectively based on its ability to create "interesting" clusters), such that the similarity between objects within a subgroup is larger than the similarity between objects belonging to different subgroups". Clustering algorithms partition data into a certain number of clusters (groups, subsets, or categories). There is no universally agreed upon definition [8].

Thus, methodologies to evaluate clusters with different levels of abstraction in order to find "interesting" patterns are useful; these methodologies could help to improve the analysis of cluster structure creating representations, facilitating the selection of clusters of interest. Methods for tree structure

[1] For instance, no extensive prior knowledge of the dataset is required, and it can detect "natural" groupings in feature space.

representation and data abstraction have been used for this task, revealing the topology and organization of clusters.

On the one hand, hierarchical methods are used to help explain the inner organization of datasets, since the hierarchical structure imposed by the data produces a separation of clusters that is mapped onto different branches. Hierarchical clustering algorithms organize data into a hierarchical structure according to a proximity matrix. The results of Hierarchical clustering are usually depicted by a binary tree or dendrogram. The root node of the dendrogram represents the whole data set and each leaf node is regarded as a data object. The intermediate nodes describe to what extent the objects are proximal among them; and the height of the dendrogram usually expresses the distance between each pair of objects or clusters, or an object and a cluster. The ultimate clustering results can be obtained by cutting the dendrogram at different levels. This representation provides very informative descriptions and visualization for the potential data clustering structures, especially when real hierarchical relations exist in the data, like the data from evolutionary research on different species of organisms. Therefore, this hierarchical organization enables us to analyze complicated structures as well as the exploration of the dataset at multiple levels of detail [23].

On the other hand, data abstraction permits the extraction of a simple and compact representation of a data set. Here, simplicity is either from the perspective of automatic processing (so that a machine can perform further processing efficiently) or is human-oriented (so that the representation obtained is easy to comprehend and intuitively appealing). In the clustering context, a typical data abstraction is a compact description of each cluster, usually in terms of cluster prototypes or representative patterns such as the centroid of the cluster [7]. Soft competitive learning methods [11] are employed on data abstraction in a self-organizing way. These algorithms attempt to distribute a number of vectors (prototype vectors) in a potentially low-dimensional space. The distribution of these vectors should reflect (in one of several possible ways) the probability distribution of the input signals which in general is not given explicitly but through sample vectors. Two principal approaches have been used for this purpose. The first is based on a fixed network dimensionality (i.e. Kohonen maps [16]). In the second approach, non fixed dimensionality is imposed on the network; hence, this network can automatically find a suitable structure and size through a controlled growth process [19].

Different approaches have been introduced in order to combine the capabilities of tree structure of the hierarchical methods and the advantages of soft competitive learning methods used for data abstraction [20, 13, 6, 22, 12, 18], obtaining networks capable of representing the structure of clusters and their prototypes in a hierarchical self-organizing way. These networks are able to grow and adapt their structure in order to represent the characteristics of clusters in the most accurate manner. Although these hybrid models provide satisfactory results, they generate crisp partitions of the datasets. The crisp segmentations tend to allocate elements of the dataset in just one branch of the tree in each level of the hierarchy and assign just one prototype to represent one cluster, so the membership to other branches or prototypes is zero. Nevertheless, in many applications crisp partitions in hierarchical structures are not the optimal representation of the clusters, since some elements of the dataset could belong to multiple clusters or branches with a certain degree of membership.

One example of this situation is presented in Geographic Information Systems (GIS) applications. One of the topics treated by GIS researchers refers to the classification of geographical zones with similar characteristics to climate, soil and terrain (conditions relevant to agricultural production) in order to create the so called agro-ecological zones (AEZ) [9]. AEZ provide the frame for various applications, such as quantification of land productivity, estimation of land's population supporting capacity, and optimization of land resource use and development. Many institutions, governments and enterprises need to know which AEZ a particular region belongs to (allocating the region to a certain AEZ cluster), in order to apply policies to invest, for instance in new cropping systems for economic viability, and sustainability. However, the geographical region of interest can vary in range of resolution depending on the application or context (i.e. countries, states, cities, parcels). In addition, the fuzzy and implicit nature of the geographic zones (in which geographical boundaries are not hard, but rather soft boundaries) transform the boundaries of the AEZ in zones of transition rather than sharp boundaries. Thus, the soft boundaries make it possible that regions in the middle of two AEZ have membership of both. The clustering method to deal with this situation has to provide views of AEZ at multiple levels, preferably in a hierarchical way. In addition, it should be capable of discovering fuzzy memberships of geographical regions to the AEZ.

For the purpose of representing degrees of membership, fuzzy logic is a feature that could be added to hierarchical self-organized hybrid models. We propose, thus Fuzzy Growing Hierarchical Self-Organizing Networks (FGHSON), with the intention of synergistically combining the advantages of Self-Organizing Networks, hierarchical structures, and fuzzy logic. FGHSON are designed to improve the analysis of datasets where it is desirable to obtain a fuzzy representation of a dataset in a hierarchical way, then discovering its structure and topology. This new model will be able to obtain a growing hierarchical structure of the dataset in a self-organizing fuzzy manner. This kind of network is based on the Fuzzy Kohonen Clustering Networks (FKCN) [4] and Hierarchical Self-Organizing Structures (HSS) [17, 21, 20, 22].

This book chapter is organized as follows: In the next section the Hierarchical Self-Organizing Structures and the Fuzzy Kohonen Clustering Networks will be explained then, our model will be described. Section 3 focuses on the application of the methodology using the Iris benchmark and an example dataset, a further example where model parameters are tuned is also presented. Finally, in Section 4 conclusions are drawn and future extensions of the work described.

2 Methods

2.1 Hierarchical Self-Organizing Structures

The ability to obtain hierarchically structured knowledge from a dataset using autonomous learning has been widely used in many areas. This is due to the fact that hierarchical self-organizing structures permit unevenly distributed real-world data to be represented in a suitable network structure, during an unsupervised training process. These networks capture the unknown data topology in terms of hierarchical relationships and cluster structures.

Different methodologies have been presented in this area with various approaches. It is possible to classify hierarchical self-organizing structures in two classes taking into account the algorithm of self-organization used. The first family of models is based on Kohonen self-organizing maps (SOM), and the second on Growing Cell Structures (GCS) [10].

With respect to approaches based on GCS, Hierarchical Growing Cell Structures (HiGCS) [5], TreeGCS [13] and the Hierarchical topological clustering (TreeGNG) [6] have been proposed. The algorithms derived from GCS are based on periodic node deletion, node activity and the volume of the input space classified by the node. This approach tends to represent examples with high occurrence rates, and therefore takes low frequency examples as outliers or noise. As a result, examples with low presence rates are not represented in the model. Nevertheless, in many cases it is desirable to discover novelties in the dataset, so taking into account the observations with low occurrence rates could allow discovery of those exceptional behaviors.

For this reason, we focused our research on approaches based on SOM [17, 21, 20], particularly the Growing Hierarchical Self-Organizing Map (GHSOM)[22] due to its ability to take into account the observations with low presence rates as part of the model. This is possible since the hierarchical structure of the GHSOM is adapted according to the requirements of the input space. Therefore, areas in the input space that require more units for appropriate data representation create deeper branches than others. This process is done without eliminating nodes that represent examples with low occurrence rates.

2.2 Fuzzy Kohonen Clustering Networks

FKCN [4] integrate the idea of fuzzy membership from Fuzzy C-Means (FCM), with the updating rules of SOM. Thus, creating a self-organizing algorithm that automatically adjusts the size of the updated neighborhood during a learning process, which usually terminates when the FCM objective function is minimized. The update rule for the FKCN algorithm can be given as:

$$W_{i,t} = W_{i,t-1} + \alpha_{ik,t}(Z_k - W_{i,t-1});\ for\ k = 1,2,...,n;\ for\ i = 1,2,...,c \quad (1)$$

where $W_{i,t}$ represents the centroid[2] of the ith cluster at iteration t , Z_k is the kth vector example from the dataset and α_{ik} is the only parameter of the algorithm and according to [14]:

$$\alpha_{ik,t} = (U_{ik,t})^{m(t)} \quad (2)$$

Where $m(t)$ is an exponent like the fuzzification index in FCM and $U_{ik,t}$ is the membership value of the compound Z_k to be part of cluster i. Both of these constants vary at each iteration t according to:

$$U_{ik} = \left(\sum_{j=1}^{c} \left(\frac{\|Z_k - W_i\|}{\|Z_k - W_j\|} \right)^{2/(m-1)} \right)^{-1} ;\ 1 \leq k \leq n\ ;\ 1 \leq i \leq c \quad (3)$$

$$m(t) = m0 - m\Delta \cdot t \quad ; \quad m\Delta = (m0 - m_f)/iterate\ limit \quad (4)$$

Where $m0$ is a constant value greater than the final value (m_f) of the fuzzification parameter m. The final value m_f should not be less than 1.1, in order to avoid a divide by zero error in equation (3). The iterative process will stop if $\left\|W_{i,(t)} - W_{(i,t-1)}\right\|^2 < \epsilon$, where ϵ is a termination criterion or after a given number of iterations. At the end of the process, a matrix U is obtained, where U_{ik} is the degree of membership of the Z_k element of the dataset to the cluster i. In addition, the centroid of each cluster will form the matrix W where W_i is the centroid of the ith cluster. The FKCN algorithm is given below:

1. Fix c, and $\epsilon > 0$ to some small positive constant.
2. Initialize $W_0 = (W_{1,0}, W_{2,0}, \cdots, W_{c,0}) \in \Re^c$.
   Choose $m_0 > 1$ and $t_{max} = max.\ number\ of\ iterations$.
3. For $t = 1, 2, \cdots, t_{max}$
   a. Compute all cn learning rates $\alpha_{ik,t}$ with equations (2) and (3).
   b. Update all c weight vectors $W_{i,t}$ with
   $W_{i,t} = W_{i,t-1} + [\sum_{k=1}^{n} \alpha_{ik,t}(Z_k - W_{i,t-1})] \,/\, \sum_{j=1}^{n} \alpha_{ij,t}$
   c. Compute $E_t = \left\|W_{i,(t)} - W_{(i,t-1)}\right\|^2 = \sum_{i=1}^{c} \left\|W_{i,(t)} - W_{(i,t-1)}\right\|^2$
   d. If $E_t < \epsilon$ stop.

[2] In the perspective of neural networks it represents a neuron or a prototype vector. So the number of neurons or prototype vectors will be equal to the number of clusters.

2.3 *Fuzzy Growing Hierarchical Self-Organizing Networks*

Fuzzy Growing Hierarchical Self-Organizing Networks (FGHSON) are based on a hierarchical fuzzy structure of multiple layers, where each layer consists of several independent growing FKCNs. This structure can grow by means of an unsupervised self-organizing process in two manners (inspired by [22]):

a. Individually, in order to find the more suitable number of prototypes (which compose a FKCN) that may represent in an accurate manner the input dataset.
b. On groups of FKCNs in a hierarchical mode, permitting the hierarchy to reveal a particular set of characteristics of data.

Both growing processes are modulated by three parameters that regulate the breadth (growth of the layers), depth (hierarchical growth) and membership degree of data to the prototype vectors.

The FGHSON works as follows:

1) Initial Setup and Global Network Control
The main motivation of the FGHSON algorithm is to properly represent a given dataset. The quality of this representation is measured in terms of the difference between a prototype vector and the example vectors represented by this. The *quantization error* qe is used for this purpose. The qe measures the dissimilarity of all input data mapped onto a particular prototype vector, hence it can be used to guide a growth process with the aim of achieving an accurate representation of the dataset reducing the qe. The qe of a prototype vector W_i is calculated according to (5) as the mean Euclidean distance between its prototype and the input vectors Z_c that are part of the set of vectors C_i mapped onto this prototype.

$$qe_i = \sum_{Z_c \in Ci} \|W_i - Z_c\| \; ; \; C_i \neq \phi \tag{5}$$

The first step of the algorithm is focused on obtaining a global measure that allows us to know the nature of the whole dataset. For this purpose the training process begins with the computation of a global measure of error qe_0. qe_0 represents the qe of the single prototype vector W_0 that forms the layer 0, see figure 1(a), calculated as shown in (6). Where, Z_k represents the input vectors from the whole data set Z and W_0 is defined as a prototype vector $W_0 = [\mu_{0_1}, \mu_{0_2}, \ldots, \mu_{0_n}]$, where μ_{0_i} for $i = 1, 2, \ldots, n$; is computed as the average of μ_{0_i} in the complete input dataset. In other words W_0 is a vector that corresponds to the mean of the input variables.

$$qe_0 = \sum_{Z_k \in Z} \|W_0 - Z_k\| \tag{6}$$

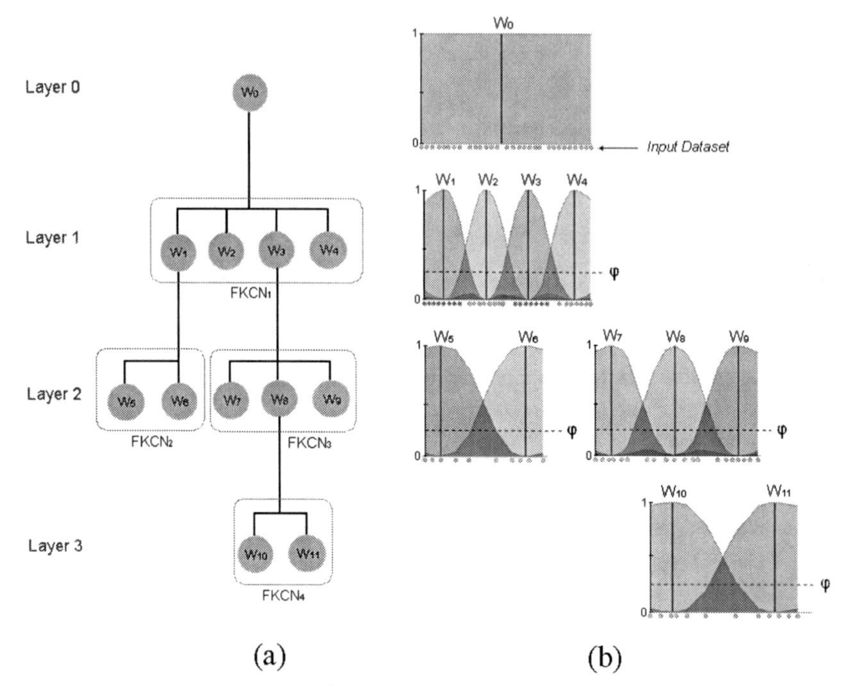

Fig. 1 (a) Hierarchical structure showing the prototype vectors and FKCNs created in each layer for a supposed case. (b) Membership degrees in each layer, corresponding to the network shown in the diagram. The parameter φ (the well known $\alpha - cut$) represents the minimal degree membership of an observation to be part of the dataset represented by a prototype vector, the group of data with a desired membership to a prototype will be used for the training of a new FCKN in the next layer (depth process). In this particular diagram the dataset is unidimensional (represented by the small circles below the membership plot) in order to simplify the example

The value of qe_0 will help to measure the minimum quality of data representation of the prototype vectors in the subsequent layers. Succeeding prototypes have the task of reducing the global representation error qe_0.

2) Breadth growth process
The construction of the first layer starts after the calculation of qe_0. This first layer consists of a FKCN ($FKCN_1$) with two initial prototype vectors. The growth process of the $FKCN_1$ begins by adding a new prototype vector and training it until a suitable representation of the dataset is achieved. Each of these prototype vectors is an n-dimensional vector W_i (with the same dimensionality as the input patterns), which is initialized with random values. The $FKCN_1$ is trained as shown in section 2.2, taking as input (in the exceptional case of the first layer) the whole dataset. More precisely, the $FKCN_1$ is allowed to grow until the qe of the prototype for its preceding

layer (qe_0 in the case of layer 1) is reduced to at least a fixed percentage τ_1. Continuing with the creation of the first layer, the number of prototypes in the $FKCN_1$ will be adapted. To achieve this, the *mean quantization error of the map* (MQE) is computed according to expression (7), where d refers to the number of prototype vectors contained in the FKCN, and qe_i represents the quantization error of the prototype W_i.

$$MQE_m = \frac{1}{d} \cdot \sum_i qe_i \tag{7}$$

The MQE is evaluated using (8) to measure the quality of data representation, and is used also as stopping criterion for the growing process of the FKCN. In (8) qe_u represents the qe of the corresponding prototype u in the upper layer. In the specific case of the first-layer, the stopping criterion is shown in (9).

$$MQE < \tau_1 \cdot qe_u \tag{8}$$

$$MQE_{layer1} < \tau_1 \cdot qe_0 \tag{9}$$

If the stopping criterion (8) is not fulfilled, it is necessary to aggregate more prototypes for a more accurate representation. For this aim, the prototype with the highest qe is selected and is denoted as the error prototype e. A new prototype is inserted in the place where e was computed. After the insertion, all the FKCN parameters are reset to the initial values (except for the values of the prototype vectors) and the training begins according to the standard training process of FKCN. Note that the same value of the parameter τ_1 is used in each layer of the FGHSON. Thus, at the end of the process, a layer 1 is obtained with a $FKCN_1$ formed by a set of prototype vectors W, see figure 1(a). In addition, a membership matrix U is obtained. This matrix contains the membership degree of the dataset elements to the prototype vectors, as explained in section 2.2.

3) Depth growth process

As soon as the breadth process of the first layer is finished, its prototypes are examined for further growth (depth growth or hierarchical growth). In particular, those prototypes with a large quantization error will indicate which clusters need a better representation by means of new FKCNs. The new FKCNs form a second layer, for instance W_1 and W_3 in figure 1(a). The selection of these prototypes is regulated by qe_0 (calculated previously in step 1) and a parameter τ_2 which is used to describe the desired level of granularity in the data representation. More precisely, each prototype W_i in the first layer that does not fulfill the criterion given in expression (10) will be subject to hierarchical expansion.

$$qe_i < \tau_2 \cdot qe_0 \tag{10}$$

After the expansion process and creation of the new FKCNs, the breadth process described in stage 2 begins with the newly established FKCNs, for instance, $FKCN_2$ and $FKCN_3$ in figure 1(a). The methodology for adding new prototypes, as well as the termination criterion of the breadth process, is essentially the same as used in the first layer. The difference between the training processes of the FKCNs in the first layer and all subsequent layers, is that only a fraction of the whole input data is selected for training. This portion of data will be selected according to a minimal membership degree (φ). This parameter φ (an $\alpha - cut$) represents the minimal degree of membership for an observation to be part of the dataset represented by a prototype vector. Hence, φ is used as a selection parameter, so all the observations represented by W_i have to fulfill expression (11), where U_{ik} is the degree of membership of the ${Z_k}^{th}$ element of the dataset to the cluster i. As an example, figure 1(b) shows the membership functions of the FKCNs in each layer, and how φ is used as a selection criteria to divide the dataset.

$$\varphi < U_{ik} \tag{11}$$

At the end of the creation of layer two, the same procedure described in step 2 is applied to build layer 3 and so forth.

The training process of the FGHSON is terminated when no prototypes require further expansion. Note that this training process does not necessarily lead to a balanced hierarchy, i.e., a hierarchy with equal depth in each branch. Rather, the specific distribution of the input data is modeled by a hierarchical structure, where some clusters require deeper branching than others.

3 Experimental Testing

3.1 Iris Data Set

In this experiment the Iris dataset[3] is used in order to show the adaptation of the FGHSON to those areas where an absolute membership to a single prototype is not obvious. Therefore, FGHSON must (in an unsupervised manner) look at the representation of the dataset on the areas where observations of the same category share similar zones. For instance in the middle of the data cloud formed by the Virginica and Versicolor observations (see figure 2(a)).

The parameters of the algorithm were set to $\tau_1 = 0.2$, $\tau_2 = 0.03$, and $\varphi = 0.2$. After training, a structure of four layers was obtained. The zero layer is used to measure the whole deviation of the dataset as was presented in section 2.3. The first layer consist of a FKCN with three prototype vectors as shown in figure 2(b), this distribution of prototypes aim to represent three Iris

[3] There are three categories in the data set : Iris Setosa, Iris Versicolor and Iris Virginical. Each having 50 observations with four features: sepal length (SL), sepal width (SW), petal length (PL), and petal width (PW).

categories. The second layer (figure 2(c)) reaches a more fine-grained description of the dataset, placing prototypes in almost all of the data distribution, adding prototypes in the zones where more representation was needed. Finally in figure 2(d), it is possible to observe an over population of prototypes in the middle of the cloud of Virginica and Versicolor observations. This occurs because this part of the dataset presents observations with ambiguous membership in the previous layer, then, several prototypes are placed in this new layer for proper representation. Hence, permitting those observations to obtain a higher membership of its new prototypes. The outcome of the process is a more accurate representation of this zone.

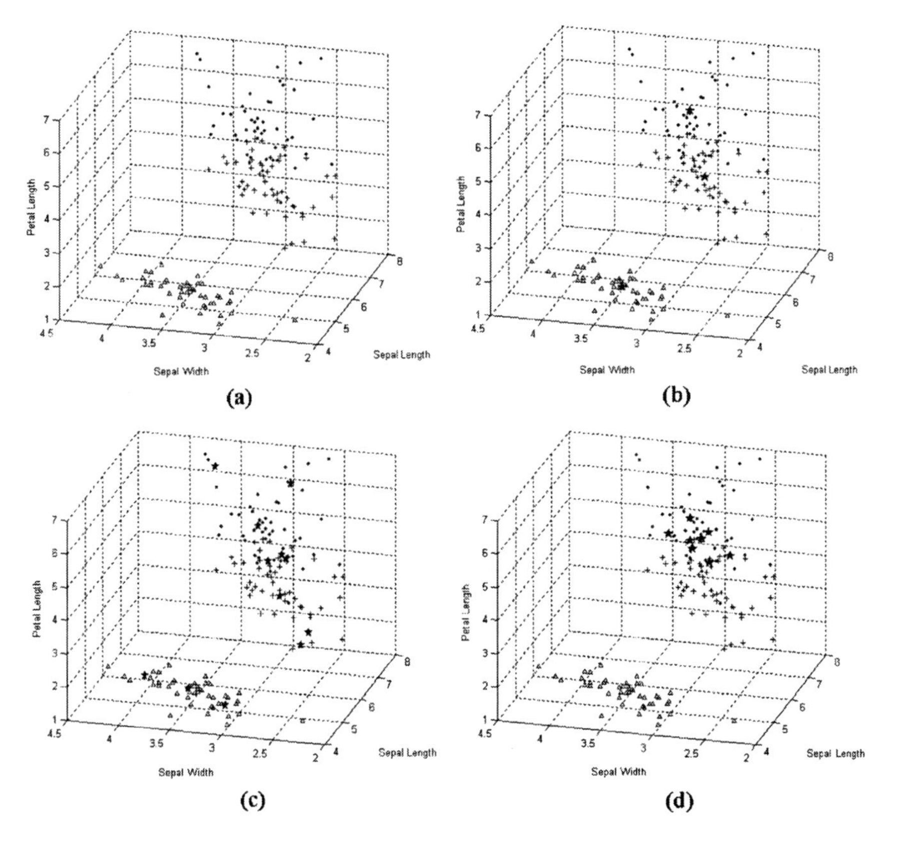

Fig. 2 Distribution of the prototype vectors, represented by stars, in each layer of the hierarchy. (a) Iris data set. There are three Iris categories: Setosa, Versicolor, and Virginica represented respectively by triangles, plus symbols, and dots. Each has 50 samples with 4 features. Here, only three features are used: PL, SW, and SL. (b) First layer (c) Second layer and (d) Third layer of the FGHSON, in this layer prototypes are presented only in the zone where observations of Virginica and Vesicolor share the same area, so the new prototypes represent each category in a more accurate manner.

3.2 Example Set

An exmaple set, as presented by Martinez et al [19] is used in order to show the capabilities of the FGHSON to represent a dataset that has multiple dimensionalities. In addition, it is possible to illustrate how the model stops the growing process in those parts where the desired representation is reached and keep growing where a low membership or poor representation is present. The parameters of the algorithm were set to $\tau_1 = 0.3$, $\tau_2 = 0.065$, and $\varphi = 0.2$. Four layers were created after training the network. In figure 3(a) the first layer is shown, in this case seven prototypes were necessary to represent the dataset at this level, one for the 1D oval, one for the 2D plane and five for the 3D parallelepiped (note that there are no prototypes clearly associated to the line).

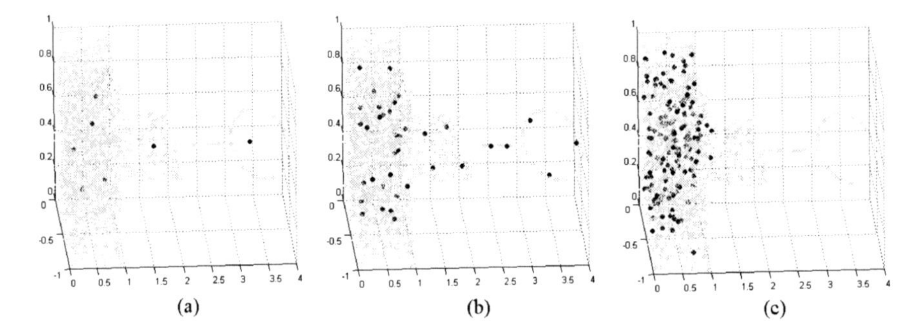

Fig. 3 Distribution of the prototype vectors (represented by black points) (a) First layer (b) Second layer (c) Third layer.

In the second layer shown in the figure 3, a more accurate distribution of prototypes is reached, so it is possible to observe prototypes adopting the form of the dataset. Additionally, in regions where the quantization error was large, the new prototypes allow a better representation (e.g., along the line). In layer three (see figure 3(c)), no more prototypes are needed to represent the circle, the line and the plane; but a third hierarchical expansion was necessary to represent the parallelepiped. In addition, due to the data density in the parallelepiped, many points are members of multiple prototypes, so several prototypes were created.

3.3 Tuning the Model Parameters

In order to explore the performance of the algorithm, different values for the parameters φ, τ_1 and τ_2 were tested using the Iris dataset. The tests were performed using ten different values of τ_1 (breadth parameter), ten of τ_2 (depth parameter) and eight of φ ($\alpha - cut$), forming 800 triplets. For each triplet (φ, τ_1 and τ_2) a FGHSON was trained using the following fixed parameters: $t_{max} = 100$ (maximum number of iterations), $\epsilon = 0.0001$ (termination criterion) and $m_0 = 2$ (fuzzification parameter).

Several variables were obtained in order to measure the quality of the networks created for every FGHSON generated; for instance the number of hierarchical levels of the obtained network, the number of FKCNs created for each level, and finally the quantization error by prototype and level. The analysis of these values will allow discovery of the relationships between the parameters (φ, τ_1 and τ_2) and the topology of the networks (represented in this experiment by the levels reached for each network and the number of FKCN created). In addition, it will be possible to observe the relationship between the quantization errors of prototypes by level and the parameters of the algorithm. This activity makes it possible for us to find values of the parameters that allow us to build the most accurate structure, based on the number of prototypes, the quantization error and the number of levels present in the network.

Due the large amount of information involved, a graphical representation of the obtained data was used in order to facilitate visualization of the results. For this, 3D plots were used as follows: the parameter τ_1 (which regulates the breadth of the networks) and the parameter τ_2 (which regulates the depth of the hierarchical architecture) are shown on the x-axis and y-axis respectively. The z-axis shows the quantity of levels in the hierarchy (see figure 4 and figure 5). Each 3D plot corresponds to one fixed value of φ. Hence, eight 3D plots represent the eight different values evaluated for φ, then each 3D plot contains 100 possible combinations of the duple (τ_1,τ_2) for a specific φ. Therefore, analysis of τ_1, τ_2 and φ and the levels of the 800 networks were generated and plotted.

Furthermore, additional information was added to the 3D plots. The number of FKCNs created for level were represented by a symbol in the 3D plots (see figure 4 and figure 5 left side). The higher quantization error of the prototypes that were expanded is shown in a new group of 3D plots; in others words this prototype is the "father" of the prototypes in that level[4]. The rounded value of the quantization error is shown as a mark in the 3D plot for each triplet of values, in each level (see figure 4 and figure 5 right side).

Examining the obtained results, there are some interesting results related to the quantization error and the topology of the network. For instance, figure 4 and figure 5 show the different networks created. It can be seen that for values of τ_2 above 0.3 the model generates networks with just one level, so an interesting area to explore lies between the values of $\tau_2 = 0.1$, 0.2 and 0.3. With respect to the quantization error (figure 4 and figure 5 right side) for almost all values of φ, the lower quantization error with the lower number of

[4] For this reason, in level one all the values are 291 (see figure 4 and figure 5 right side) because the prototype "father" that is expanded has the same quantization error for all networks; in the case of the first level this error is called qe_0, as is described in section 2.3.

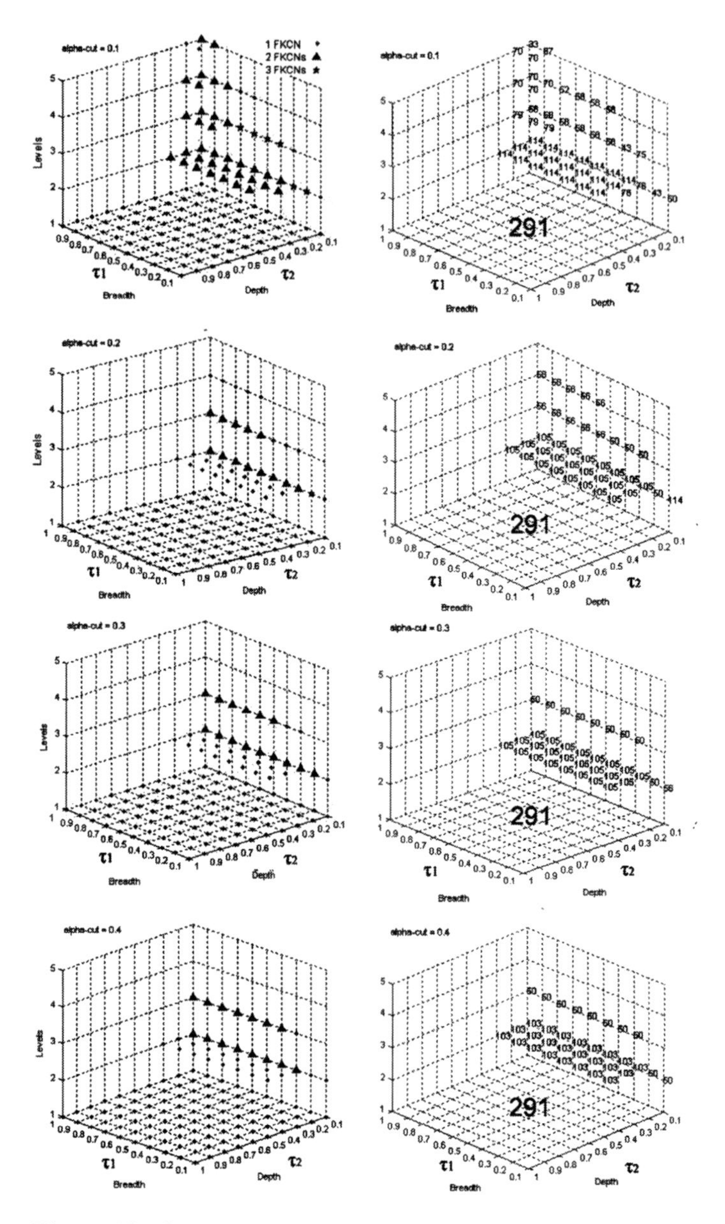

Fig. 4 The figure has 3D plots showing the results obtained using $\varphi = 0.1$, 0.2, 0,3 and 0.4. On the left side it is possible observe the levels obtained for each triplet $(\varphi, \tau_1, \tau_2)$, in addition the number of FKCN created for each level are represented by a symbol. On the right side the higher quantization error of the prototypes that were expanded is shown; in others words this prototype is the "father" (with higher quantization error of the prototypes in that level). In the special case of the first level all the values are 291, because the prototype "father" that is expanded has the same quantization error for all the networks; in the case of the first level this error is called qe_0, as described in section 2.3.

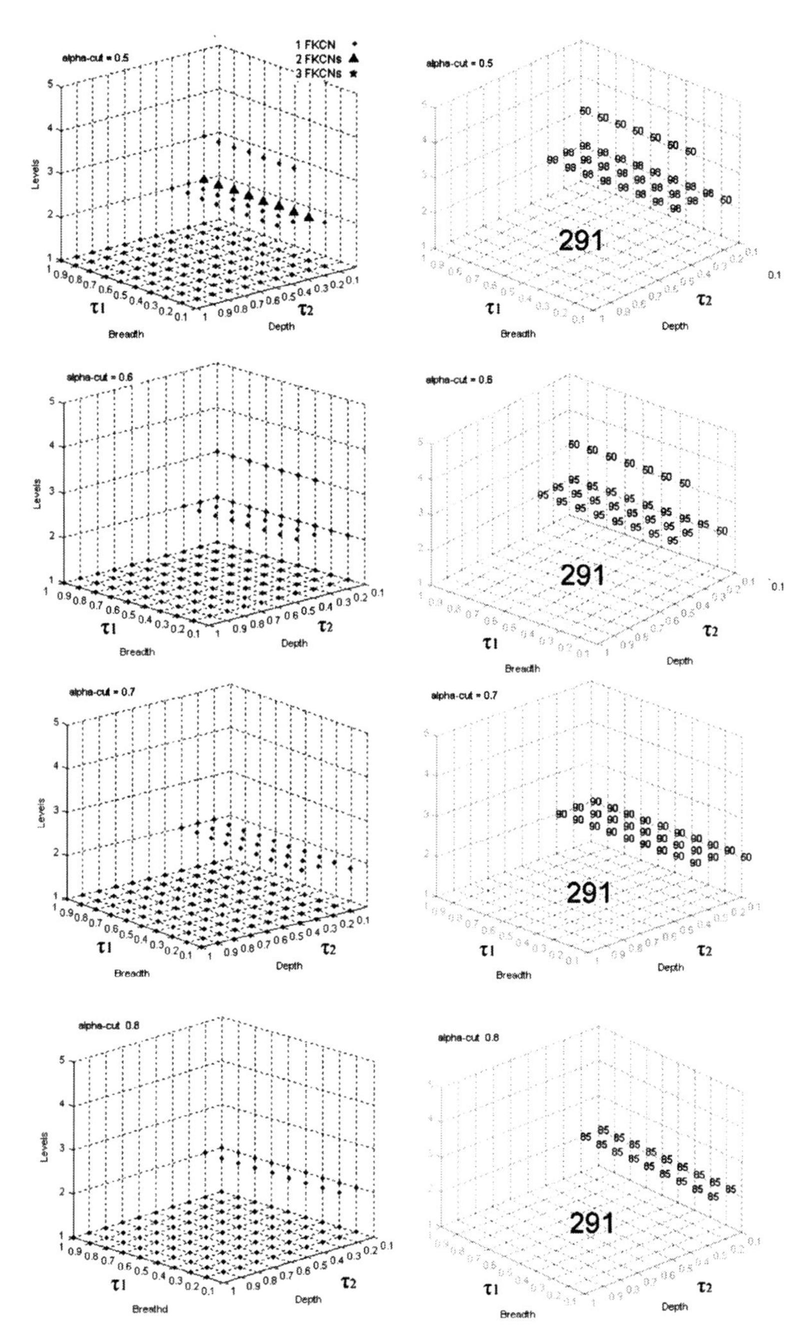

Fig. 5 The figure shows 3D plots of the results obtained using φ = 0.5, 0.6, 0.7 and 0.8, in addition the number of FKCN created for each level are represented by a symbol. On the right side the higher quantization error of the prototypes that were expanded is shown; in others words this prototype is the "father" (with higher quantization error) of the prototypes in that level.

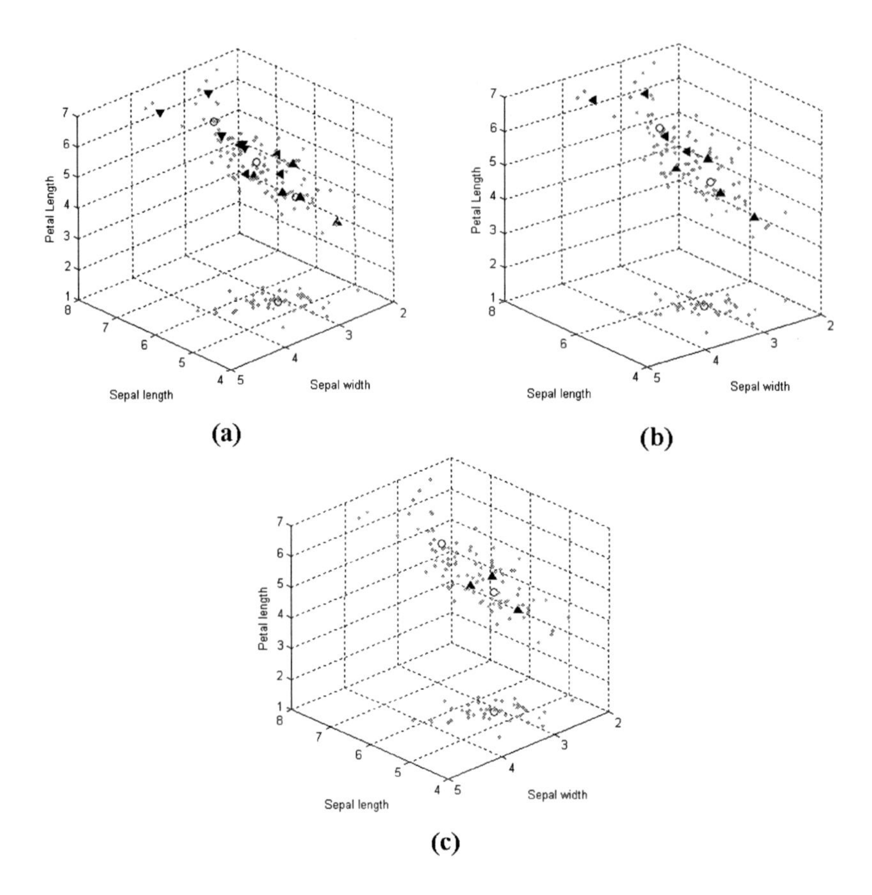

Fig. 6 Structures obtained tuning the model with the values (a) $\varphi = 0.1$, $\tau_1 = 0.2$, $\tau_2 = 0.1$ (b) $\varphi = 0.3$, $\tau_1 = 0.2$, $\tau_2 = 0.1$, and (c) $\varphi = 0.3$, $\tau_1 = 0.2$, $\tau_2 = 0.1$. In this figure it is possible to observe the distribution of the prototype vectors; the prototypes of the first level are represented by circles and the prototypes of the second level are represented by triangles.

levels was presented in the points ($\tau_1 = 0.1$, $\tau_2 = 0.2$) and(or) ($\tau_1 = 0.2$, $\tau_2 = 0.1$) (see figure 4 and figure 5).

In the next step of the analysis, the value of the parameters which generated the best networks so far were selected, based on the premise that the most accurate network has to present a lower number of levels, FKCNs, and a lower quantization error. For a selected group of three networks (see table 1), the distribution of the prototypes on the dataset were plotted in order to analyze how the prototypes of these selected networks had been adapted to the dataset (see figure 6).

Some remarks could be made about the plots obtained. In the first example ($\varphi = 0.1$, $\tau_1 = 0.2$, $\tau_2 = 0.1$) shown in figure 6(a), there are four prototypes in

Table 1 Parameters and results of the best networks selected

φ	τ_1	τ_2	Levels	qe[a]	FKCNs
0.1	0.2	0.1	2	43	3
0.3	0.2	0.1	2	50	2
0.6	0.2	0.1	2	50	1

[a] The higher quantization error of the "father" prototype of the prototypes in that level.

the first level of the hierarchy; these prototypes represent four classes[5]. Then, the prototypes of this layer represent the three classes of iris, in addition they also take the problematic region between Versicolor and Virginica as a fourth class. Furthermore, new prototypes are created in the second layer in order to obtain a more accurate representation of the dataset, creating a proliferation of prototypes. This phenomena is due to the low φ (0.1) being selected. This is because the quantity of elements represented for each prototype is large (due to low membership, a lot of data can be a member of one prototype) so, many prototypes are necessary to reach a low quantization error.

In the next example with $\varphi = 0.3$, $\tau_1 = 0.2$, $\tau_2 = 0.1$, it is possible observe (figure 6(b)) the three prototypes created in the first level. In this case the number of the prototypes matches the number of classes in the iris dataset. Nevertheless, there is (as in the previous example) an abundance of prototypes in the Virginica-Versicolor group. But in this case the number of prototypes is lower compared with the preceding example, showing how φ affects the quantity of prototypes created.

Finally, in the last example with $\varphi = 0.6$, $\tau_1 = 0.2$, $\tau_2 = 0.1$. Three prototypes are created in the first level of the network matching the classes of the Iris dataset (figure 6(c)); additionally, in the second layer one of the previous prototypes is expanded to three prototypes in order to represent the fuzzy areas of the data set. This last network presents the lower values of vector quantization, levels of hierarchy, and FKCNs; so it is possible to select this as the more accurate topology. Consider the previously defined premise which said that the most accurate network had to present lower number of levels, number of FKCNs, and the lower quantization error.

4 Conclusion

The Fuzzy Growing Hierarchical Self-organizing Networks are fully adaptive networks able to hierarchically represent complex datasets. Moreover, they allow a fuzzy clustering of the data, allocating more prototype vectors or

[5] It knows that there are three classes (iris Setosa, Virginica and Versicolor) but the fourth exists in an area where Versicolor and Virginica present similar characteristics.

branches to heterogeneous areas or to regions where the data have similar membership degree to several clusters. This property can help to better describe the structure of the dataset and the inner data relationships.

In this book chapter the effects of using different values for the parameters of the algorithm, have been presented using the Iris dataset as an example. It was shown how the different parameters affect the topology and quantization error of the networks created. In addition, some of the better networks created were examined in order to show how different representations of the same dataset can be obtained with similar accuracy.

Acknowledgements. This work is part of a cooperation project between BIOTEC, CIAT, CENICAÑA (Colombia) and HEIG-VD (Switzerland) named Precision Agriculture and the Construction of Field-crop Models for Tropical Fruits. The financial support is given by several institutions in Colombia (MADR, COLCIENCIAS, ACCI) and the State Secretariat for Education and Research (SER) in Switzerland.

References

1. Anderberg, M.: Cluster Analysis for Applications. Academic, New York (1973)
2. Backer, E., Jain, A.: A clustering performance measure based on fuzzy set decomposition. IEEE Trans. Pattern Anal. Mach. Intell. PAMI-3(1), 66–75 (1981)
3. Baraldi, A., Alpaydin, E.: Constructive feedforward ART clustering networks-Part I and II. IEEE Trans. Neural Netw. 13(3), 645–677 (2002)
4. Bezdek, J., Tsao, K., Pal, R.: Fuzzy Kohonen clustering networks. In: IEEE Int. Conf. on Fuzzy Systems, pp. 1035–1043 (1992)
5. Burzevski, V., Mohan, C.: Hierarchical Growing Cell Structures. Tech Report: Syracuse University (1996)
6. Doherty, J., Adams, G., Davey, N.: TreeGNG - Hierarchical topological clustering. In: Proc. Euro. Symp. Artificial Neural Networks, pp. 19–24 (2005)
7. Diday, E., Simon, J.C.: Clustering analysis. In: Digital Pattern Recognition, pp. 47–94. Springer, Heidelberg (1976)
8. Everitt, B., Landau, S., Leese, M.: Cluster Analysis. Arnold, London (2001)
9. Fischer, G., van Velthuizen, H.T., Nachtergaele, F.O.: Global agro-ecological zones assessment: methodology and results. Interim Report IR-00-064. IIASA, Laxenburg, Austria and FAO, Rome (2000)
10. Fritzke, B.: Growing cell structures: a self-organizing network for unsupervised and supervised learning. Neural Networks 7(9), 1441–1460 (1994)
11. Fritzke, B.: Some competitive learning methods, Draft Doc. (1998)
12. Taniichi, H., Kamiura, N., Isokawa, T., Matsui, N.: On hierarchical self-organizing networks visualizing data classification processes. In: Annual Conference, SICE 2007, pp. 1958–196 (2007)
13. Hodge, V., Austin, J.: Hierarchical growing cell structures: TreeGCS. IEEE Transactions on Knowledge and Data Engineering 13(2), 207–218 (2001)
14. Huntsberger, T., Ajjimarangsee, P.: Parallel Self-organizing Feature Maps for Unsupervised Pattern Recognition. Int. Jo. General Sys. 16, 357–372 (1989)

15. Jain, A., Murty, M., Flynn, P.: Data clustering: a review. ACM Comput. Surv. 31(3), 264–323 (1999)
16. Kohonen, T.: Self-organized formation of topologically correct feature maps. Biological Cybernetics 43(1), 59–69 (1982)
17. Lampinen, J., Oja, E.: Clustering properties of hierarchical self-organizing maps. J. Math. Imag. Vis. 2(2–3), 261–272 (1992)
18. Luttrell, S.: Hierarchical self-organizing networks. In: Proceedings of the 1st IEE Conference on Artificial Neural Networks, London, UK, pp. 2–6. British Neural Network Society (1989)
19. Martinez, T., Schulten, J.: Topology representing networks. Neural Networks 7(3), 507–522 (1994)
20. Merkl, D., He, H., Dittenbach, M., Rauber, A.: Adaptive hierarchical incremental grid growing: An architecture for high-dimensional data visualization. In: Proc. Workshop on SOM, Advances in SOM, pp. 293–298 (2003)
21. Miikkulainen, R.: Script recognition with hierarchical feature maps. Connection Science 2, 83–101 (1990)
22. Rauber, A., Merkl, D., Dittenbach, M.: The growing hierarchical self-organizing map: Exploratory analysis of high-dimensional data. IEEE Transactions on Neural Networks 13(6), 1331–1341 (2002)
23. Xu, R., Wunsch, D.: Survey of clustering algorithms. IEEE Transactions on Neural Networks 16(3), 645–678 (2005)

Self-Organizing Neural Grove: Efficient Multiple Classifier System with Pruned Self-Generating Neural Trees

Hirotaka Inoue

Abstract. Multiple classifier systems (MCS) have become popular during the last decade. Self-generating neural tree (SGNT) is a suitable base-classifier for MCS because of the simple setting and fast learning capability. However, the computation cost of the MCS increases in proportion to the number of SGNTs. In an earlier paper, we proposed a pruning method for the structure of the SGNT in the MCS to reduce the computational cost. In this paper, we propose a novel pruning method for more effective processing and we call this model self-organizing neural grove (SONG). The pruning method is constructed from both an on-line and an off-line pruning method. Experiments have been conducted to compare the SONG with an unpruned MCS based on SGNT, an MCS based on C4.5, and the *k*-nearest neighbor method. The results show that the SONG can improve its classification accuracy as well as reducing the computation cost.

1 Introduction

Classifiers need to find hidden information in the large amount of given data effectively and must classify unknown data as accurately as possible [1]. Recently, to improve the classification accuracy, multiple classifier systems (MCS) such as neural network ensembles, bagging, and boosting have been used for practical data mining applications [2, 3, 4, 5]. In general, the base classifiers of the MCS use traditional models such as neural networks (backpropagation network and radial basis function network) [6] and decision trees (CART and C4.5) [7].

Neural networks have great advantages of adaptability, flexibility, and universal nonlinear input-output mapping capability. However, to apply these neural

Hirotaka Inoue
Kure National College of Technology, 2-2-11 Agaminami, Kure,
Hiroshima 737-8506, Japan
e-mail: hiro@kure-nct.ac.jp

L. Franco et al. (Eds.): Constructive Neural Networks, SCI 258, pp. 281–291.
springerlink.com

networks, it is necessary that human experts determine the network structure and some parameters, and it may be quite difficult to choose the right network structure suitable for a particular application at hand. Moreover, a long training time is required to learn the input-output relation of the given data. These drawbacks prevent neural networks being the base classifier of the MCS for practical applications.

Self-generating neural trees (SGNTs) [8] have simple network design and high speed learning. SGNTs are an extension of the self-organizing maps (SOM) of Kohonen [9] and utilize competitive learning. The SGNT capabilities make it a suitable base classifier for the MCS. In order to improve the accuracy of SGNN, we propose ensemble self-generating neural networks (ESGNN) for classification [10] as one of the MCS. Although the accuracy of ESGNN improves by using various SGNTs, the computational cost, that is, the computation time and the memory capacity increases in proportion to the increasing number of SGNNs in the MCS.

In an earlier paper [11], we proposed a pruning method for the structure of the SGNN in the MCS to reduce the computational cost. In this paper, we propose a novel MCS pruning method for more effective processing and we call this model a self-organizing neural grove (SONG). This pruning method is comprised of two stages. At the first stage, we introduce an on-line pruning method to reduce the computational cost by using class labels in learning. At the second stage, we optimize the structure of the SGNT in the MCS to improve the generalization capability by pruning the redundant leaves after learning. In the optimization stage, we introduce a threshold value as a pruning parameter to decide which subtree's leaves to prune and estimate using 10-fold cross-validation [12]. After the optimization, the SONG can improve its classification accuracy as well as reducing the computational cost. Bagging [2] is used as a resampling technique for the SONG.

In this work, we investigate the improvement performance of the SONG by comparing it with an MCS based on C4.5 [13] using ten problems in a UCI machine learning repository [14]. Moreover, we compare the SONG with *k*-nearest neighbor (*k*-NN) [15] to investigate the computational cost and the classification accuracy. The SONG demonstrates higher classification accuracy and faster processing speed than *k*-NN on average.

The rest of the paper is organized as follows: the next section shows how to construct the SONG. Then Section 3 is devoted to some experiments to investigate its performance. Finally we present some conclusions, and outline plans for future work.

2 Constructing Self-Organizing Neural Grove

In this section, we describe how to prune redundant leaves in the SONG. First, the on-line pruning method used in learning the SGNT is outlined. Second, we show the optimization method in constructing the SONG. Finally, we show a simple example of the pruning method for a two dimensional classification problem.

2.1 On-Line Pruning of Self-Generating Neural Tree

SGNT is based on SOM and implemented as a competitive learning algorithm. The SGNT can be constructed directly from the given training data without any human intervention required. The SGNT algorithm is defined as a tree construction problem of how to construct a tree structure from the given data, which consist of multiple attributes, under the condition that the final leaves correspond to the given data.

Before we describe the SGNT algorithm, we explain some notations used.

- input data vector: $e_i \in \mathbb{R}^m$.
- root, leaf, and node in the SGNT: n_j.
- weight vector of n_j: $w_j \in \mathbb{R}^m$.
- the number of the leaves in n_j: c_j.
- distance measure: $d(e_i, w_j)$.
- winner leaf for e_i in the SGNT: n_{win}.

The SGNT algorithm is a hierarchical clustering algorithm. The pseudo C code of the SGNT algorithm is given in Figure 1 where several sub procedures are used. Table 1 shows the sub procedures of the SGNT algorithm and their specifications.

In order to decide the winning leaf n_{win} in the sub procedure `choose(e_i,n_1)`, competitive learning is used. If an n_j includes the n_{win} as its descendant in the SGNT, the weight w_{jk} $(k = 1, 2, \ldots, m)$ of the n_j is updated as follows:

$$w_{jk} \leftarrow w_{jk} + \frac{1}{c_j} \cdot (e_{ik} - w_{jk}), \quad 1 \leq k \leq m. \tag{1}$$

```
Input:
  A set of training examples E = {e_i}, i = 1, ... , N.
  A distance measure d(e_i,w_j).
Program Code:
  copy(n_1,e_1);
  for (i = 2, j = 2; i <= N; i++) {
    n_win = choose(e_i, n_1);
    if (leaf(n_win)) {
      copy(n_j, w_win);
      connect(n_j, n_win);
      j++;
    }
    copy(n_j, e_i);
    connect(n_j, n_win);
    j++;
    prune(n_win);
  }
Output:
  Constructed SGNT by E.
```

Fig. 1 SGNT algorithm

Table 1 Sub procedures of the SGNT algorithm

Sub procedure	Specification
$copy(n_j, e_i/w_{win})$	Create n_j, copy e_i/w_{win} as w_j in n_j.
$choose(e_i, n_1)$	Decide n_{win} for e_i.
$leaf(n_{win})$	Check n_{win} whether n_{win} is a leaf.
$connect(n_j, n_{win})$	Connect n_j as a child leaf of n_{win}.
$prune(n_{win})$	Prune leaves if they have the same class.

After all training data are inserted into the SGNT as the leaves, each one has a class label as the outputs and the weights of each node are the averages of the corresponding weights of all its leaves. The topology of the whole SGNT network reflects the given feature space. For more details concerning how to construct and perform the SGNT, see [8]. Note, to optimize the structure of the SGNT effectively, we remove the threshold value of the original SGNT algorithm in [8] to control the number of leaves based on the distance because of the trade-off between the memory capacity and the classification accuracy. In order to avoid the above problem, we introduce a new pruning method in the sub procedure `prune(n_win)`. We use the class label to prune leaves. For leaves that have the n_{win} parent node, if all leaves belong to the same class, then these leaves are pruned and the parent node is given the class.

2.2 Optimization of the SONG

The SGNT has a high speed processing capability. However, the accuracy of the SGNT is inferior to the conventional approaches, such as nearest neighbor, because the SGNT cannot guarantee reaching the nearest leaf for unknown data. Hence, we construct the SONG by taking the majority of plural SGNT outputs to improve the accuracy.

Although the accuracy of the SONG is superior or comparable to the accuracy of conventional approaches, the computational cost increases in proportion to the increase in the number of SGNTs in the SONG. In particular, the huge memory requirement prevents the use of the SONG for large datasets even with the most advance computers.

In order to improve the classification accuracy, we propose an optimization method of SONG for classification. This method has two parts, the merge phase and the evaluation phase. The merge phase is performed as a pruning algorithm to reduce dense leaves (Figure 2). This phase uses the class information and a threshold value α to decide which subtree's leaves to prune or not. For leaves that have the same parent node, if the proportion of the most common class is greater than or equal to the threshold value α, then these leaves are pruned and the parent node is given the most common class.

The optimum threshold values α of the given problems are different from each other. The evaluation phase is performed to choose the best threshold value by introducing 10-fold cross validation (Figure 3).

```
1 begin    initialize j = the height of the SGNT
2   do for each subtree's leaves in the height j
3      if the ratio of the most class ≥ the threshold value α,
4      then merge all leaves to parent node
5      if all subtrees are traversed in the height j,
6      then j ← j − 1
7   until j = 0
8 end.
```

Fig. 2 The merge phase

```
1 begin initialize α = 0.5
2   do for each α
3      evaluate the merge phase with 10-fold cross validation
4      if the best classification accuracy is obtained,
5      then record the α as the optimal threshold value
6      α ← α + 0.05
7   until α = 1
8 end.
```

Fig. 3 The evaluation phase

2.3 An Example of the Pruning Method for SONG

We show an example of the pruning method for SONG in Figure 4. This is a two-dimensional classification problem with two equal circular Gaussian distributions that have an overlap. The shaded plane is the decision region of class 0 and the other plane is the decision region of class 1 by the SGNT. The dotted line is the ideal decision boundary. The number of training samples is 200 (class0: 100,class1: 100) (Figure 4(a)).

The unpruned SGNT is given in Figure 4(b). In this case, 200 leaves and 120 nodes are automatically generated by the SGNT algorithm. In this unpruned SGNT, the height is 7 and the number of units is 320. In this, we define the unit to count the sum of the root, nodes, and leaves of the SGNT. The root is the node which is of height 0. The unit is used as a measure of the memory requirement in the next section. Figure 4(c) shows the pruned SGNT after the optimization stage in $\alpha = 1$. In this case, 159 leaves and 107 nodes are pruned away and 48 units remain. The decision boundary is the same as the unpruned SGNT. Figure 4(d) shows the pruned SGNT after the optimization stage in $\alpha = 0.6$. In this case, 182 leaves and 115 nodes are pruned away and only 21 units remain. Moreover, the decision boundary is improved more than the unpruned SGNT because this case can reduce the effect of the overlapping class by pruning the SGNT.

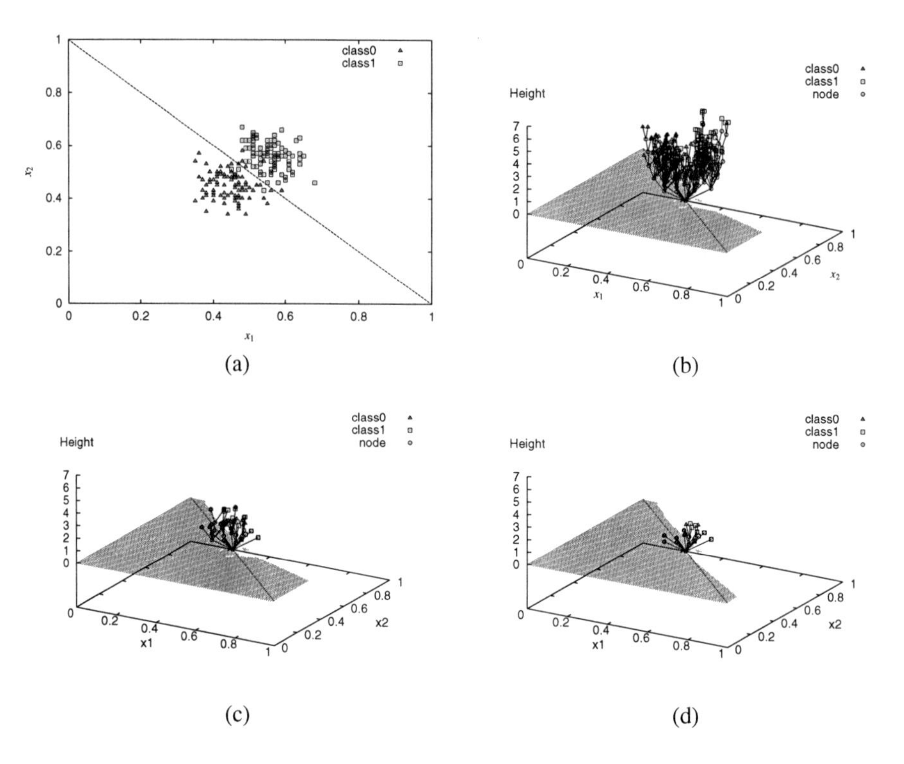

Fig. 4 An example of the SONG pruning algorithm, (a) a two dimensional classification problem with two equal circular Gaussian distribution, (b) the structure of the unpruned SGNT, (c) the structure of the pruned SGNT ($\alpha = 1$), and (d) the structure of the pruned SGNT ($\alpha = 0.6$). The shaded plane is the decision region of class 0 by the SGNT and the dotted line shows the ideal decision boundary

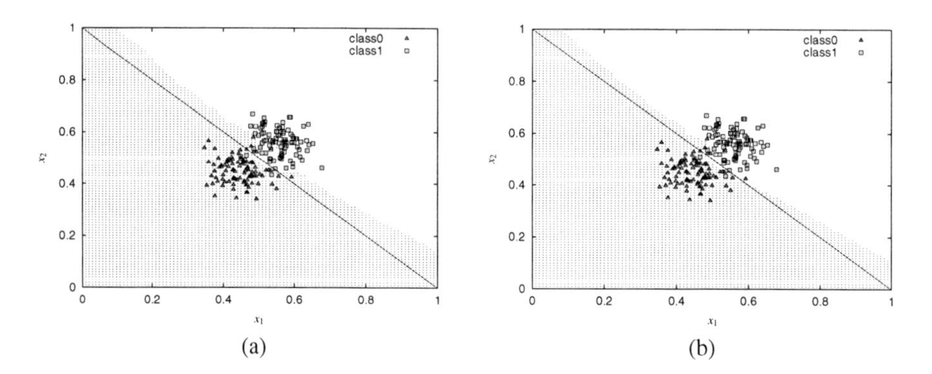

Fig. 5 An example of the SONG's decision boundary ($K = 25$), (a) $\alpha = 1$, and (b) $\alpha = 0.6$. The shaded plane is the decision region of class 0 by the SONG and the dotted line shows the ideal decision boundary

In the above example, we use all training data to construct the SGNT. The structure of the SGNT is changed by the order of the training data. Hence, we can construct the SONG from the same training data by changing the input order.

To show how well the SONG is optimized by the pruning algorithm, we show an example of the SONG in the same problem used above. Figure 5(a) and Figure 5(b) show the decision region of the SONG in $\alpha = 1$ and $\alpha = 0.6$, respectively. We set the number of SGNTs K to 25. The result of Figure 5(b) is a better estimation of the ideal decision region than the result of Figure 5(a). We investigate the pruning method for more complex problems in the next section.

3 Experimental Results

We investigate the computational cost (the memory capacity and the computation time) and the classification accuracy of the SONG with bagging for ten benchmark problems in the UCI machine learning repository [14]. Table 2 presents the abstract of the datasets.

We evaluate how SONG is pruned using 10-fold cross-validation for the ten benchmark problems. In this experiment, we use a modified Euclidean distance measure for the SONG and k-NN. Since the performance of the SONG is not sensitive in the threshold value α, we set the different threshold values α which are moved from 0.5 to 1; $\alpha = [0.5, 0.55, 0.6, \ldots, 1]$. We set the number of SGNTs K in the SONG to 25 and execute 100 trials by changing the sampling order of each training set. All experiments in this section were performed on an UltraSPARC workstation with a 900MHz CPU, 1GB RAM, and Solaris 8.

Table 3 shows the average memory requirement and classification accuracy of 100 trials for the SONG. As the memory requirement, we count the number of units which is the sum of the root, nodes, and leaves of the SGNT. The average memory requirement is reduced from between 65% to 96.6% and the classification accuracy is improved by 0.1% to 2.9% by optimizing the SONG. This confirms that the

Table 2 Brief summary of the datasets. N is the number of instances, m is the number of attributes

Dataset	N	m	classes
balance-scale	625	4	3
breast-cancer-w	699	9	2
glass	214	9	6
ionosphere	351	34	2
iris	150	4	3
letter	20000	16	26
liver-disorders	345	6	2
new-thyroid	215	5	3
pima-diabetes	768	8	2
wine	178	13	3

Table 3 The average memory requirement and classification accuracy of 100 trials for the bagged SGNT in the SONG. The standard deviation is given inside the bracket on classification accuracy ($\times 10^{-3}$)

	memory requirement			classification accuracy		
Dataset	pruned	unpruned	ratio	pruned	unpruned	ratio
balance-scale	107.68	861.18	12.5	0.866(6.36)	0.837(7.83)	+2.9
breast-cancer-w	30.88	897.37	3.4	0.97(2.41)	0.966(2.71)	+0.4
glass	104.33	297.75	35	0.714(13.01)	0.709(14.86)	+0.5
ionosphere	50.75	472.39	10.7	0.891(6.75)	0.862(7.33)	+2.9
iris	15.64	208.56	7.4	0.962(6.04)	0.955(5.45)	+0.7
letter	6197.5	27028.56	22.9	0.956(0.77)	0.955(0.72)	+0.1
liver-disorders	163.12	471.6	34.5	0.648(12.89)	0.636(13.36)	+1.2
new-thyroid	49.45	298.21	16.5	0.958(7.5)	0.957(7.49)	+0.1
pima-diabetes	204.4	1045.03	19.5	0.749(7.05)	0.728(7.83)	+2.1
wine	15	238.95	6.2	0.976(4.41)	0.972(5.57)	+0.4
Average	693.88	3181.96	16.9	0.869	0.858	+1.1

Table 4 The improved performance of the pruned MCS and the MCS based on C4.5 with bagging

	MCS based on SGNT			MCS based on C4.5		
Dataset	SGNT	MCS	ratio	C4.5	MCS	ratio
balance-scale	0.779	**0.866**	+8.7	0.795	0.827	+3.2
breast-cancer-w	0.956	**0.97**	+1.4	0.946	0.963	+1.7
glass	0.642	0.714	+7.2	0.664	**0.757**	+9.3
ionosphere	0.852	0.891	+3.9	0.897	**0.92**	+2.3
iris	0.943	**0.962**	+1.9	0.953	0.947	−0.6
letter	0.879	**0.956**	+7.7	0.880	0.938	+5.8
liver-disorders	0.59	0.648	+5.8	0.635	**0.736**	+10.1
new-thyroid	0.939	**0.958**	+1.9	0.93	0.94	+1
pima-diabetes	0.695	0.749	+5.4	0.749	**0.767**	+1.8
wine	0.955	**0.976**	+2.1	0.927	0.949	+2.2
Average	0.823	0.869	+4.6	0.837	**0.874**	+3

SONG can be effectively used for all datasets with regard to both the computational cost and the classification accuracy.

To evaluate SONG's performance, we compare it with an MCS based on C4.5. We set the number of classifiers K in the MCS to 25 and we construct both MCSs by bagging. Table 4 shows the improved performance of the SONG and the MCS based on C4.5. The results of the SGNT and the SONG are the average of 100 trials. The SONG performs better than the MCS based on C4.5 for 6 of the 10 datasets. Although the MCS based on C4.5 degrades the classification accuracy for iris, SONG can improve the classification accuracy for all problems. Therefore, SONG is an

Table 5 The classification accuracy, the memory requirement, and the computation time of ten trials for the best pruned SONG and k-NN

	classification acc.		memory requirement		computation time (s)	
Dataset	SONG	k-NN	SONG	k-NN	SONG	k-NN
balance-scale	0.878	**0.888**	**109.93**	562.5	**0.82**	1.14
breast-cancer-w	**0.974**	0.969	**26.8**	629.1	**1.18**	1.25
glass	**0.758**	0.701	**91.33**	192.6	0.36	**0.08**
ionosphere	**0.912**	0.866	**51.38**	315.9	1.93	**0.2**
iris	**0.973**	0.96	**11.34**	135	0.13	**0.05**
letter	0.958	**0.96**	**6208.03**	18000	**208.52**	503.14
liver-disorders	**0.685**	0.653	**134.17**	310.5	**0.54**	0.56
new-thyroid	**0.972**	**0.972**	**45.74**	193.5	0.23	**0.05**
pima-diabetes	**0.764**	0.751	**183.57**	691.2	**1.72**	2.49
wine	**0.983**	0.977	**11.8**	160.2	0.31	**0.15**
Average	**0.885**	0.869	**687.41**	2119.1	**21.57**	50.91

efficient MCS on the basis of both the scalability for large scale datasets and the robust improving generalization capability for the noisy datasets comparable to the MCS with C4.5.

To show the advantages of SONG, we compare it with k-NN on the same problems. The best classification accuracy of 100 trials with bagging were chosen. In k-NN, we choose the best accuracy where k is 1,3,5,7,9,11,13,15, and 25 with 10-fold cross-validation. All methods are compiled using gcc with the optimization level `-O2` on the same workstation.

Table 5 shows the classification accuracy, the memory requirement, and the computation time achieved by the SONG and k-NN. Although there are compression methods available for k-NN [16], they take enormous computation time to construct an effective model. We use the exhaustive k-NN in this experiment. Since k-NN does not discard any training sample, the size of this classifier corresponds to the training set size. The results of k-NN correspond to the average measures obtained by 10-fold cross-validation, the same experimental procedure adapted in SONG. Next, we show the results for each category.

First, with regard to the classification accuracy, SONG is superior to k-NN for 8 of the 10 datasets and gives 1.6% improvement on average. Second, in terms of the memory requirement, even though the SONG includes the root and the nodes which are generated by the SGNT generation algorithm, this is less than k-NN for all problems. Although the memory requirement of the SONG is totally used K times in Table 5, we release the memory of SGNT for each trial and reuse the memory for effective computation. Therefore, the memory requirement is suppressed by the size of the single SGNT. Finally, in view of the computation time, although the SONG consumes the cost of K times the SGNT to construct the model and test for the unknown dataset, the average computation time is faster than k-NN. The SONG is slower than k-NN for small datasets such as glass, ionosphere, and iris. However, it is faster than k-NN for large datasets such as balance-scale, letter, and pima-diabetes. In the case of letter, in particular, the computation time of the SONG is faster than k-NN by about 2.4 times. We need to repeat 10-fold cross validation many times to select the optimum parameters for α and k. This evaluation consumes much computation time for large datasets such as letter. Therefore, the SONG based on the fast and compact SGNT is useful and practical for large datasets. Moreover, the SONG is capable parallel computation because each classifier behaves independently. In conclusion, the SONG is a practical method for large-scale data mining compared with k-NN.

4 Conclusions

In this paper, we proposed a new pruning method for the MCS based on SGNT, which is called SONG, and evaluated the computation cost and the accuracy.We introduced an on-line and off-line pruning method and evaluated the SONG by 10-fold cross-validation. Experimental results showed that the memory requirement is significant reduce, and by using the pruned SGNT as the base classifier of the SONG, accuracy is increased. The SONG is a useful and practical MCS to classify large datasets. In future work, we will study an incremental learning and a parallel and distributed processing of the SONG for large scale data mining.

References

1. Han, J., Kamber, M.: Data Mining: Concepts and Techniques. Morgan Kaufmann Publishers, San Francisco (2000)
2. Breiman, L.: Bagging predictors. Machine Learning 24, 123–140 (1996)
3. Schapire, R.E.: The strength of weak learnability. Machine Learning 5(2), 197–227 (1990)
4. Quinlan, J.R.: Bagging, Boosting, and C4.5. In: Proceedings of the Thirteenth National Conference on Artificial Intelligence, Portland, OR, August 4-8, 1996, pp. 725–730. AAAI Press, The MIT Press (1996)
5. Rätsch, G., Onoda, T., Müller, K.R.: Soft margins for AdaBoost. Machine Learning 42(3), 287–320 (2001)
6. Bishop, C.M.: Neural Networks for Pattern Recognition. Oxford University Press, New York (1995)
7. Duda, R.O., Hart, P.E., Stork, D.G.: Pattern Classification, 2nd edn. John Wiley & Sons Inc., New York (2000)
8. Wen, W.X., Jennings, A., Liu, H.: Learning a neural tree. In: the International Joint Conference on Neural Networks, Beijing, China, November 3-6, 1992, vol. 2, pp. 751–756 (1992)
9. Kohonen, T.: Self-Organizing Maps. Springer, Berlin (1995)
10. Inoue, H., Narihisa, H.: Improving generalization ability of self-generating neural networks through ensemble averaging. In: Terano, T., Liu, H., Chen, A.L.P. (eds.) PAKDD 2000. LNCS, vol. 1805, pp. 177–180. Springer, Heidelberg (2000)

11. Inoue, H., Narihisa, H.: Optimizing a multiple classifier system. In: Ishizuka, M., Sattar, A. (eds.) PRICAI 2002. LNCS (LNAI), vol. 2417, pp. 285–294. Springer, Heidelberg (2002)
12. Stone, M.: Cross-validation: A review. Math. Operationsforsch. Statist. Ser. Statistics 9(1), 127–139 (1978)
13. Quinlan, J.R.: C4.5: Programs for Machine Learning. Morgan Kaufmann, San Mateo (1993)
14. Blake, C., Merz, C.: UCI repository of machine learning databases (1998)
15. Patrick, E.A., Frederick, P., Fischer, I.: A generalized k-nearest neighbor rule. Information and Control 16(2), 128–152 (1970)
16. Zhang, B., Srihari, S.N.: Fast k-nearest neighbor classification using cluster-based trees. IEEE Transactions on Pattern and Machine Intelligence 26(4), 525–528 (2004)

Author Index

LaVergne, TN USA
25 November 2009
165305LV00005B/14/P

9 783642 045110

Overhauling Learning for Multilingual Students